Fundamentals of High Frequency CMOS Analog Integrated Circuits

Duran Leblebici • Yusuf Leblebici

Fundamentals of High Frequency CMOS Analog Integrated Circuits

Second Edition

 Springer

Duran Leblebici
Istanbul Technical University
Istanbul, Turkey

Yusuf Leblebici
Sabancı University
Istanbul, Turkey

Additional material to this book can be downloaded from (https://www.springer.com/us/book/9783030636579)

ISBN 978-3-030-63660-9   ISBN 978-3-030-63658-6 (eBook)
DOI 10.1007/978-3-030-63658-6

This Springer imprint is published by the registered company Springer Nature Switzerland AGThe registered company address is: Gewerbestrasse 11, 6330 Cham, Switzerland

To Yıldız, Anıl, and Deniz Ebru,
the ladies of our family

Preface to the First Edition

In the first half of the twentieth century the radio was the main activity area of the electronics industry, and correspondingly, the RF circuits occupied a considerable part in the electronic engineering curriculum and books published in this period. Properties of the resonance circuits and electronic circuits using them: the single-tuned amplifiers as the input stages of receivers, the double-tuned circuits as the IF amplifiers, the RF sinusoidal oscillators and high power class-C amplifiers have been investigated in depth. It must be kept in mind that the upper limit of the radio frequencies of those days was several tens of MHz, the inductors used in tuned circuits were air-core or ferrite-core coils with inductance values in the micro-henries to milli-henries range, having considerably high quality factors, ranging from 100 to 1000, and the tuning capacitors were practically lossless.

The knowledge developed for the vacuum-tube circuits easily adapted to the transistors with some modifications related to the differences of the input and output resistances of the devices. In the meantime, the upper limit of the frequency increased to about 100 MHz for FM radio and to hundreds of MHz for UHF-TV. The values of the inductors used in these circuits correspondingly decreased to hundreds to tens of nano-henries. But these inductors were still wound, high-Q discrete components.

In the second half of the twentieth century, the emergence of the integrated circuits drastically increased the reach of electronic engineering. Digital electronics on one side and the analog electronics using the potentials of the operational amplifiers on the other side, forced the curricula and the textbooks to skip certain old and "already known" subjects (among them the resonance circuits and the tuned amplifiers), to open room to these new subjects. The development of the (inductorless) active filters that replaced the conventional passive LC filters extensively used in telecommunication systems, even decreased the importance of inductors.

Rapid development of CMOS technology and the steady decrease of the dimensions of the devices according to "Moore's Law", led to the increase in complexity of ICs (from now on, the VLSI circuits) and the operating frequencies as well. In the digital realm this helped to improve the performances of digital computers and digital telecommunication systems. In the analog realm the operation frequencies of the circuits increased to the GHz region and correspondingly the inductance values decreased to below

20 nH so that now it was possible to realize them as on-chip components. Hence the freedom from external bulky discrete inductors opened a new horizon towards small and lightweight mobile systems: the mobile telephone, GPS systems, Bluetooth, etc.

But there is a problem related to this development: the quality factors of the on-chip inductors are very low, usually around 10 - 100 and the tuning capacitors are not lossless anymore. Most of the earlier theory (and design practice) that was developed with very high Q discrete components does not easily translate to such integrated high-frequency circuits and on-chip components with less than ideal characteristics. Instead of relying on a comprehensive theory and analytical design of high-frequency analog circuits that was in place many decades earlier, most of the new-generation designers were tempted to adopt rather ad-hoc design strategies not grounded in sound theory, and to explain away the inevitable inconsistencies as "secondary effects". To make matters even more complicated, systematic treatment of subjects such as high-frequency circuit behavior and resonance / tuned circuits have been missing from electrical engineering curricula for several decades, and analog designers entering the field of RF / high-frequency design had to re-learn these subjects.

One of the objectives of this book is to fill this gap: to introduce the fundamental aspects of high-frequency circuit operation, to systematically discuss the behavior of key components (in particular, submicron MOSFETs and on-chip passive components), summarize the behavior of the series and the parallel resonance circuits in detail and investigate the effects of the losses of the on-chip inductors and capacitors, that are usually not taken into account in formulae derived for high-Q resonance circuits.

Since all these circuits are being developed and used usually in the GHz range, sometimes close to the physical limits of the devices, it becomes necessary to recapitulate the behavior of the MOS transistors together with their important parasitics and the frequency related secondary effects[1]. In Chap. 1 the basic current-voltage relations of MOS transistors are derived, taking into account the parabolic (not linear) shape of the inversion charge profile that lead to a different approach to understand the channel shortening effect and the gate-source capacitance of the transistor. The velocity saturation effect and the behavior of a MOS transistor under a velocity saturation regime are also investigated as an important issue for small geometry devices. Although it is not used extensively for H-F applications, the sub-threshold regime is also investigated in brief.

In addition to the intrinsic behavior of MOS transistors the parasitics –that are inevitable and have severe effects on the overall behavior especially at high frequencies–have been discussed, mostly in connection with the BSIM3 parameters. The properties, limits and parasitics of the passive on-chip components, namely resistors, fixed value and variable capacitors (varactors)

---

[1] This part of the book must not be considered as an alternative to the existing sound and comprehensive models such as EKV and others, but rather as an attempt to explain the behavior of MOS transistors, based on the basic laws of the electrostatics and circuit theory, that all electronics students already know.

and inductors are also summarized in this chapter from a realistic and design-oriented point of view.

The subject of Chap. 2 is the DC properties of the basic analog MOS circuits that will be investigated in the following chapters. The interactive use of the analytic expressions – that provide interpretable knowledge about the basic behavior of the circuit- and SPICE simulations – that gives the designer a possibility to "experiment", to fine tune and optimize the circuit, all secondary and parasitic effects included, have been illustrated throughout the chapter. It is believed that the ability of using together the analytical expressions and the power of SPICE is a "must" for an analog designer.

In Chap. 3 the frequency dependent behavior of the basic circuits are given, not only limited to the frequency characteristics of the gain but the input and output impedances. Their important properties, usually not dealt in books, are investigated and their effects on the performance of the wide-band circuits are underlined. The basics of the techniques used to enhance the gain: the additive approach (distributed amplifiers) and the cascading strategies to reach the wide-band amplifiers (not only voltage amplifiers, but the current amplifiers, the trans-admittance amplifiers and the trans-impedance amplifiers) are systematically investigated.

In the Chap. 4 first the resonance circuits are recapitulated with this approach and the behavioral differences of the high-Q and the low-Q resonance circuits are underlined. Afterwards, the tuned amplifiers are systematically investigated taking into account the low-Q effects, not only for the single tuned amplifiers but the double tuned and the staggered tuned amplifiers, that are not covered in many new (even older) books in detail. The LNA, which is one of the most important classes of tuned amplifiers, is also investigated in this chapter together with the noise behavior of the MOS transistors, which is developed with a different approach.

The LC sinusoidal oscillators are given in Chap. 5 with the negative resistance approach and the classical positive feedback approach as well, with the emphasis to the effects of the low-Q components. The problems related to the frequency stability of the LC oscillators are discussed and the phase noise in LC oscillators is investigated with a different approach.

The last chapter is devoted to a summary of the higher-level system view of HF analog circuits, especially in the context that virtually all such high-frequency circuits are eventually integrated with considerable digital circuitry for interface, post-processing, and calibration purposes – and that such integration is increasingly done on the same silicon substrate. The traditional system-level view of high-frequency components and circuits is strongly influenced by conventional (all-analog) modulation and transmission systems modeling, which is based almost exclusively on the frequency domain. The behavior of all digital systems, on the other hand, is preferably described in time domain. While the translation between these two domains is (in theory) quite straightforward, the designers must develop a sense of how some of their choices in the analog realm eventually influence the behavior of the digital part, and vice versa. The data converters (analog-to-digital and digital-to-analog converters) naturally play an important role in this translation between domains, and Chap. 6 attempts to summarize the key criteria and parameters that are used to describe system-level performance.

The target audience of this book includes advanced undergraduate and graduate-level students who choose analog / mixed-signal microelectronics as their area of specialization, as well as practicing design engineers. The required background that is needed to follow the material is consistent with the typical physics, math and circuits background that is acquired by the third (junior) year of a regular electrical and computer engineering (ECE) curriculum.

Solved design examples are provided to guide the reader through the decision process that accompanies each design task, emphasizing key trade-offs and eventual approximations.

A number of individuals have contributed their time and their efforts, to the creation of this textbook. In particular, both authors would like to thank Mrs. Yıldız Leblebici who read the entire manuscript, carefully checked the analytical derivations throughout all chapters, and provided valuable insight as an experienced electronics teacher. The authors also acknowledge the generous support of Mr. Giovanni Chiappano from austriamicrosystems A.G. for offering the use of transistor parameters in numerous examples.

The idea of this book was originally launched with the enthusiastic encouragement of Dr. Philip Meyler of Cambridge University Press, who saw the need for a design-oriented text in this field and patiently followed through its early development. We are deeply grateful to Dr. Julie Lanca-shire, our publisher, for her guidance, support, and encouragement over the years, leading up to the final stages of production. The editorial staff of Cambridge University Press has been wonderfully supportive throughout this project. We especially like to thank Ms. Sarah Matthews for her valuable assistance.

Last but not least, the authors also like to thank all reviewers who read all or parts of the manuscript, and provided very valuable comments.

Preface to the Second Edition

The first edition of this book, published in 2009, was received very positively by a wide audience wishing to gain a deeper understanding of high-frequency circuit design fundamentals and to explore beyond the widely used automated design procedures. During the ten-plus years since its first publication, the contents have further matured and been enriched with new subjects reflecting the feedback received through teaching the material in senior- and graduate-level classes. The text has also been thoroughly updated to improve the understandability of the subjects.

This second edition has been revised extensively to expand and clarify some of the key topics and to provide a wide range of design examples and problems. New material has been added for basic coverage of core topics, such as wide-band LNAs, noise feedback concept and noise cancellation, inductive-compensated band-widening techniques for flat-gain or flat-delay characteristics, and basic communication system concepts that exploit the convergence and co-existence of analog and digital building blocks in RF systems. A new chapter (Chap. 5) has been added on noise and linearity, addressing key topics in a comprehensive manner. All of the other chapters have also been revised and largely re-written, with the addition of numerous solved design examples and exercise problems.

The authors would especially like to thank Mr. Charles B. Glaser, Editorial Director at Springer, who gave his very enthusiastic and strong support for the project. The entire production staff at Springer, particularly Ms. Olivia Ramya Chitranjan, was extremely helpful in all stages of the production. The authors would also like to give their sincere thanks to Dr. Alper Çabuk of the MKR-IC Design House, Istanbul, for the rigorous review of the text and to Ms. Deniz Ebru Leblebici for the final proofreading of the manuscript.

Istanbul, Turkey

Duran Leblebici
Yusuf Leblebici

Contents

About the Authors

Duran Leblebici is professor emeritus of electrical and electronics engineering at Istanbul Technical University (ITU). He has been teaching a range of undergraduate and graduate courses, from device electronics and fabrication technologies to integrated electronic circuits and RF IC design, for more than 50 years. He is the author of three textbooks in the field of electronics. He also established the first microelectronics laboratory and the first VLSI design house at ITU. He received the Distinguished Service Award of the Turkish Scientific and Technological Research Council (TUBITAK) in 1992, in recognition of his services to microelectronics education.

Yusuf Leblebici is president of Sabancı University and former director and chair professor of the Microelectronic Systems Laboratory at the Swiss Federal Institute of Technology in Lausanne (EPFL). He has previously worked as a faculty member at the University of Illinois at Urbana-Champaign, at Istanbul Technical University, and at Worcester Polytechnic Institute (WPI), where he established and directed the VLSI Design Laboratory, and also served as a project director at the New England Center for Analog and Mixed-Signal IC Design. He is the co-author of more than 400 scientific articles and 6 textbooks.

1.1 MOS Transistors

The basic structure of an n-channel Metal-Oxide-Semiconductor (NMOS) transistor built on a p-type substrate is shown in Fig. 1.1. The MOS transistor consists of two disjoint p-n junctions (source and drain), bridged by a MOS capacitor composed of the thin gate oxide and the polysilicon gate electrode. If a positive voltage with a sufficiently large magnitude is applied to the gate electrode, the resulting vertical electric field between the substrate and the gate attracts the negatively charged electrons to the surface. Once the electron concentration on the surface exceeds the majority hole concentration of the p-type substrate, the surface is said to be inverted, i.e., a conducting channel is formed between the source and the drain. The carriers, i.e., the electrons in an NMOS transistor, enter the channel region underneath the gate through the source contact and leave the channel region through the drain contact; their movement in the channel region is subject to the control of the gate voltage.

To ensure that both p-n junctions are continuously reverse biased, the substrate potential is kept lower than the source and drain terminal potentials. Note that the device structure is symmetrical with respect to the drain and source regions; the different roles of these two regions are defined only in conjunction with the applied terminal voltages and the direction of the drain current. In an n-channel MOS (NMOS) transistor, the source is defined as the n+ region which has a lower potential than the other n+ region, the drain. This means that the current flow direction is from the drain to the source. By convention, all terminal voltages of the device are defined with respect to the source potential.

The value of the gate-to-source voltage (V_{GS}) necessary to cause surface inversion (to create the conducting channel) is called the threshold voltage V_{TH}. This quantity depends on various device and process parameters such as the work function difference between the gate and the substrate, the intrinsic substrate (surface) Fermi potential, the depletion region charge concentration, the interface charge concentration, the gate oxide thickness and oxide (dielectric) permittivity, as well as the concentration of the channel implantation that is used to adjust the threshold voltage level.

For a gate-to-source voltage exceeding the threshold voltage, an n-type conducting channel is formed between the source and the drain, which is capable of carrying the drain (channel) current. If a small, positive voltage is applied to the drain, a current proportional to this voltage will start to flow from the drain to the source through the conducting channel. The effective resistivity of the continuous inversion

---

The original version of this chapter was revised. The correction to this chapter is available at https://doi.org/10.1007/978-3-030-63658-6_8

© The Author(s), under exclusive license to Springer Nature Switzerland AG 2021
D. Leblebici, Y. Leblebici, *Fundamentals of High Frequency CMOS Analog Integrated Circuits*,
https://doi.org/10.1007/978-3-030-63658-6_1, corrected publication 2021

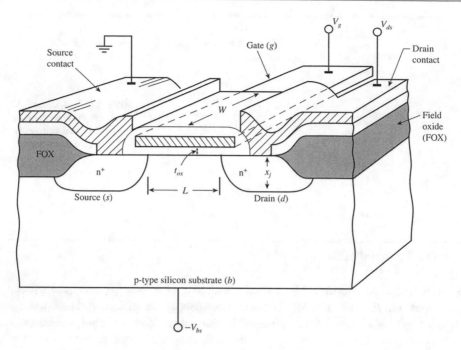

Fig. 1.1 Simplified cross-section view of an n-channel MOS (NMOS) transistor (after Taur and Ning)

layer between the source and the drain depends on the gate voltage. This operating mode is called the linear (or triode) mode, where the channel region acts as a voltage-controlled resistor. During this operating mode, the electron velocity in the channel is usually much lower than the drift velocity limit.

As the applied drain voltage is increased, the inversion layer charge and the channel depth at the drain end start to decrease. Eventually, when the drain voltage reaches a limit value called the saturation voltage (V_{Dsat}), the inversion charge at the drain is reduced – theoretically – to zero, and the velocity of electrons – theoretically – reaches very high values, as discussed in the following sections. This event is named as the "pinch-off" of the channel. Beyond the pinch-off point, i.e., for drain voltage values larger than the saturation voltage, electrons travel in a very shallow pinched-off channel with a very high velocity, which is called the "saturation velocity." This operating regime is known as the saturation mode.

If the transistor is formed on an n-type substrate, using two p+ regions as source and drain, this structure is called a p-channel MOS (PMOS) transistor. In a PMOS transistor, the fundamental mechanisms of surface inversion and channel conduction are exactly the same as in NMOS transistors, although the majority carriers consist of holes, not electrons. Thus, the gate-to-source voltage applied to the gate electrode to achieve surface inversion must be negative. Also, it should be taken into account that the hole mobility is considerably smaller than the electron mobility at room temperature, which leads to a smaller effective channel conductance for the PMOS transistor with the same channel dimensions. Nevertheless, the complementary nature of NMOS / PMOS biasing and operating conditions offers very useful circuit implementation possibilities, which underlines the importance and the wide-spread use of Complementary MOS (CMOS) circuits in a very large range of applications. The schematic cross-section of a CMOS pair realized on a p-type substrate is shown in Fig.1.2. The NMOS transistor is formed directly on the substrate.[1] The PMOS transistor is realized in an n-well that acts as the bulk of the PMOS transistor.

---

[1] Alternatively, to optimize the doping concentrations of different regions for better control of the device parameters, both transistors can be realized in separate wells.

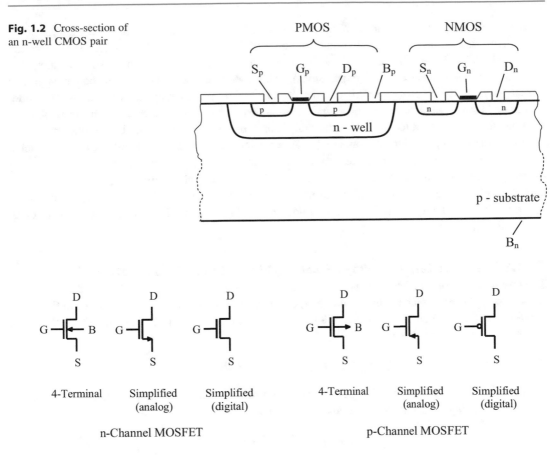

Fig. 1.2 Cross-section of an n-well CMOS pair

Fig. 1.3 Commonly used circuit symbols for NMOS and PMOS transistors

Commonly used circuit symbols for n-channel and p-channel MOS transistors are shown in Fig. 1.3. While the four-terminal representation shows all external terminals of the device, the three-terminal symbol is usually preferred for simplicity. Unless noted otherwise, the substrate terminals are always assumed to be connected to the lowest potential for NMOS devices, and to the highest potential for PMOS devices.

1.1.1   Current-Voltage Relations of MOS Transistors

The basic (so-called Level-1) current-voltage relations of a MOS transistor are given in most basic electronics textbooks. Since these relations contain a small number of parameters, they are convenient for hand calculations. The parameters of these expressions are:

- The mobility of electrons (or holes), μ
- The gate capacitance per unit area, C_{ox}
- The threshold voltage of the transistor, V_{TH}
- The gate-length modulation coefficient, λ
- The aspect ratio of the transistor, (W/L)

In the following, these relations are derived with a different approach, to remind the reader of the fundamentals, and also to clarify the understanding of device behavior. In addition, the derivation presented here is based on a realistic profile of the channel-region inversion charge (as calculated from the fundamental electric field expressions), as opposed to the classical gradual channel approach which assumes linear charge profiles in the channel.[2] This model represents the transistor under moderate to strong inversion conditions with reasonable accuracy for hand calculations, provided that the channel length is not too short, and the transistor is not in the velocity saturation region.

For short channel MOS transistors in which the carrier velocities reach saturation, i.e., approach a limit velocity, this model is no longer valid. Since a great majority of transistors realized in analog MOS integrated circuits today have channel lengths in the sub-half-micron range, they may easily enter the velocity saturation region. Therefore, it is necessary to derive rules to check if a transistor is operating in the velocity saturation region or not, and to obtain expressions that are valid for velocity saturated transistors.

1.1.1.1 The Basic Current-Voltage Relations Without Velocity Saturation

The cross-section of an NMOS transistor having an inversion layer along the channel due to a gate-source voltage greater than the threshold voltage, and zero source-drain voltage, is shown in Fig. 1.4a. Since there is no surface inversion for gate voltages smaller than the threshold voltage V_{TH}, the value of the inversion charge density is

$$Q_i = -C_{ox}(V_{GS} - V_{TH}) \quad [\text{coulomb/cm}^2]$$

where $(V_{GS} - V_{TH})$ is the "effective gate voltage" for this case. The amount of the inversion charge of a transistor having a channel length L and a channel width W is

$$\overline{Q}_i = -C_{ox}WL(V_{GS} - V_{TH})$$

The minus signs in front of these expressions denote that this is a negative charge, since the carriers in the inversion layer of an NMOS transistor are electrons.

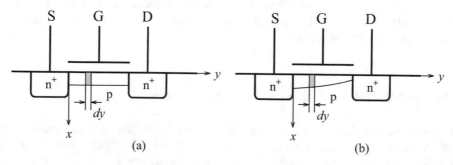

Fig. 1.4 Inversion channel profiles of a NMOS transistor for (**a**) $V_{GS} > V_{TH}$, $V_{DS} = 0$, (**b**) $V_{GS} > V_{TH}$, $V_{DS} > 0$

---

[2] It must be noted that the classical gradual channel approximation is physically impossible, because a linear decrease of the channel voltage is only possible for constant channel resistance (constant charge profile), which contradicts the original assumption.

When we apply a positive drain voltage with respect to the source, a drain current (I_D) flows in the $-y$ direction and is constant along the channel (Fig. 1.4b). However, due to the voltage drop on the channel resistance, the voltage along the channel is not constant. Although the effective gate voltage – inducing the inversion charge – at the source end of the channel is ($V_{GS} - V_{TH}$), it decreases along the channel and becomes equal to ($V_{GD} - V_{TH}$) = ($V_{GS} - V_{DS} - V_{TH}$) at the drain end. If the channel voltage with respect to the source is denoted by $V_c(y)$, the effective gate voltage as a function of y can be written as

$$V_{eff}(y) = (V_{GS} - V_{TH}) - V_c(y) \tag{1.1}$$

and the amount of the inversion charge in an infinitesimal channel segment dy is

$$d\overline{Q}_i(y) = -C_{ox}W[(V_{GS} - V_{TH}) - V_c(y)]dy \tag{1.2}$$

Since the drain current is constant along the channel, for any y position the current can be expressed as

$$I_D = \frac{d\overline{Q}_i(y)}{dt} = \frac{d\overline{Q}_i(y)}{dy/v(y)} \tag{1.3}$$

where $v(y)$ is the velocity of electrons at position y, and can be expressed in terms of the electron mobility and the electric field strength at y:

$$v(y) = \mu_n E(y) = -\mu_n \frac{dV_c(y)}{dy} \tag{1.4}$$

Using (1.2), (1.3), and (1.4), we obtain

$$I_D = \mu_n C_{ox} W[(V_{GS} - V_{TH}) - V_c(y)]\frac{dV_c(y)}{dy}$$

which gives the electric field strength as

$$E(y) = \frac{dV_c(y)}{dy} = \frac{I_D}{\mu_n C_{ox}W[(V_{GS} - V_{TH}) - V_c(y)]} \tag{1.4a}$$

and

$$\frac{I_D}{\mu_n C_{ox}W}dy = [(V_{GS} - V_{TH}) - V_c(y)]dV_c(y)$$

After integration from the source end ($y = 0$) to y we obtain

$$\frac{I_D}{\mu_n C_{ox}W}y = (V_{GS} - V_{TH})V_c(y) - \frac{1}{2}V_c^2(y) \tag{1.5}$$

From (1.5), the channel voltage $V_c(y)$ corresponding to a certain gate voltage and drain current can be deduced as

$$V_c(y) = (V_{GS} - V_{TH}) \mp \sqrt{(V_{GS} - V_{TH})^2 - \frac{2I_D}{\mu_n C_{ox}W}y} \tag{1.6}$$

To satisfy the obvious physical condition $V_c(0) = 0$, the sign before the square root term has to be negative. The effective gate voltage is found from (1.1) and (1.6):

$$V_{eff}(y) = (V_{GS} - V_{TH}) - V_c(y) = (V_{GS} - V_{TH})\sqrt{1 - \frac{2I_D}{\mu_n C_{ox} W (V_{GS} - V_{TH})^2} y} \qquad (1.7)$$

It is useful to interpret (1.7) for certain cases:

(a) For $y = 0$ (at the source end of the channel) the effective channel voltage is $V_{eff}(0) = (V_{GS} - V_{TH})$, as expected.

(b) If $V_c(L) = V_{DS} = (V_{GS} - V_{TH})$, the effective channel voltage at $y = L$ is equal to zero and the channel is pinched-off at the drain end of the channel. For this case, the value of the drain current can be found as

$$I_D = I_{Dsat} = \frac{1}{2} \mu_n C_{ox} \frac{W}{L} (V_{GS} - V_{TH})^2 \qquad (1.8)$$

In this expression, $\mu_n C_{ox}$ is a technology-dependent parameter (KP_n) and has the same value for all NMOS transistors on a chip.[3] For a given technology, (1.8) can be written as

$$I_D = I_{Dsat} = \frac{1}{2} KP_n \frac{W}{L} (V_{GS} - V_{TH})^2 \qquad (1.8a)$$

I_{Dsat} is called the "saturation current"[4] corresponding to a given gate voltage. The variations of the effective channel voltage along the channel and the corresponding inversion charge for a "saturated" transistor are plotted in Fig. 1.5a and b, respectively, based on (1.2) and (1.7). Note that the inversion

Fig. 1.5 The variation of (**a**) the effective channel voltage, (**b**) the corresponding inversion charge, and (**c**) the velocity of electrons along the channel, for a transistor pinched-off at the drain end of the channel

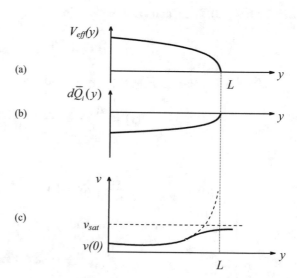

---

[3] Similarly, $KP_p = \mu_p C_{ox}$ is a parameter common to all PMOS transistors on a chip.

[4] The phrase "saturation current" is also used to express the value of the maximum drain current per 1-micron channel width for a certain technology, when the gate and the drain both are connected to the maximum permissible voltage for this technology. This "saturation current" takes into account all secondary effects discussed in the following sections, affecting the drain current. For example, for AMS 035 micron, 3.3 V technology the typical value of the maximum saturation current is given as 540 μA/micron for NMOS transistors and 240 μA/micron for PMOS transistors, respectively. This is not to be confused with the saturation current that is given as a function of the gate voltage, according to (1.8a).

charge profile in Fig. 1.5b is found to be a second-order function of the distance y, and not a linear profile as usually assumed in the conventional gradual channel approach. While this does not influence the current-voltage relationship, the realistic charge profile will later be used for a more straightforward calculation of the channel capacitance.

It is useful to note and to interpret an important fact: the current remains constant along the channel, but the electron density is decreasing. To maintain the current constant along the channel, i.e., to carry the same amount of charge in a certain time interval along the channel, the electron velocity has to increase from the source end to the drain end of the channel. Even more dramatically, the electron velocity theoretically has to reach infinity in case of pinch-off of the channel, since the inversion charge decreases to zero. But it is known that the velocity of electrons (and holes) cannot exceed a certain limit value and approach asymptotically this "saturation velocity," v_{sat}. In Fig. 1.5c, the velocity of electrons along the channel is plotted. The dashed curve corresponds to the theoretical behavior, without any velocity limitation. The solid curve takes into account the velocity limitation. It is obvious that due to this limitation, the electron density does not decrease to zero but has a finite value to maintain the drain current with the limit velocity.

(c) If $V_c(L) = V_{DS} < (V_{GS} - V_{TH})$, the effective channel voltage is always positive along the channel. In other words, the channel does not pinch-off. From another point of view this can be interpreted such that the distance of the pinch-off point (L') is longer than the channel length. The drain current in this case can be solved from (1.6), for $y = L$ and $V_c(y) = V_{DS}$ as

$$I_D = \mu_n C_{ox} \frac{W}{L} \left[(V_{GS} - V_{TH}).V_{DS} - \frac{1}{2} V_{DS}^2 \right] \qquad (1.9)$$

that reduces to (1.8) for $V_{DS} = (V_{GS} - V_{TH})$, as expected. The variations of the effective channel voltage, the inversion (electron) charge density and the velocities of electrons along the channel are plotted, qualitatively, in Fig. 1.6.

Fig. 1.6 The variation of (a) the effective channel voltage, (b) the corresponding inversion charge, and (c) the velocity of electrons along the channel, for a transistor operating in the resistive (no pinch-off) region

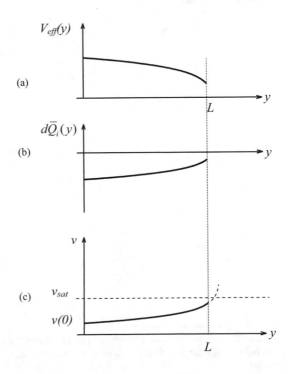

For this "pre-saturation" region the variation of the drain current for small V_{DS} values can be written as

$$I_D \cong \mu_n C_{ox} \frac{W}{L}(V_{GS} - V_{TH})V_{DS} \text{ for } V_{DS} \ll (V_{GS} - V_{TH}) \tag{1.9a}$$

This means that the drain current is proportional to the drain-source voltage. In other words, the transistor acts as a resistor in this region. That is why this region is also known as the "resistive region."

(d) For $V_c(L) = V_{DS} > (V_{GS} - V_{TH})$, the transistor is in saturation. Assume that the transistor is pinched-off at the drain end of the channel and then the drain-source voltage increases by ΔV_{DS}. The effective channel voltage becomes equal to zero (the channel voltage becomes equal to $(V_{GS}$-$V_{TH})$) at a distance L' smaller than L. Since the current (I_D) and the electron velocity (v_{sat}) are constant in the interval L'- L, the electron charge density has to remain constant from L' to L which implies a linear variation of the potential along the pinched-off portion of the channel. Corresponding variations of the effective channel voltage and the charge density along the channel are shown in Fig. 1.7a and b. As a result of the linear potential variation, the electric field strength along this region is constant. In addition, it must have a value corresponding to the saturation velocity, that is around $E = 10^5$ [V/cm] for silicon.[5] Correspondingly, the velocity of electrons is equal to the saturation velocity along the pinched-off region of the channel as shown in Fig.1.7c.

Fig. 1.7 The variation of (a) the effective channel voltage, (b) the corresponding inversion charge, and (c) the velocity of electrons along the channel, for a transistor pinched-off before the drain end of the channel

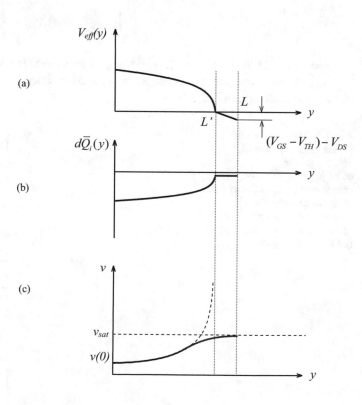

Fig. 1.8 The inversion charge profiles corresponding to (**a**) $V_{GS} > V_{TH}, V_{DS} = 0$, (**b**) the pre- pinch-off (resistive) regime, (**c**) the onset of the pinch-off or saturation, (**d**) deep saturation

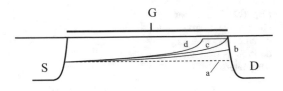

In Fig. 1.8, the inversion charge profiles for four different operating regimes are given for comparison.

From these considerations, we can conclude that

$$\Delta L = (L - L') = \frac{V_{DS} - (V_{GS} - V_{TH})}{E_{sat}} \tag{1.10}$$

The influence of the drain-source voltage on the drain current at a certain channel length can now be calculated:

$$I_{Dsat} = \frac{1}{2}\mu_n C_{ox} \frac{W}{L} (V_{GS} - V_{TH})^2$$

$$\frac{dI_D}{dV_{DS}} = \frac{dI_D}{dL} \times \frac{dL}{dV_{DS}} = \left(-I_D \frac{1}{L}\right) \times \left(-\frac{1}{E_{sat}}\right) = \Lambda . I_D \tag{1.11}$$

where

$$\Lambda = \frac{1}{L.E_{sat}} \tag{1.12}$$

(1.11) gives the slope of the output characteristic curve corresponding to a certain V_{GS}, at the beginning of the saturation region, and is equal to the output conductance of the transistor (g_{ds}) for this point. Thus, the drain current corresponding to any drain-source voltage for the same gate-source voltage can be calculated as

$$I_D = I_{Dsat} + g_{ds} (V_{DS} - V_{DS(sat)}) = I_{Dsat} + g_{ds}[V_{DS} - (V_{GS} - V_{TH})] \tag{1.13}$$

From (1.12) and (1.13) the drain current can be written as

$$I_D = I_{Dsat} \frac{1}{1 - \Lambda[V_{DS} - (V_{GS} - V_{TH})]} \cong I_{Dsat}\{1 + \Lambda[V_{DS} - (V_{GS} - V_{TH})]\} \tag{1.14}$$

Using simple linear relations shown in Fig.1.8, it is possible to express the drain current in terms of the conventional "lambda parameter" as

$$I_D = I_{Dsat} \frac{1 + \lambda V_{DS}}{1 + \lambda(V_{GS} - V_{TH})}$$

$$= \frac{1}{2}\mu_n C_{ox} \frac{W}{L} (V_{GS} - V_{TH})^2 \frac{1 + \lambda V_{DS}}{1 + \lambda(V_{GS} - V_{TH})} \tag{1.14a}$$

For $V_{DS} \gg (V_{GS} - V_{TH})$, (1.14a) can be simplified as

Fig. 1.9 The output characteristic curve for a certain V_{GS} value and definitions of Λ and λ parameters

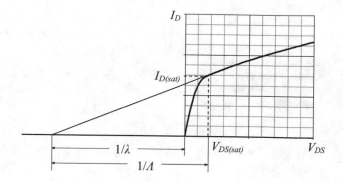

Fig. 1.10 The output characteristic curves of a typical NMOS transistor. The border between the resistive region and the saturation region corresponding to $V_{DS} = (V_{GS} - V_{TH})$ is plotted as a dashed line

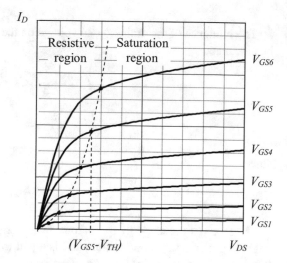

$$I_D = I_{Dsat}(1 + \lambda.V_{DS})$$
$$= \frac{1}{2}\mu_n C_{ox}\frac{W}{L}(V_{GS} - V_{TH})^2(1 + \lambda.V_{DS}) \tag{1.15}$$

This expression is commonly used to model the "channel length modulation effect." However, it should be kept in mind that this approximation is not very accurate, especially for short-channel devices (Fig. 1.9).

In Fig. 1.10, the output characteristic curves of an NMOS transistor covering these three characteristic features, namely the pre-saturation (or resistive) region, the onset of the pinch-off, and the saturation region, are given. For PMOS transistors, the characteristic curves have a similar shape. But since in PMOS transistors the gate-source voltage must be negative to induce p-type inversion layer, and consequently, the polarities of the drain-source voltage and the drain current are negative, all voltages and currents on the characteristics must be marked as "negative."

The initial slope of the characteristic curve corresponding to a certain gate-source voltage can be calculated from (1.9a) as

$$\frac{dI_D}{dV_{DS}} \cong \mu_n C_{ox}\frac{W}{L}(V_{GS} - V_{TH})$$

and the inverse of this conductance is called the "on resistance, r_{on}" of the transistor:

$$r_{on} = \frac{1}{\mu_n C_{ox} \frac{W}{L} (V_{GS} - V_{TH})} \qquad (1.16)$$

which expresses the series resistance exhibited by the transistor when it is used as a switch.

For analog applications, a MOS transistor is – almost – always used in the saturation region. Therefore, the parameters corresponding to this region have prime importance and will be investigated in detail, later on.

1.1.1.2 Current-Voltage Relations Under Velocity Saturation

In Sect. 1.1.1.1, we observed that the velocity of electrons reaches the "saturation velocity" at the drain end of the channel. In Fig. 1.11, the velocity of carriers in silicon is shown as a function of the lateral electric field strength. The saturation velocities of electrons and holes are approximately 10^7 cm/s and 8×10^6 cm/s, respectively, in bulk silicon. The initial slope of the velocity curve corresponds to the low field mobility of electrons and holes. It has been shown that the saturation velocities of electrons and holes in the inversion layer of a MOS structure are considerably lower than those in bulk silicon and are given as 6.5×10^6 cm/s and 5.85×10^6 cm/s, respectively [1].

We have also seen that velocity saturation can extend towards the source end of the channel under certain bias conditions. To understand this behavior, it is useful to consider the plot given in Fig. 1.11, which displays the generic behavior of carrier velocity (electrons and holes) as a function of the lateral electric field strength in the channel. This plot indicates that:

- The velocity of electrons (and holes) increases proportionally with the lateral electric field strength (parallel to the direction of current) until it reaches the vicinity of a so-called "critical field strength" (F_{cr}).

- The proportionality factor is called the "low-field mobility" of electrons and holes:

$$v_n = -\mu_n.E \qquad v_p = \mu_p.E \qquad (1.17)$$

The minus sign in the first expression indicates that the velocity of electrons is in the opposite direction of the electric field.

- As already mentioned, the saturation velocities of electrons and holes in the inversion layer of a MOS transistor are 6.5×10^6 cm/s and 5.85×10^6 cm/s, respectively.
- Carrier mobility, critical field strength, and the saturation velocity are related as

Fig. 1.11 Velocity of carriers versus the lateral electric field strength in silicon

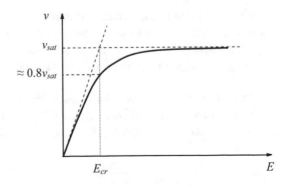

$$E_{cr(n)} \cong \frac{v_{sat}}{\mu_n} \qquad E_{cr(p)} \cong \frac{v_{sat}}{\mu_p} \qquad\qquad (1.18)$$

- The carrier velocities corresponding to the critical field strengths are approximately equal to $0.8 \times v_{sat}$.
- A final remark: the mobility of electrons (and holes) in the channel of a MOS transistor also depends on the transversal (perpendicular to the direction of the current) electric field strength [2–4]. This secondary effect must not be neglected for thin gate oxide (small geometry) devices (See Appendix A).

In a transistor operating in the pinched-off (saturated) regime, electrons travel under velocity saturation conditions at the drain end of the channel, as seen in Fig. 1.7c. If the length of this region is only a small fraction of the channel length, the effects of the velocity saturation can be neglected. Especially for short channel devices, electrons can travel under velocity saturation conditions in a major part of the channel, even along the whole channel. In these cases, the expressions derived under the assumption that the mobility of carriers is constant along the channel are no longer valid, and new expressions must be derived.

Now consider an NMOS transistor in which velocity saturation conditions are observed along the entire length of the channel, and consequently all electrons in the channel travel with saturation velocity. Since the current is constant and the velocity of the electrons is equal to the saturation velocity, the charge density also must be constant along the channel. This is equal to the charge density at the source end of the channel (1.3), and the total charge in the channel region is

$$\overline{Q}_i = WLC_{ox}(V_{GS} - V_{TH}) \qquad\qquad (1.19)$$

If this charge is being swept in t seconds, which is $t = L/v_{sat}$, the drain current under velocity saturation conditions becomes

$$I_{D(v-sat)} = WC_{ox}(V_{GS} - V_{TH})v_{sat} \qquad\qquad (1.20)$$

From (1.18) and (1.20), we can write

$$I_{D(v-sat)} = WC_{ox}(V_{GS} - V_{TH})\left(\mu_n E_{crit(n)}\right) \qquad\qquad (1.21)$$

For smaller values of the field strength, this expression reduces to

$$I_D = \mu_n C_{ox} \frac{W}{L}(V_{GS} - V_{TH})V_{DS} \qquad\qquad (1.22)$$

which is identical to the current expression in the resistive region (1.9a) for small values of V_{DS}, plotted in Fig.1.12 with a dashed line. For the rest of this curve, there are two alternatives:

- If the velocity saturation does not exist, the transistor enters into the normal saturation region at $V_{DS(sat)} = (V_{GS} - V_{TH})$ and the current saturates as shown with curve-A, to a value given in (1.8).

- If the velocity of the electrons reaches saturation velocity, the current saturates to the value given in (1.20), as shown with curve-B. For this case, the drain-source voltage corresponding to the onset of the velocity saturation can be found from (1.20) and (1.21) as

Fig. 1.12 Comparison of
normal saturation and
velocity saturation

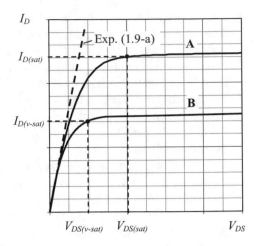

$$V_{DS(v-sat)} = \frac{L v_{sat}}{\mu_n} \tag{1.23}$$

Until now, we have assumed that the velocities of electrons in the channel of a velocity saturated NMOS transistor are equal to the saturation velocity that corresponds to the extreme velocity saturation. From the inspection of Fig. 1.11, we see that the saturation starts at a lower field strength and the velocity of electrons asymptotically approaches v_{sat}. Therefore, it is more realistic to modify (1.23) as

$$V_{DS(v-sat)} = \frac{L k v_{sat}}{\mu_n} \tag{1.23a}$$

where k is a constant smaller than unity. $k = 0.8$ corresponding to the critical field strength can be used as an appropriate value. Similarly, the drain current under velocity saturation conditions starts to saturate at approximately $k v_{sat}$ that gradually increases to v_{sat}. Therefore, to modify (1.20) as

$$I_{D(v-sat)} = k W C_{ox}(V_{GS} - V_{TH}) v_{sat} \tag{1.20a}$$

is more realistic.

Now we can interpret the results related to the velocity saturation:

- If $V_{DS(v-sat)}$ is smaller than $V_{DS(sat)}$, the transistor is in the velocity saturation regime. This leads us to an expression to check if a transistor is in velocity saturation:

$$\frac{L k v_{sat}}{\mu_n} < (V_{GS} - V_{TH}) \rightarrow L < \frac{\mu(V_{GS} - V_{TH})}{k v_{sat}} \tag{1.24}$$

- According to (1.23a), for a transistor operating in the velocity saturation region, the saturation voltage is not only smaller than the normal saturation voltage, but also independent of the gate-source voltage. This is a valuable property for circuits that have a tight supply voltage budget. For example, for a $L = 0.13$ μm transistor with $V_{TH} = 0.25$ V and $\mu_n = 200$ cm^2/V.s, the calculated value of $V_{DS(v-sat)}$ is 0.34 V.

- The velocity saturation is a small geometry phenomenon. For example, assuming 0.2 V gate overdrive and $\mu_n = 400$ cm^2/V.s electron mobility, an NMOS transistor is subject to velocity saturation if the channel length is smaller than 0.15 μm.
- Since the hole mobility is always smaller than the electron mobility, velocity saturation occurs only for extremely short channel PMOS transistors. For example, assuming 0.2 V gate overdrive and $\mu_p = 150$ cm^2/V.s hole mobility value, a PMOS transistor can enter into the velocity saturation regime only if the channel length is smaller than 85 nm.
- For a certain gate length, the transistor may enter the velocity saturation regime if the gate overdrive voltage increases. For $L = 0.13$ μm transistor with $V_{TH} = 0.25$ V and $\mu_n = 200$ cm^2/V.s, the velocity saturation occurs for $(V_{GS} - V_{TH}) > 0.34$ V, or for $V_{GS} > 0.59$ V.
- According to (1.21), for a velocity saturated transistor the drain current is linearly related to the gate voltage. Figure 1.13a shows the gate-source voltage to drain current transfer characteristics of a $W/L = 13$ μm/0.13 μm and a $W/L = 130$ μm/1.3 μm transistor. The long channel transistor that is not subject to velocity saturation has an obvious quadratic characteristic. For the short channel transistor, the characteristic is linear after the onset of the velocity saturation.
- In Fig. 1.13b, the output characteristics of a 0.13 μm NMOS transistor are shown. Approximately equal intervals between the curves indicate the linear relation of the drain current to the gate-source voltage. The small and almost constant value of the drain-source saturation voltage is about 0.4 V (which approximately matches the calculated value). The nature of velocity saturation in small-geometry devices can be best appreciated when these curves are compared with the characteristics of a long channel (non-velocity saturated) transistor shown in Fig. 1.11c.

1.1.1.3 The Subthreshold Regime

Until now, it has been assumed that for gate-source voltages smaller than the threshold voltage V_{TH}, the drain current of the transistor is zero. But it is known that a very small drain current flows for gate-source voltages considerably smaller than the threshold voltage. This is called the "subthreshold" current and it can be controlled by the gate-source voltage. In this section, the subthreshold regime will be explained with a different approach based on the basic behaviors of the bipolar transistors.

Figure 1.14a shows the simplified cross-section of an NMOS transistor. When the channel is not inverted, this structure can also be interpreted as an NPN bipolar junction transistor (BJT) such that the source, substrate, and the drain of the MOS transistor correspond to the emitter, base, and collector of the bipolar transistor, respectively. The base of the BJT is connected to its emitter as shown in Fig. 1.14b.

From the BJT theory, it is known that the basic Ebers-Moll or Gummel-Poon expressions can be reduced to

$$I_C = -I_{CBS}\left(e^{-V_{CB}/V_T} - 1\right) \tag{1.25}$$

for $V_{BE} = 0$, where $V_T = kT/q \simeq 26$ mV, I_{CBS} the reverse saturation current of the collector-base junction when the emitter is short-circuited to the base [5]. I_{CBS} can be expressed as

$$I_{CBS} = qAn_i^2\left(\frac{D_n}{p_{p0}L_n} + \frac{D_p}{n_{n0}L_p}\right) \tag{1.26}$$

where q is the unit charge, A is the cross-section of the collector junction, n_i is the intrinsic carrier density of silicon, D_n, D_p are the diffusion coefficients, L_n, L_p are the diffusion lengths of electrons and holes. p_{p0} and n_{n0} are the majority carrier concentrations in the p-type and n-type regions that

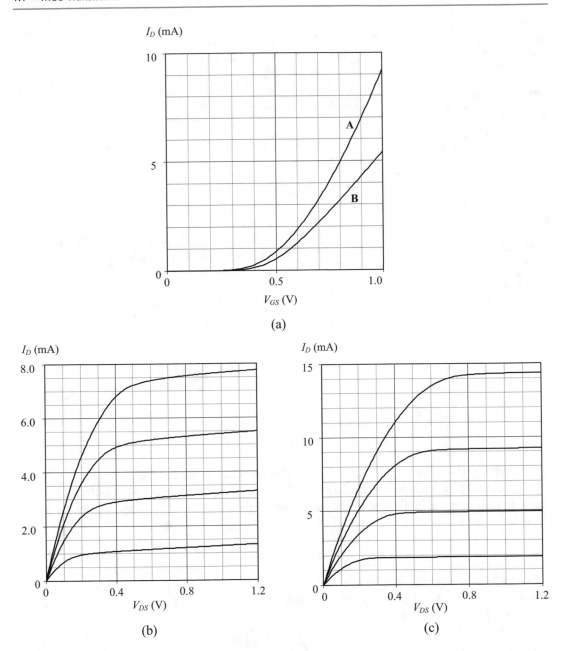

Fig. 1.13 (a) $I_D = f(V_{GS})$ characteristic curves of two 0.13 micron technology transistors; (A) non-velocity saturated 130 μm/1.3 μm transistor, (B) velocity saturated 13 μm/0.13 μm transistor. (b) The output characteristics of a 13 μm/ 0.13 μm transistor. Note the effects of the velocity saturation. (c) The output characteristics of a 130 μm/1.3 μm transistor. ($V_{GS} = 0.6$ V to 1.2 V, with 0.2 V intervals)

correspond to the base and collector regions of the bipolar transistor, and also to the substrate and source/drain regions of the MOS transistor. Since the source and drain doping densities (corresponding to the majority carrier concentrations) are always much higher than those of the substrate, (1.26) can be simplified as

Fig. 1.14 (a) Cross-section of an NMOS transistor, with its source connected to the bulk (or to the appropriate well). (b) Transistor is represented as a BJT, with its base region (bulk) connected to the emitter (the source region of the NMOS transistor)

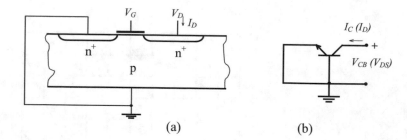

Fig. 1.15 The variation of the surface potential as a function of the gate voltage for an NMOS structure ($C_{ox} = 5 \times 10^{-7}$ F/cm^2). The upper curve (A) was calculated for $N_A = 10^{16}$ (cm^{-3}), and the lower curve (B) for $N_A = 10^{17}$ (cm^{-3})

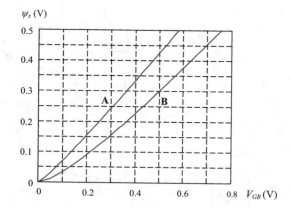

$$I_{CBS} \cong qAn_i^2 \frac{D_n}{p_{p0}L_n} = M\frac{1}{p_{p0}} \tag{1.26a}$$

which corresponds to the drain current of the NMOS transistor in subthreshold.

It is known that the carrier concentrations in the channel region depend on the surface potential. The hole concentration on the surface of the channel region in terms of the surface potential (ψ_s) is given as

$$p'_{p0} = p_{p0}e^{(-\psi_s/V_T)} \tag{1.27}$$

For gate voltage values considerably smaller than the threshold voltage of the transistor, it can be shown that the surface potential is approximately proportional with the gate voltage as shown in Fig. 1.15 [6]. Hence (1.27) can be written as

$$p'_{p0} = p_{p0}e^{(-nV_{GS}/V_T)} \tag{1.28}$$

where n has a value between 0.6 to 0.9. Equation (1.21) shows us that the gate voltage controls the hole concentration in the channel and the change of the gate voltage in the positive direction decreases the hole density in the "base region" of the NPN bipolar transistor. In other words, p_{p0} in (1.26a) decreases to p'_{p0} as a function of the gate voltage. Now (1.26a) can be written as

$$I_{CBS} = \frac{M}{p_{po}}e^{nV_{GS}/V_T} = I_{CBS0}.e^{nV_{GS}/V_T} \tag{1.29}$$

and from (1.25)

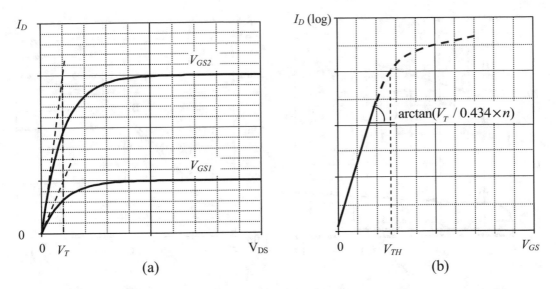

Fig. 1.16 (a) The subthreshold drain current as a function of the drain-source voltage, for two different gate voltage values ($V_{GS2} > V_{GS1}$). (b) The drain current as a function of the gate-source voltage plotted on a logarithmic I_D scale

$$I_D = -I_{CBS0} \cdot e^{nV_{GS}/V_T}\left(e^{-V_{DS}/V_T} - 1\right) \tag{1.30}$$

where I_{CBS0} is a structural parameter (constant) of the MOS transistor and will be shown with I_0.

This expression can be evaluated as follows:

- For a given value of V_{GS}, the drain current can be plotted as a function of V_{DS} as shown in Fig. 1.16a. The "corner" of this curve corresponds to $V_T = 26$ mV. The drain current starts to "saturate" after this point.

- The "family" of curves corresponding to different values of V_{GS} forms the "output characteristics" of the MOS transistor operating in the subthreshold regime. The separations between these curves vary exponentially, similar to the output characteristics of a bipolar transistor driven with the base-emitter voltage (not base current).

- In Fig. 1.16b, the variation of the drain current as a function of the gate-source voltage on a logarithmic axis is shown. For $V_{DS} \gg V_T$,

$$I_D \cong I_0 e^{nV_{BE}/V_T}$$

$$\log I_D = \log I_0 + \frac{nV_{GS}}{V_T}\log e = \log I_0 + \frac{nV_{GS}}{V_T} \times 0.434$$

and the inverse slope of the plot, which is usually expressed as the "subthreshold slope" in [mV/decade]

$$S[\text{mV/dec.}] = \frac{V_T[\text{mV}]}{0.434 \times n} \tag{1.31}$$

The numerical value of the subthreshold slope for $n = 0.7$ and 0.8 are $S = 85.2$ and 74.9 [mV/dec.], respectively.

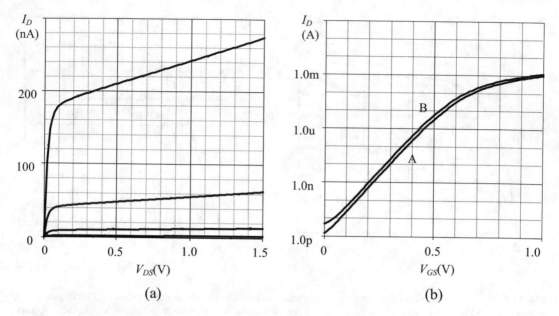

Fig. 1.17 The subthreshold characteristics of an AMS035, NMOS transistor ($W = 35$ μm, $L = 0.35$ μm). (a) The output characteristics ($V_{GS} = 0.25$ V to 0.4 V, steps: 0.05 V). (b) $I_D = f(V_{GS})$ (A: $V_{DS} = 0.5$ V, B: $V_{DS} = 3$ V). The subthreshold slope is 75 mV/dec.

- This slope is constant up to the vicinity of the threshold voltage. Then the transistor leaves the subthreshold regime and enters the strong inversion regime. This part of the curve is plotted with a dashed line.

In Fig. 1.17a and b, the output characteristic curves and the drain current-gate voltage characteristics of a 0.35 μm NMOS transistor are given for operating conditions in the subthreshold regime, as obtained from PSpice simulations. These curves are (in principle) in good agreement with curves given in Fig. 1.16. The width of the pre-saturation region on the output characteristic curves is in the order of V_T, as expected. This is valuable information for low supply voltage budget cases.

The drain current of a given transistor under subthreshold conditions is always considerably lower than that of the normal (in inversion) operation. Since the parasitic capacitances of the device are the same for both of these cases, the high-frequency performance of a MOS transistor operating in the subthreshold regime is always inferior compared to normal operation. Therefore, the subthreshold operation must be considered as suitable only for low frequency and low power applications.

1.1.2   Determination of Model Parameters and Related Secondary Effects

The current-voltage relations given in the previous sections are the most basic relations of an MOS transistor. They are simple enough for hand calculations and suitable to help understand the basic behavior of devices and basic circuits containing MOS transistors. To use these expressions for hand calculations, the model parameters in them, namely μ, C_{ox}, V_{TH}, Λ (or λ) and certainly the gate dimensions (W and L), must be known. In this section, we will discuss how to determine the numerical values of these basic parameters, and the related secondary effects that must be taken into account.

1.1.2.1 Mobility, μ

As seen from (1.9a), the drain current of an NMOS transistor is directly influenced by the mobility of electrons (similarly, by the mobility of holes in the case of a PMOS transistor) in the inversion channel. But the mobility is not a "constant." It depends on the manufacturing process and the bias conditions of the transistor. The low-field mobility values for a certain process are given by the manufacturer among the high-level parameter sets, for example, BSIM-3. The numerical value of the low-field mobility for electrons and holes are usually in the range of 200...500 cm^2/V.s and 70...150 cm^2/V.s, respectively.[6] Reliable average values of mobility for a specific process can be found in the high-level parameter sets (for example BSIM-3) supplied by the manufacturers.

Another important issue is the transversal electrical field dependence of the mobility. It has been shown that carrier mobilities decrease with the transversal field strength, which depends on the thickness of the gate oxide, the value of the threshold voltage, and the gate voltage[7] [3, 4]. This effect is more pronounced for small geometry (thin gate oxide) processes. For example, for $T_{ox} = 5$ nm, $V_{TH} = 0.5$ V, and $V_{GS} = 1$ V, the electron mobility decreases to approximately 75% of its low-field value. The obvious result of this important secondary effect is a considerable discrepancy between the drain current and the value calculated from (1.8), assuming constant mobility. In Fig. 1.18, the $I_D = f(V_{GS})$ curve (the transfer characteristic) of a MOS transistor is given with and without the influence of the transversal field. From the comparison of these curves, it can be seen that:

- Due to the transversal field, the drain current decreases considerably for high gate voltage values
- Due to this effect, the transfer characteristic appears to be more "linear" than quadratic

Fig. 1.18 The transfer characteristic of an NMOS transistor ($\mu_{no} = 500$ cm^2/V.s, $t_{ox} = 5$ nm, $V_{TH} = 0.5$ V and $W/L = 10$) when the transversal field effect is neglected (A) and not neglected (B)

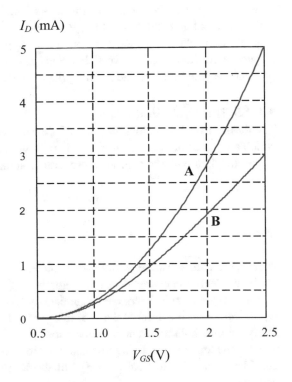

I_D (mA)

V_{GS}(V)

---

[6] In the model parameters lists released by the manufacturers, the mobility is given in [cm^2/V.s] or [m^2/V.s]. Care is necessary to use the same "dimension for length" in calculations.

[7] See Appendix-A

- The slope of the transfer characteristic at a certain operating point (i.e., the measure of gate-source voltage dependence of the drain current, which is called the "transconductance parameter" of the transistor) decreases considerably

This effect is obviously a small geometry problem and must not be overlooked when the gate oxide thickness is less than 10 nm. For hand calculations, it is convenient to use the μ value corresponding to the actual gate-source bias voltage of the transistor, in the circuit.

1.1.2.2 Gate Capacitance, C_{ox}

The gate capacitance corresponds to the maximum value (i.e., for accumulation or strong inversion) of the gate-bulk capacitance per unit area. Its value can be calculated as

$$C_{ox} = \frac{\varepsilon_0 \varepsilon_{ox}}{T_{ox}} \tag{1.32}$$

If we insert the values of ε_0 and ε_{ox} (8.85×10^{-14} F/cm and 3.9 for silicon dioxide as the gate insulator) and express the gate oxide thickness in nm for convenience, (1.32) can be arranged as

$$C_{ox} = \frac{34.5}{T_{ox}[\mathrm{nm}]} \times 10^{-7} \ \ [\mathrm{F/cm^2}] \tag{1.32a}$$

The value of the gate capacitance is one of the most accurately determined parameters of a MOS transistor and depends only on the thickness of the gate oxide, T_{ox}. The value of T_{ox} is usually in the range of 2...8 [nm] for submicron transistors.[8] An important issue related to the gate oxide thickness is the electrostatic breakdown of the gate dielectric. The breakdown field strength for silicon dioxide is given as 10 MV/cm, which corresponds to 1 V per nm. This high sensitivity of the gate oxide with respect to breakdown (an irreversible device failure mechanism) usually requires special measures to prevent the gate-bulk voltage reaching the breakdown value.

1.1.2.3 Threshold Voltage, V_{TH}

The threshold voltage, from the point of view of the circuit designer, is the gate-to-source voltage at which an appreciable drain current starts to flow. It is well-known from device physics that the value of the threshold voltage depends on several structural parameters and can be expressed as

$$V_{TH} = \Phi_{MS} - \frac{Q_{tot}}{C_{ox}} - \frac{Q_s}{C_{ox}} - 2\Phi_F - q\frac{D_I}{C_{ox}}$$

where:

Φ_{MS}: is the gate-substrate work function difference [V]
Q_s: is the depletion charge density [coulomb/cm^2]
Q_{tot}: is the total oxide-interface charge density [coulomb/cm^2]
Φ_F: is the Fermi potential of the substrate (or well) [V]
D_I: is the threshold adjustment implant dose [cm^{-2}]
and they depend on materials and processes. However, material properties and process parameters are considered the "intellectual property" of the IC manufacturer, and hence, not available to circuit designers.

---

[8] The production tolerance of the gate oxide thickness is usually within $\pm10\%$. Consequently, as the "most robust" parameter of a MOSFET, C_{ox} can have a value 10% higher or lower than the given value.

Fig. 1.19 (a) Example where transistors (M1 and M2) are not subject to substrate bias effect. (b) Example where M2 and M3 are subject to substrate bias effect

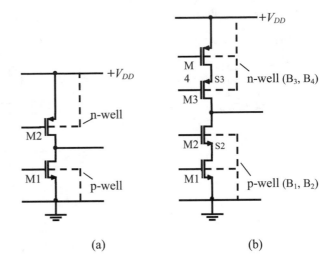

(a)

(b)

The "basic" threshold voltage is given in the parameters list of advanced simulation models as (VTH0). It is the threshold voltage of a large transistor when the source is connected to the bulk (or to the appropriate well). Several effects influencing the value of the threshold voltage have been theoretically investigated in detail [7, 8] and also included in the advanced models. These secondary effects can be ignored for the hand calculations, with one exception: the substrate bias effect.

The MOS transistors in a CMOS circuit are usually operated with the source connected to the n-well for NMOS transistors and to the p-well for PMOS transistors, as shown in Fig. 1.19a. But in some cases, a series connection of same-type transistors is needed, as shown in Fig. 1.19b. In this case, the source regions of M1 and M4 can be connected to the appropriate wells. However, the source of M2 is connected to the drain of M1 and has to be separated from the well. This means that there is a positive voltage between the source of M2 and the p-well. Similarly, there is a negative voltage between the source of M3 and the n-well. This bias (V_{SB}) increases the magnitude of the threshold voltage. It has been shown that the increase of the threshold voltage depends on a number of structural parameters such as the gate oxide thickness, the doping concentration of the substrate, and the doping properties of the channel region, and modeled in detail in advanced models.

For hand calculations, the substrate bias effect can be ignored and left to the fine-tuning of the circuit as a whole with SPICE. If the accurate values of the threshold voltages are needed in the hand-calculation stage, it is possible to use SPICE simulations to obtain the actual threshold voltage of a transistor under a specified substrate bias condition. Assume that the drain-source voltage of M1 in Fig. 1.19b is 1 V. It means that the source of M2 is 1 V positive with respect to ground. In Fig. 1.20 the $I_D = f(V_{GS})$ curves of a transistor for $V_{SB} = 0$ and $V_{SB} = 1$ V are shown, obtained from a simple SPICE simulation. From these curves, the threshold voltages for no substrate bias and for a 1 V substrate bias can be found as 0.5 V and 0.7 V, respectively. This example shows that the substrate bias can significantly change the value of the threshold voltage, and must not be ignored.

1.1.2.4 Channel Length Modulation Factor, λ

It is known that the drain current does not remain constant in the saturation region but exhibits a gradual increase with the drain-source voltage. This effect is modeled with the λ parameter for low-level models. In these models, it is assumed that the saturation region tangents of the output

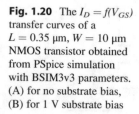

Fig. 1.20 The $I_D = f(V_{GS})$ transfer curves of a $L = 0.35\ \mu m$, $W = 10\ \mu m$ NMOS transistor obtained from PSpice simulation with BSIM3v3 parameters. (A) for no substrate bias, (B) for 1 V substrate bias

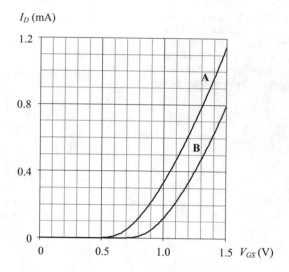

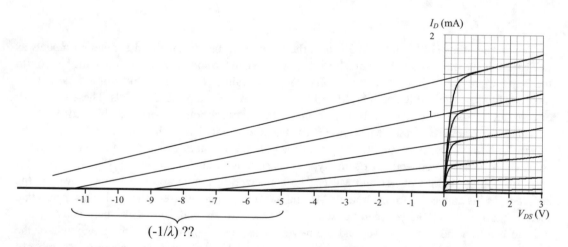

Fig. 1.21 Determination of the λ parameter from the output characteristic curves obtained from SPICE simulation with BSIM-3v3 parameters. Note that it is not possible to find a unique λ parameter value that is valid for all operating points

characteristic curves intersect the horizontal axis at the same point and that the voltage corresponding to this point is $(-1/\lambda)$. This effect is modeled in detail in advanced high-level models. In Fig. 1.21, the output characteristics of an NMOS transistor obtained from SPICE simulation with BSIM-3v3 parameters are shown. The attempt to determine the λ parameter (or similarly the Λ parameter defined in Sect. 1.1.1.) from these curves shows that for a realistic transistor the tangents do not intersect the horizontal axis at the same point, and therefore it is not possible to find a single λ (or Λ) parameter that is valid for all operating conditions.

On the other hand, λ (or Λ) is a useful parameter for hand calculations to determine the DC operating points with better accuracy and to calculate the small-signal parameters corresponding to a certain operating point. One solution would be to determine the value of λ corresponding to the actual operating point from the characteristic curves.

1.1.2.5 Gate Length (*L*) and Gate Width (*W*)

The device parameters (such as mobility, threshold voltage, etc.) reviewed in the previous sections are usually not considered as design parameters; their values are mostly dictated by the process technology, and the circuit designer has little or no influence to set their values. The only significant design freedom is found in choosing the dimensions of the gate, namely the gate length (*L*) and the gate width (*W*). These dimensions are usually called the "drawn" geometries, i.e., the dimensions on the mask layout. Note that the actual dimensions on the chip are always somewhat different from these values, as a result of the lithography and etching steps of fabrication.

It will be shown in the following sections that the high-frequency performance of a MOS transistor strongly depends on the gate length of the transistor; shorter channel lengths provide better high-frequency performance. Consequently, it is wise to use the minimum possible channel length value[9] for a certain technology, if the high-frequency performance has prime importance. It is also known that the small-signal output resistance of a MOS transistor is smaller for shorter channel lengths. But a high output resistance is desired for certain types of circuits, e.g., for current sources. Therefore, a trade-off must be made, and the designer has to decide on the appropriate channel length for each transistor, sometimes following several iterations guided by hand calculations and/or by simulations.

It is known that the saturation drain current is proportional to the aspect ratio (*W/L*) for a non-velocity-saturated MOS transistor and proportional to the channel width (*W*) for a velocity saturated transistor:

$$I_D = I_{Dsat} = \frac{1}{2}\mu_n C_{ox} \frac{W}{L}(V_{GS} - V_{TH})^2$$
$$= \frac{1}{2}KP_n \frac{W}{L}(V_{GS} - V_{TH})^2 \tag{1.8}$$

$$I_{D(v-sat)} = kWC_{ox}(V_{GS} - V_{TH})v_{sat} \tag{1.20a}$$

The gate-source voltage dependence of the drain current, also called the transconductance parameter, is also proportional to *W/L* for a non-velocity-saturated transistor and proportional to the channel width *W* for a velocity saturated transistor:

$$g_{m(sat)} = \frac{dI_D}{dV_{GS}} = \mu_n C_{ox} \frac{W}{L}(V_{GS} - V_{TH}) = \sqrt{2\mu C_{ox}\frac{W}{L}I_D} \tag{1.33}$$

$$g_{m(v-sat)} = kWC_{ox}v_{sat} \tag{1.34}$$

These four expressions indicate that a wider channel provides higher drain current and higher transconductance. But a wide channel means higher area consumption on the chip and higher parasitic capacitances that affect the high-frequency performance of the device. Therefore, there is another design trade-off related to the dimensions of the transistors in a circuit.

For a given process technology (i.e., for a given $KP = \mu C_{ox}$ value), (1.8) can be normalized as

$$K = \frac{I_D}{(KP/2)} = \frac{W}{L}(V_{GS} - V_{TH})^2 \tag{1.35}$$

---

[9] The usable minimum channel length is usually a part of the name of the technology. For example, TSMC018 indicates that for this technology offered by TSMC, the minimum "drawn" channel length is 0.18 micrometers.

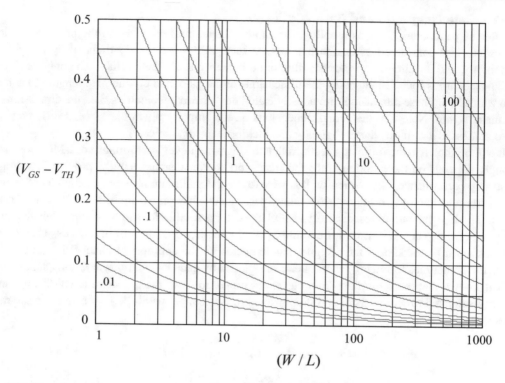

Fig. 1.22 The relation between *(V$_{GS}$ –V$_{TH}$)* and the aspect ratio for different values of *I$_D$*. Parameter *K = I$_D$ / (KP/2)* is plotted from *K = 0.01* to *K = 100* with 1–2-5 intervals. Note that for PMOS transistors the absolute values of *(V$_{GS}$ -V$_{TH}$)* and *K* must be used

In Fig. 1.22, the gate overdrive is plotted against the aspect ratio (*W/L*) for different values of the *K* parameter. From this figure, the necessary aspect ratio corresponding to an "acceptable" gate overdrive can be easily found, for a certain drain current (or vice versa). Note that the gate overdrive voltage is also the saturation voltage of the transistor, and must be "acceptably" small, especially if there are a number of transistors sharing the total DC supply voltage, as in Fig. 1.19b.

1.1.3   Parasitics of MOS Transistors

The "parasitics" of a MOS transistor correspond to all non-intentional and non-avoidable passive or active devices that exist around the MOS transistor. Namely, they are the parasitic capacitances between different regions of the MOS transistor, the resistances associated with the numerous terminals of the device, the p-n junctions (for example the drain-well junction) that are integral parts of the MOS transistor, and similar bipolar structures.[10] These parasitic p-n junctions and bipolar transistors are normally reverse-biased such that their currents are negligibly small. But for VLSI circuits containing millions of MOS transistors, the totality of these reverse-biased junction currents, also called substrate currents, becomes a severe problem.

---

[10] It is known that the so-called latch-up effect in CMOS inverters is a result of the parasitic bipolar transistors that are integral parts of the structure.

1.1.3.1 Parasitic Capacitances

In Fig. 1.23, the parasitic capacitances and resistances of a MOS transistor are shown. These parasitic elements are all geometry dependent. Some of them, for example the junction capacitances, are also technology and bias dependent. In this section, the parasitic capacitances of MOS transistors will be examined under saturation conditions.

The Total Gate-Source Capacitance, C_{gs}:

The gate-source capacitance of a MOS transistor biased in the saturation region is the sum of two components: the gate capacitance (C_g) corresponding to the carrier charge in the inversion layer, which is induced and controlled by the gate-source voltage, and the gate-source overlap capacitance (C_{gso}).

The incremental inversion layer charge of a transistor operating in the saturation region as a function of the channel voltage $V_c(y)$ was obtained as

$$d\overline{Q}_i(y) = -C_{ox}W[(V_{GS} - V_{TH}) - V_c(y)]dy \qquad (1.2)$$

If we insert the channel voltage that was given with (1.6) to (1.2), we obtain

$$d\overline{Q}_i(y) = -C_{ox}W(V_{GS} - V_{TH})\sqrt{1 - \frac{2I_D}{\mu_n C_{ox}W(V_{GS} - V_{TH})^2}y}.dy \qquad (1.36)$$

At the onset of saturation, i.e., $y = L$, (1.36) can be simplified as

$$d\overline{Q}_i(y) = -C_{ox}W(V_{GS} - V_{TH})\sqrt{1 - \frac{y}{L}}.dy \qquad (1.37)$$

If we integrate (1.37) along the channel, we obtain the total inversion charge in the channel as

$$\overline{Q}_i = \frac{2}{3}C_{ox}WL(V_{GS} - V_{TH}) \qquad (1.38)$$

and the corresponding capacitance

$$C_g = \frac{d\overline{Q}_i}{dV_{GS}} = \frac{2}{3}C_{ox}WL \qquad (1.39)$$

Fig. 1.23 The parasitic capacitances and resistances of a MOS transistor

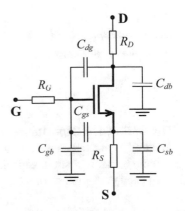

The physical parameters necessary to calculate the value of the gate-source overlap capacitance C_{gso} are usually not readily available. But the value of this capacitance per unit width of the channel (CGSO) is given among the model parameters, usually in [F/m]. Hence the total gate-source capacitance can be written as

$$C_{gs} = \frac{2}{3}C_{ox}WL + C_{gso} = \frac{2}{3}C_{ox}WL + (\text{CGSO} \times W)$$

or for a certain channel length L,

$$C_{gs} = W\left(\frac{2}{3}C_{ox}L + \text{CGSO}\right) = C_{ox}WL\left(\frac{2}{3} + \frac{\text{CGSO}}{C_{ox}L}\right)$$

$$= C_{ox}WL \times k_{ol}$$

(1.40)

Example 1.1

The capacitance-related parameters of typical 0.18-micron technology are:

Gate oxide thickness:	TOX = 4.2E-9	[m]
Gate-source overlap capacitance:	CGSO = 1.21E-10	[F/m]

Calculate the gate-source capacitance of a 20 μm/0.18 μm transistor.

Let us first find the value of the gate capacitance from (1.32a).

$$C_{ox} = \frac{34.5}{4.2} \times 10^{-7} = 8.2 \times 10^{-7} \ \left[\text{F/cm}^2\right]$$

Since TOX and CGSO are given in terms of [m], C_{ox} must be converted to [F/m²]:

$$C_{ox} = 8.2 \times 10^{-3} \ \left[\text{F/m}^2\right]$$

From (1.40)

$$C_{gs} = 8.2 \times 10^{-3} \times \left(20 \times 10^{-6}\right) \times \left(0.18 \times 10^{-6}\right)\left(\frac{2}{3} + \frac{1.21 \times 10^{-10}}{\left(8.2 \times 10^{-3}\right) \times \left(0.18 \times 10^{-6}\right)}\right)$$

$$= 29.5 \times 10^{-15}(0.667 + 0.082) = 22.1 \times 10^{-15} \ \text{F} = 22.1 \ \text{fF}$$

An important piece of information obtained from this example is that the overlap capacitance is more than 10% of the total. This means that the overlap capacitance for short-channel transistors is not a secondary component of C_{gs}, and must not be ignored.

Problem 1.1

Derive an expression for the gate-source capacitance of a MOS transistor operating in the velocity saturation region.

$$\text{Answer}: C_{gs} = C_{ox}WL + (\text{CGSO} \times W)$$

The Drain-Gate Capacitance, C_{dg}

The drain-gate capacitance of a MOS transistor consists of only the drain-gate overlap capacitance. Although this is a small capacitance in magnitude, it has a very important influence on the high-frequency performance of the MOS transistors and will be explained later. Similar to the gate-source overlap capacitance, its value per unit width (usually per meter) is given in the model parameter lists as (CGDO). Therefore, the value of the drain-gate capacitance of a MOS transistor can be found as

$$C_{dg} = (\text{CDGO}) \times W [\text{m}] [\text{farad}] \tag{1.41}$$

The Gate-Substrate Capacitance, C_{gb}

Its value is given as CGBO, i.e., the gate-substrate capacitance per unit length of the channel. Since the channel lengths of transistors are usually small, the gate-substrate capacitance always has a small numerical value compared to the other parasitic capacitances and can be readily neglected for hand calculations.

The Drain-Substrate and Source-Substrate Capacitances, C_{db} and C_{sb}

C_{db} (and similarly C_{sb}) is the total junction capacitance of the drain-substrate (or drain-well) junction, which must be biased in the reverse direction. These junctions have two parts that are different in nature: the bottom junction and the side-wall junctions. It is known from basic p-n junction theory that the capacitance of a reverse-biased junction decreases with the magnitude of the bias voltage. The variation of the capacitance depends on the doping properties of the junction:

$$C_j(V) = C_j(0) \left(1 - \frac{V}{\phi_B} \right)^{-m}$$

where $C_j(0)$ is the value of the capacitance for zero bias, ϕ_B is the built-in junction potential and m is the grading coefficient of the junction, whose value varies between 1/3 to 1/2, depending on the doping profiles.

In the model parameter lists, the value of the junction capacitance for zero bias is given in two parts: (CJ) is the bottom junction capacitance per unit area (usually in farad per square meter) and (CJSW) is the side-wall junction capacitance per unit length (usually in farad/meter). The grading coefficients for these junctions are also given separately as (MJ) and (MJSW). The built-in voltage is represented with (PB).

For hand calculations, this capacitance can be taken into account with some simplifying assumptions. Although somewhat pessimistic, using the zero-bias value of the junction capacitance is a simple but convenient approach.

Example 1.2

The simplified plan view of a MOS transistor fabricated with the AMS 035 micron geometry is given below. The dimensions of the transistor are $L = 0.35$ μm, $W = 20$ μm. The widths of the source and drain regions are typically $X = 0.85$ μm. The related model parameters are given next to the figure. Calculate the source and drain junction capacitances (a) for $V = 0$, (b) for $V = -1$ V.

CJ	9.4 e-4
CJSW	2.5 e-10
MJ	3.4e-1
MJSW	2.3 e-1
PB	6.9 e-1

(a) The bottom area of the junction:

$$A = X \times W = (0.85 \times 10^{-6}) \times (20 \times 10^{-6}) = 17 \times 10^{-12} \, [\text{m}^2]$$

The bottom junction zero bias capacitance:

$$C_j(0) = (\text{CJ}) \times A = (9.4 \times 10^{-4})(17 \times 10^{-12}) = 159.8 \times 10^{-16} [\text{F}] = 15.98 \, [\text{fF}]$$

The total length of the side-wall junction:

$$L_{SW} = 2 \times (X + W) = 2 \times (0.85 \times 10^{-6} + 20 \times 10^{-6}) = 41.7 \times 10^{-6} [\text{m}]$$

The total side-wall capacitance:

$$C_{jsw}(0) = (\text{CJSW}) \times L_{SW} = (2.5 \times 10^{-10}) \times (41.7 \times 10^{-6}) = 10.425 \, [\text{fF}]$$

The total zero-bias junction capacitance:

$$C_{jT}(0) = C_j(0) + C_{jsw}(0) = 26.4 \, [\text{fF}]$$

(b) The bias-dependent factor for the bottom capacitance for 1 V reverse bias:

$$\left(1 - \frac{(-1)}{0.69}\right)^{-0.34} = 0.737$$

The bottom junction capacitance for 1 V reverse bias:

$$C_j(-1) = 15.98 \times 0.737 = 11.78 \, [\text{fF}]$$

The bias-dependent factor for the side-wall capacitance for 1 V reverse bias:

$$\left(1 - \frac{(-1)}{0.69}\right)^{-0.23} = 0.814$$

The side-wall junction capacitance for 1 V reverse bias:

$$C_{jsw}(-1) = 10.425 \times 0.814 = 8.48 \, [\text{fF}]$$

The total junction capacitance for 1 V reverse bias:

$$C_{jT}(-1) = 11.78 + 8.48 = 20.26 \, [\text{fF}]$$

That is 6.14 fF (23.3%) smaller than the zero-bias value.

1.1.3.2 The High-Frequency Figure of Merit: f_T

An important definition for a MOS transistor that is related to the parasitic capacitances of the device is the "high-frequency figure of merit," f_T. It is known that the low-frequency input current of a MOS transistor is practically zero, and consequently, the low-frequency current gain is infinite. At higher frequencies, on the other hand, the capacitive current that flows into the gate terminal becomes non-negligible. Hence, the current gain decreases at high frequencies. f_T is defined as the frequency for which the magnitude of the current gain is equal to unity.

The capacitive (small signal) input current of a MOS transistor can be expressed as

$$i_i = v_{gs}.j\omega C_{gs}$$

where C_{gs} is the total gate-source capacitance of the transistor.

The maximum value of the output signal current can be written from the definition of the transconductance, as $i_o = g_m v_{gs}$. From these expressions, the current gain can be written as

$$A_i = \frac{i_o}{i_i} = \frac{g_m}{j\omega C_{gs}}, \quad |A_i| = \frac{g_m}{\omega C_{gs}}$$

Consequently, f_T is found as

$$f_T = \frac{1}{2\pi} \frac{g_m}{C_{gs}} \tag{1.42}$$

It is instructive to compare this expression for a non-velocity saturated and a velocity saturated transistor.

For a non-velocity saturated transistor, the input capacitance that was given in (1.40) can be arranged as

$$C_{gs} = WLC_{ox} \left(\frac{2}{3} + \frac{CGSO}{LC_{ox}} \right)$$

The transconductance of a non-velocity saturated transistor was given as (1.33). From (1.40) and (1.33) f_T can be arranged as

$$f_T = \frac{1}{2\pi} \frac{\mu(V_{GS} - V_{TH})}{L^2 \left(\frac{2}{3} + \frac{CGSO}{LC_{ox}} \right)} \tag{1.43}$$

and in terms of the drain current:

$$f_T = \frac{1}{2\pi} A \sqrt{\frac{I_D}{W}} \tag{1.43a}$$

where A is a technology-dependent parameter and can be calculated as

$$A = \sqrt{\frac{2\mu}{k_{ol}^2 C_{ox} L^3}} \tag{1.43b}$$

This expression indicates that for a non-velocity saturated transistor

- f_T increases with mobility. Therefore, NMOS transistors exhibit better high-frequency performance.
- f_T increases with the gate bias voltage (gate overdrive) and consequently with the DC current of the transistor.
- f_T strongly depends on the gate length. Therefore, short-channel devices are better for high-frequency applications.

For a velocity saturated transistor, using the transconductance expression given in (1.34) and the gate-source capacitance found in Problem 1.1, f_T can be expressed as

$$f_T = \frac{1}{2\pi} \frac{k v_{sat}}{L\left(1 + \frac{CGSO}{LC_{ox}}\right)} \tag{1.44}$$

From this expression, we conclude that:

- Since the saturation velocity of holes is only slightly smaller than that of electrons, the high-frequency performance of velocity saturated PMOS transistors is comparable to that of NMOS transistors.
- For a velocity saturated transistor, f_T is independent of the DC operating conditions.

1.1.3.3 The Parasitic Resistances

The effects of the series source (or drain) resistances of small geometry transistors can be examined in two parts: the intrinsic resistance and the extrinsic resistance. The intrinsic resistance is the sum of (i) the resistance of the accumulation layer in the source region induced by the gate voltage, and (ii) the spreading resistance from the accumulation layer to the bulk of the source [9–11]. This component (spreading resistance) varies with the gate voltage. The extrinsic source (or drain) resistance is composed of the source region resistance and the resistances of the related silicon-to-metal contacts.

All these components are modeled in high-level simulation models and the associated model parameters are provided by the manufacturers. In these models, the source and drain series resistance are usually evaluated and modeled together as a total series resistance [12], assuming that the structure is symmetrical and the source (or drain) series resistance is equal to one half of the total series resistance. It must be stressed that for analog design, these two parasitics have to be considered as separate resistors. Although the magnitudes of these resistors are equal, their effects on the behavior of the device in a circuit are not the same. The obvious effect of the series drain resistance of a MOS transistor is the increase in the saturation voltage, due to the voltage drop on this resistance. The different and important effects of the series source resistance are the decrease of the drain current corresponding to a certain gate-source voltage and the decrease of the transconductance, as investigated in Chap. 2.

The plan view of a MOS transistor is shown in Fig. 1.24a. The parasitic series resistance of the source (or drain) of a transistor is the sum of three components:

The intrinsic resistance (R_{Si}), the extrinsic resistance (R_{Se}), and the contact resistance (R_{cont}), which are shown in Fig. 1.24b. It must be kept in mind that the total contact resistance is the parallel equivalent of all individual contact resistances.

The numerical value of the source (or drain) resistance of a MOS transistor can be obtained using the model parameters given in BSIM3v3 parameter lists. In BSIM3v3, the parameter related to the

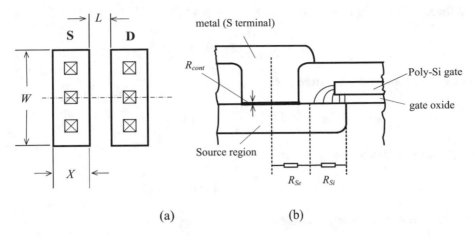

Fig. 1.24 (a) Plan view of the source and drain regions of a MOS transistor. (b) The components of the source resistance: $R_S = R_{Si} + R_{Se} + R_{cont}$

intrinsic part of the series resistance (RDSW) is given as the total series resistance per micron width of the transistor. For hand calculation purposes, the gate voltage dependence of this resistor can be ignored and the approximate value of the series intrinsic source (or drain) resistance "per 1-micron width" of the source (or drain) can be found with

$$R_{Si} = \frac{1}{2} \frac{(\text{RDSW})}{W[\mu\text{m}]} \tag{1.45}$$

The total extrinsic resistance is modeled with (RSH), the sheet resistances of the source and drain regions. If the contacts are placed at a distance of (X/2) from the source (Fig. 1.24a), the resistance from the edge of the source to the line of contacts can be calculated with

$$R_{Se} = \frac{1}{2} (\text{RSH}) \times (\text{nrs}) \tag{1.46}$$

where (nrs) is defined by geometry and is equal to $X/(N.\,W)$, where N is the number of "fingers" of the transistor, which will be explained later on.

The other component of the extrinsic source resistance is the equivalent contact resistance (R_{cont}), which is inversely proportional to the number of parallel-connected contacts. The value of a typical contact resistance is given by the manufacturer.

Example 1.3
Let us calculate the source series resistance of an AMS035 PMOS transistor as shown in Fig. 1.24. The dimensions are $L = 0.35$ μm, $W = 5$ μm, and $X = 0.85$ μm.

The available parameter values from the datasheets of AMS are as follows:

RDSW	1.033e+03
RSH	1.290e+02
Contact res.	60 ohm per 0.4 μm × 0.4 μm contact from p-diff. to metal-1.

The intrinsic component of the source resistance from (1.45)

$$R_{Si} = \frac{1}{2} \frac{1033}{5[\mu m]} = 103.3 \text{ ohm}$$

Source region body resistance from (1.46)

$$R_{Se} = \frac{1}{2} 129 \times \left(\frac{0.85}{5}\right) = 10.96 \text{ ohm}$$

The equivalent contact resistance

$$R_{Sc} = \frac{60}{3} = 20 \text{ ohm}$$

Hence the total series source resistance is

$$R_S = 103.3 + 10.96 + 20 = 134.26 \text{ ohm}$$

Problem 1.2 *The parasitic resistance-related parameters for AMS035 NMOS transistors are given with their typical and maximum values as follows:*

RDSW   345
RSH   75 typical, 85 maximum
Contact res.   *30 ohm per 0.4 μm × 0.4 μm contact, typical, 100 ohm maximum.*

The dimension of the transistor is 20 μm/0.35 μm. The number of contacts on source and drain regions is 10.

Calculate the "worst-case" source and drain parasitic series resistances.

Since the DC gate current is negligible, the gate resistance of a MOS transistor has no effect on the DC (and low frequency) performance. But at high frequencies, due to the gate capacitance, the gate resistance affects the high-frequency performance of the device. The thermal noise of the gate series resistance is another factor that must not be overlooked.

It must be noted that the gate capacitance and the resistance of the gate conductor form a distributed R-C line along the width of the channel (See Fig.1.25). The gate signal applied from one end of the gate electrode propagates along the channel width, until it reaches the other end of the gate R-C transmission line [13].

For analog applications, it is interesting to investigate the gate (input) signal attenuation along the channel width, as a function of the frequency. The [y] parameters of a passive two-port R-C line are given in [14] as hyperbolic functions of $(l\sqrt{s.rc})$, where s is the complex frequency, l the length of the line, and r and c are the resistance and capacitance per unit length. From these expressions, the voltage transfer coefficient of the line from its input to its unloaded output port can be found as

$$A = \frac{v_o}{v_i} = \frac{1}{\cosh\left(l\sqrt{s.rc}\right)} \quad (1.47)$$

The magnitude of A is obviously equal to unity for low frequencies along the line. This means that the voltage at the end of the line is equal to the input voltage at low frequencies, but the transfer

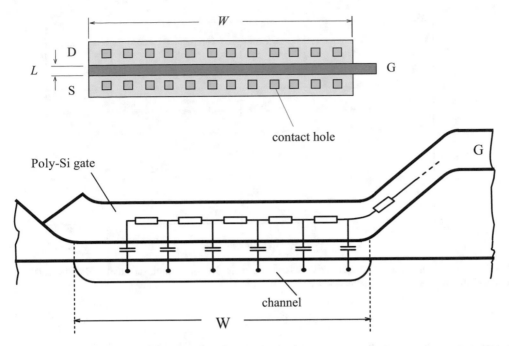

Fig. 1.25 The plan (top) view of a MOS transistor and the schematic representation of the distributed R-C delay line composed of the gate resistance and the gate capacitance

coefficient decreases with the frequency. We can calculate the frequency[11] at which the magnitude of A decreases to $\left(1/\sqrt{2}\right)$.

From

$$\cosh\left(l\sqrt{s.rc}\right) = \sqrt{2}$$

which corresponds to

$$\left(l\sqrt{s.rc}\right) = 0.882 \text{ or } \left(s.l^2rc\right) = 0.78 \tag{1.48}$$

This expression can be applied to the gate *R-C* line of a MOS transistor, provided that the conductivity of the channel is sufficiently high, i.e., the channel and the bulk can be considered as the "ground" with respect to the gate conductive strip. For this case, the total resistance and the total capacitance of the gate electrode are $R = r \times W$ and $C = c \times W$, respectively. R and C can be expressed, in terms of the sheet resistance of the gate electrode and C_{ox}, as

$$R = R_{sh}\frac{W}{L} \quad \text{and} \quad C = C_{ox}WL \tag{1.49}$$

Therefore,

---

[11] This frequency will be called the maximum usable frequency. Note that it is necessary to evaluate this frequency together with the high-frequency figure of merit f_T of the transistor.

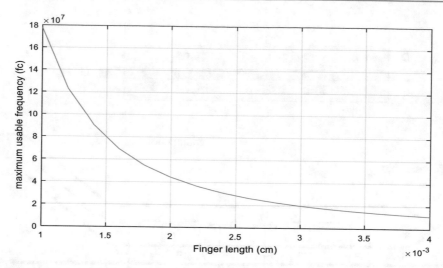

Fig. 1.26 The maximum usable frequency as a function of the RC line length (channel width) for a 0.18-micron technology having $R_{sh} = 8$ ohm/square and 8.625×10^{-7} F/cm^2

$$rc = R_{sh}C_{ox} \tag{1.50}$$

From these expressions, the complex frequency corresponding to $A = \left(1/\sqrt{2}\right)$ can be calculated as

$$s_c = \frac{0.78}{R_{sh}C_{ox}W^2}$$

and the corresponding frequency as

$$f_c = \frac{0.78}{2\pi \times R_{sh}C_{ox}W^2} = \frac{0.124}{R_{sh}C_{ox}W^2} \tag{1.51}$$

The maximum usable frequency for UMC 0.18 micron technology is calculated from (1.51) and plotted in Fig. 1.26 as a function of frequency. For example, if the channel width is more than 20 μm and the gate signal is driven from one end, the usable frequency does not exceed 4 GHz. It is obvious from these observations that the gate width has a strong adverse effect on the usable maximum frequency of a transistor. In addition to the decreasing magnitude of the line transfer function, there is certainly an associated phase shift and corresponding signal delay. This means that the control efficiency of the channel decreases from the near-end of the gate (connected to the gate contact) toward the far end, which results in the decrease of the effective g_m of the transistor.

It is also instructive to investigate the step response of the gate R-C line. The response to a unity voltage step at its open end of an R-C transmission line is given as [15]

$$v(x,t) = \mathrm{erfc}\left[\left(\frac{x}{l}\right) / \sqrt{\frac{4t}{rcl^2}} \right]$$

where r and c are the parameters of the line and l represents the length of the R-C line. This expression can be simplified as

$$v(x,t) = \mathrm{erfc}\left[\frac{x\sqrt{rc}}{2\sqrt{t}} \right] = 1 - \mathrm{erf}\left[\frac{x\sqrt{rc}}{2\sqrt{t}} \right] \tag{1.52}$$

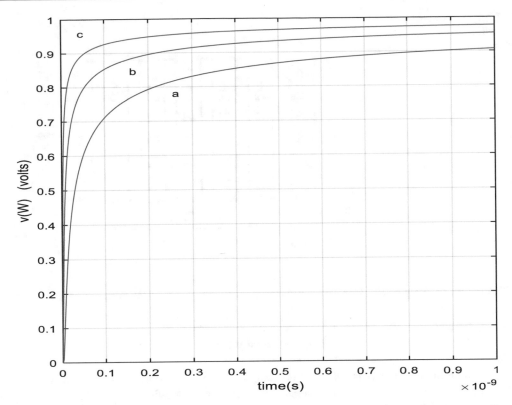

Fig. 1.27 The voltage at the end of the gate strip as a function of time for a 1 V voltage step applied between the gate and the source contacts. (**a**) corresponds to $W = 20\ \mu m$, (**b**) corresponds to $W = 10\ \mu m$, and (**c**) corresponds to $W = 5\ \mu m$

For the case of the gate R-C line of a MOS transistor, l corresponds to the channel width (W) and using (1.50), (1.52) can be arranged as

$$v(x,t) = 1 - \text{erf}\left[\frac{x\sqrt{R_{sh}C_{ox}}}{2\sqrt{t}}\right] \tag{1.53}$$

As an example for the UMC 0.18 micron technology, the voltages appearing at the end of 20 μm, 10 μm, and 5 μm gate strips for unity amplitude step inputs as functions of time are plotted in Fig. 1.27.

This example shows that the gate input signal takes a considerable time to reach the far end of the gate strip, depending on the width of the finger. This means that the channel will not be turned on uniformly, and consequently, the drain current will reach its expected value with a considerable delay. For the examined transistor, the gate voltage at the end of the finger reaches the 90% level of its final value in 0.9 ns for $W = 20\ \mu m$, in 0.2 ns for $W = 10\ \mu m$, and in 0.05 ns for $W = 5\ \mu m$. Also note that the slope of the voltage ramp decreases with the length of the gate strip, which may result in uncertainty of the turn-on behavior of certain classes of circuits (see 6.3.3).

The well-known solution to overcome these problems is to use a finger-gate structure as shown in Fig. 1.28, resulting in a transistor with a large effective channel width W which consists of multiple channel sections (so-called fingers).[12] The advantages of the finger structure can be summarized as follows:

---

[12] In this case the "channel width" axis in Fig. 1.26 must be understood as "finger length".

Fig. 1.28 Plan view of a
multi-finger MOS
transistor

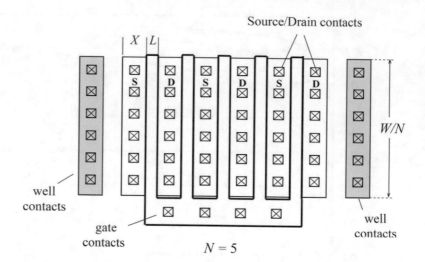

- The length of the gate RC line on each of the fingers is equal to W/N, where N is the number of fingers
- The f_c frequency of the transistor increases with the square of N
- The finger structure is "layout friendly," especially for high W values
- An odd number of fingers is better for symmetry
- An even number of fingers is better for reducing the drain (and source) parasitic capacitances
- The total source (or drain) area is smaller than that of a one-finger structure, decreasing the junction parasitic capacitances
- The source (or drain) area reduction factor is $\delta = \frac{A_F}{A} = \frac{1}{2} + \frac{1}{2N}$ which approaches ½ for a higher number of fingers
- The total source (or drain) parasitic series resistance is smaller

1.2   Passive On-chip Components

To enable the integration of a complete analog circuit or system on a single chip, the passive components – namely resistors, capacitors, and inductors compatible with the standard MOS transistor fabrication technologies – must be available. To improve the quality of passive components, it may become necessary to add extra steps to the standard fabrication process, which certainly increases the complexity of the process and consequently, the price. Figure 1.29 shows the cross-section of a typical CMOS chip with some of the passive components: a poly 2 resistor, a poly 1-poly 2 capacitor, and a metal-insulator-metal (MIM) capacitor. The on-chip inductors are usually formed using the uppermost thick metal layers.

1.2.1   On-chip Resistors

On-chip resistors can be realized by using thin conductive films or doped silicon layers. One possibility for the thin films is to use the metal (usually aluminum) films that are normally used for interconnections. The sheet resistances of thin metal films are in the order of 50...100 mΩ/□. It is

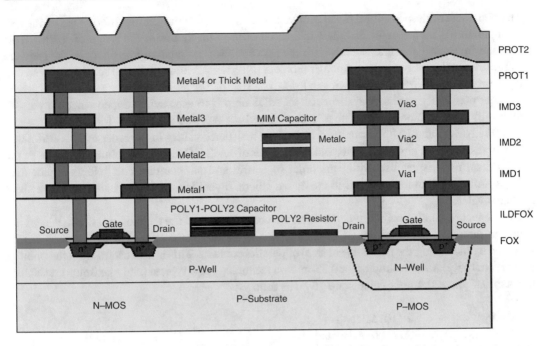

Fig. 1.29 The partial cross-section of a 0.35 micron, four metal layer CMOS chip. Some of the passive components are shown together with the NMOS and the PMOS transistors. (Courtesy of AMS)

Fig. 1.30 Plan view of a serpentine shaped polysilicon resistor. Note that $N_{sq} = 22$, $N_{cor} = 4$ and $N_c = 2$

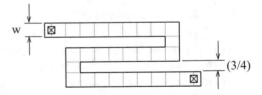

obvious that the metal films are only suitable for very low-value resistors, and are rarely used in ICs. The more important and frequently used option is the highly doped poly-silicon films that are normally used as the gate material of MOS transistors. For small geometries, they are usually silicided to decrease the resistance. The sheet resistance of the gate poly-silicon films is usually in the range of $5 \ldots 10 \, \Omega/\square$. The plan view of a thin-film resistor is shown in Fig. 1.30. The value of such a resistor can be calculated as

$$R \cong R_{sh}\left(N_{sq} + 0.6 \times N_{cor}\right) + 2R_c$$

where R_{sh} is the sheet resistance of the conductive film in ohms per square, N_{sq} is the count of normal squares along the resistor, N_{cor} is the count of corner squares and R_c is the resistance of the end contacts.

Using the gate polysilicon, resistor values of up to a few kilo-ohms can be realized. For a higher resistor value, the gate polysilicon is not convenient due to the adverse effects of the distributed parasitic capacitance of the resistor and excessive area consumption. For analog applications, IC manufacturers usually provide an additional high resistivity polysilicon film. The sheet resistance of this polysilicon layer is usually in the range of $1..2 \, k\Omega/\square$.

There are several important practical problems associated with the film resistors. One of them is the manufacturing tolerance of the sheet resistance of the film that is in the range of $\pm20\%$. This corresponds to $\pm20\%$ absolute tolerance for the resistors. But the relative tolerances of the resistors on a chip are considerably smaller. Another problem is the temperature coefficient of the resistors that is in the order of $+10^{-3}$/K for gate-poly and -0.5×10^{-3}/K for high-resistance poly.

Another possibility to realize resistors on a CMOS chip is to use several doped silicon layers, as were usually done in bipolar ICs. The n^+ and p^+ doped regions that are used to form the source and drain of the NMOS and PMOS transistors have sheet resistance values in the order of $50 \ldots 150 \, \Omega/\square$. The sheet resistance of a well is higher and in the order of $1 \ldots 2 \, k\Omega/\square$. The absolute tolerances of the doped silicon regions are usually in the order of $\pm10\%$ and temperature coefficients higher than $+10^{-3}$/K. Another drawback related to the doped silicon layers is the bias dependence of the sheet resistance values.[13]

The contact resistance certainly depends on the contact size and the technology. Contact sizes are usually in the order of the minimum geometry. The series resistance related to the contacts can be decreased by increasing the number (not size) of parallel contacts. This is also useful from the point of view of reliability. As an example, for 0.35-micron technology, the typical (and maximum) resistance values of 0.4 µm $\times$ 0.4 µm contacts are given as follows:

Metal-poly	2 (10) Ω/contact
Metal-S/D (n-type)	30 (100) Ω/contact
Metal-S/D (p-type)	60 (150) Ω/contact

1.2.2 On-chip Capacitors

There are two classes of on-chip capacitors: passive, fixed value capacitors and the variable capacitors (varactors). The passive capacitors are – in principle – basic parallel plate capacitors. Varactors are solid-state structures such as reverse biased p-n junctions and MOS capacitors. A varactor obviously can be used as a fixed value capacitor if it is biased with a fixed bias voltage.

1.2.2.1 Passive On-chip Capacitors

Two options to form parallel plate capacitors can be seen in Fig. 1.29. These options are the poly1-poly2 capacitor and the metal-insulator-metal (MIM) capacitor. The poly1-poly2 capacitors have to be realized at the bottom level of the structure. The MIM capacitors can be realized at any metal level in principle, but the practically possible level (or levels) for a certain technology is usually declared by the manufacturer.

The capacitance of a parallel plate capacitor is given as

$$C = \frac{A}{t} \varepsilon_0 \varepsilon_{ox}$$

where ε_0 and ε_{ox} are the permittivity of the vacuum and dielectric constant of the insulating layer, respectively, A is the area of the parallel plates and t is the distance between the conducting plates that

---

[13] Note that the diffusion region used as a resistor must not have a silicide layer on top, which would reduce the overall resistance to unusable levels. Many layout design tools provide the "silicide block" feature to prevent the formation of a silicide layer on top of the resistor areas.

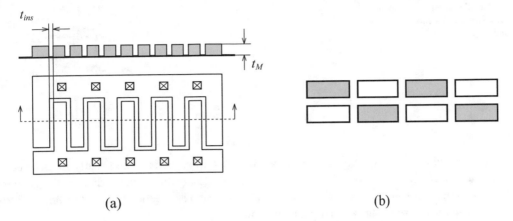

Fig. 1.31 (**a**) The plan view and the cross-section of an interdigitated capacitor. (**b**) The cross-section of a two-layer finger structure

is filled with the insulating material, usually silicon dioxide. Inserting the numerical values of ε_0 and ε_{ox}, and expressing the area in square microns and t in nanometers, the capacitance can be found as

$$C = \frac{34.5}{t[nm]} \ [\text{fF}] \tag{1.54}$$

Due to the fringe-field effect at the periphery of the capacitor, the accuracy of this expression decreases for smaller dimensions.

In poly and MIM capacitors, the electric field between the electrodes is perpendicular to the surface. To increase the surface usage efficiency, structures using the lateral electric field are developed. The simplest example for a lateral field (or interdigitated) capacitor is shown in Fig. 1.31. The capacitance of this structure is proportional to the thickness of the metal layer (t_M) and the length of the serpentine-shaped insulating region (dielectric) between the two metal electrodes, and inversely proportional to the width of the insulator (t_{ins}). It is obvious that the most suitable metal layer is the thick metal layer. Several structures using both the lateral and the perpendicular field can be developed to further increase the surface usage efficiency [16, 17].

The Parasitics of Passive On-chip Capacitors

The dominant parasitics of an on-chip capacitor are shown in Fig. 1.32, where C is the intended capacitance, C_{par1} and C_{par2} are the parasitic capacitances between the plates of the capacitor and ground, and R_s is the equivalent series resistance representing the total losses. For vertical field capacitors, for example the poly1-poly2 capacitor, the parasitic capacitance of the bottom plate is considerably bigger than the parasitic of the upper plate, and in some cases not negligibly small compared to C. In horizontal field capacitors, for example the capacitor shown in Fig. 1.31, the parasitic capacitances associated with the two electrodes are approximately equal, which is an advantage in certain cases.

The equivalent parasitic series resistance comprises the resistances of the contacts and of the plates. The latter is a distributed resistor in two dimensions; it is frequency-dependent and complicated to calculate. The overall effect of this resistance is usually represented with the quality factor

Fig. 1.32 The dominant parasitics of an on-chip capacitor

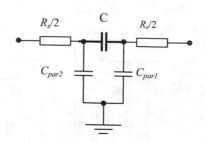

(Q) of the capacitor.[14] The quality factors of passive on-chip capacitors are usually in the range of 30...100 @ 1 GHz. It must be noted that at higher frequencies, certainly depending on the geometry of the structure, the self-inductances of the plates and associated connections start to become effective and the capacitor can resonate with its parasitic inductance.

1.2.2.2 Varactors

The varactor is a voltage-tunable capacitor which is mostly used in high-frequency applications, to tune the frequency characteristics of circuits. One of the important classes of varactors is a reverse-biased p-n junction.[15] It is known from basic semiconductor device physics that the depth (and consequently the electric charge) of the depleted regions of a p-n junction varies with the applied voltage. This means that the amount of charge in the depletion regions can be controlled with the voltage, which, by definition, corresponds to an electrostatic capacitance. The value of this capacitance can be expressed, as already mentioned in Sect. 1.1.3.1, as

$$C_j(V) = C_j(0)\left(1 - \frac{V}{\phi_B}\right)^{-m} \tag{1.55}$$

where $C_j(0)$ is the value of the capacitance for zero bias, ϕ_B is the built-in junction potential, and m is the grading coefficient of the junction whose value varies between 1/3 to 1/2, depending on the doping profiles. $C_j(0)$ depends on the junction area and the doping concentrations of the p-n junction. The junction barrier height ϕ_B is in the order of 0.65 ... 0.75 V and is given as (PB) in parameter lists, as already mentioned. There are several options to form a p-n varactor in a CMOS IC. The cross-section of one of them and the typical voltage-capacitance curve are shown in Fig. 1.33. The tuning range[16] depends on the range of the control voltage; its minimum value is zero and the maximum value is limited with the breakdown voltage of the junction and the available DC voltage value on the chip.

It is known from MOS theory that the gate-bulk capacitance of a MOS structure exhibits a certain variation. In Fig. 1.34, the variation of the capacitance of a PMOS transistor as a function of the DC voltage applied between the gate electrode and the bulk is shown. In one extreme case, where the gate voltage is sufficiently more negative than the threshold voltage V_{TH} (which is negative for a PMOS

---

[14] The quality factor of a capacitor at a certain frequency ω, can be defined as $Q = 1/R_s C\omega$, if the capacitor is in resonance at ω with an ideal (lossless) inductance (See Chap. 4).

[15] The p-n junction type of variable capacitors are extensively used, not only as an integral part of bipolar and MOS ICs, but also as discrete components acting as the tuning elements of radio and TV receivers and telecommunication systems.

[16] The tuning range is defined as $\gamma = (C_{max} - C_{min})/(C_{max} + C_{min})$

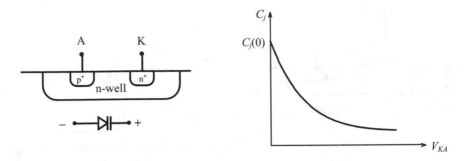

Fig. 1.33 The cross-section and the typical control voltage versus capacitance curve of a p-n junction varactor

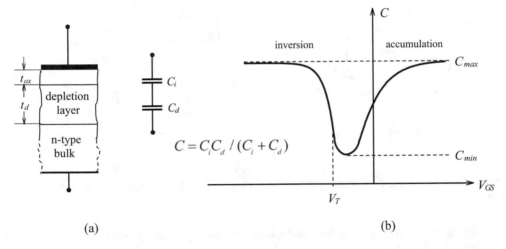

Fig. 1.34 (a) The oxide and depletion capacitance components of a MOS capacitor. (b) Typical C-V curve of a PMOS transistor

transistor), there is a strong inversion layer that forms on the lower "plate" of the MOS capacitor. The capacitance is equal to $C_i = WLC_{ox}$. In the other extreme, if the gate voltage is sufficiently more positive than the threshold voltage, the attracted majority carriers (electrons in this case) of the n-type bulk forms a conductive n-type accumulation layer. The capacitance corresponding to this case is again equal to C_i. In between these two extreme cases, for a narrow interval of the gate voltage, the semiconductor surface just below the gate oxide is – practically – depleted from the electrons and holes. The thickness of this depletion layer and consequently the fixed ion charge is subject to the control of the gate voltage, which corresponds to the capacitance $C_d = dQ_d/dV_G$. For this interval the total gate-bulk capacitance is the series equivalent of C_i and C_d: $C = C_iC_d/(C_i + C_d)$. To change the capacitance characteristic given in Fig. 1.34 to a more useful form that permits the control of capacitance monotonically in one direction, some modifications are possible.

One of the solutions compatible with standard CMOS technologies is shown in Fig. 1.35a. In this structure, the n-well (the bulk of the transistor) is connected to the highest positive potential in the circuit. Consequently, it is not possible to drive the device into the accumulation region. The source and drain regions are connected to each other. If the gate voltage is more negative than the threshold voltage, the device is in inversion mode and the capacitance is equal to C_i. This is the maximum value of the capacitance. For gate voltages that are sufficiently higher than the threshold voltage, inversion

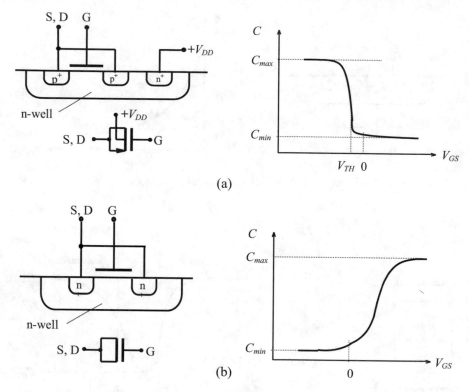

Fig. 1.35 Cross-section and the *C-V* curve of (**a**) an inversion type MOS varactor, (**b**) an accumulation type MOS varactor

is not possible, and the device is in depletion mode. The corresponding capacitance is at its minimum value and is equal to $C = C_i C_d/(C_i + C_d)$.

The second solution is shown in Fig. 1.35b. The bulk (well) again is n-type. For gate voltages higher than the threshold voltage, an electron-rich accumulation layer is formed and connects the two high concentration n-type regions. Since this high conductivity layer is just below the oxide layer, the capacitance is equal to the oxide capacitance, $C_i = WLC_{ox}$. When the gate voltage decreases toward negative values, the accumulated electrons and then the majority electrons of the n-type bulk are repelled to form a depletion layer. For sufficiently low gate voltages, the capacitance is equal to $C = C_i C_d/(C_i + C_d)$, and varies monotonically between these two extreme values.

For a certain technology – with a certain gate oxide thickness and doping concentration of the n-well – the minimum and maximum values of the capacitance are same for the inversion type and accumulation type MOS varactor. Since the surface consumption is smaller, usually the accumulation type is preferred. The tuning ranges and the quality factors of the MOS varactors are usually in the range of 40...60% and 50...80 @ 1GHz, respectively.

An important problem associated with the varactors is the nonlinearity of the *C-V* curve. For small signals superimposed on the bias voltage, the capacitance of the device has a value corresponding to this bias. It has been shown that for large signal amplitudes – in addition to a nonlinear distortion – the effective capacitance of the device decreases [18].

In Fig. 1.36, several examples are given related to the biasing of varactors. Figure 1.36a shows the biasing of a p-n junction varactor that forms a variable capacitor whose one end is grounded. The

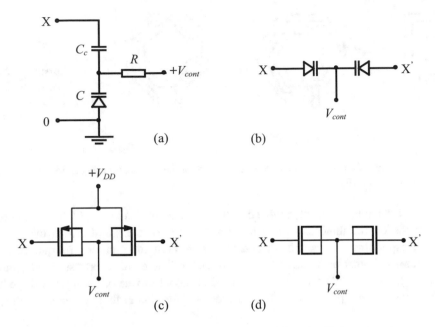

Fig. 1.36 Biasing of varactors. Biasing of: (**a**) a one-end grounded varactor, (**b**) a p-n junction type symmetrical varactor pair, (**c**) an inversion type MOS symmetrical varactor pair, (**d**) an accumulation type of MOS symmetrical varactor pair

coupling capacitor C_c must be as high as possible to profit from the full tuning range of the varactor. The bias resistor R must be high enough not to affect the Q of the varactor.

In many applications, varactors are used as floating tuning elements. In the examples shown in Fig. 1.36b, c, and d [19], the nodes X and X' have the same DC voltage. For proper operation of the varactors in Fig. 1.36 (b), the control voltage V_{cont} must remain positive with respect to V_X. The control voltage in Fig. 1.36c must be positive with respect to V_X. In Fig. 1.36d, V_{cont} can be changed from negative to positive voltages with respect to V_X.

1.2.3   On-chip Inductors

Throughout the history of electronic and telecommunication systems, inductors have always been among the most bulky and expensive passive components of the circuits. In the case of integrated circuits, the situation was even worse: it was completely impossible to integrate larger-valued inductors as on-chip components. The goal of most of the active filters developed during the last five decades was to eliminate the inductors and to replace them with capacitors and active circuit combinations. Thanks to earlier developments in microelectronics technology, which resulted in operational amplifiers and OTAs, active filters have effectively eliminated the inductors from many application areas. Eventually, the maximum operation frequencies of systems increased up to the "gigahertz" region, where it was not possible to apply the conventional active filter techniques. But in parallel to the increase of the frequency, the necessary inductance values decreased down to the range of "nanohenries," which is small enough to be integrated on-chip.

On-chip inductors are usually shaped by etching the uppermost thick metal layer that is usually aluminum in standard technologies and has a thickness of around one micron. Copper is an alternative with a higher conductivity, and is provided by some foundries.

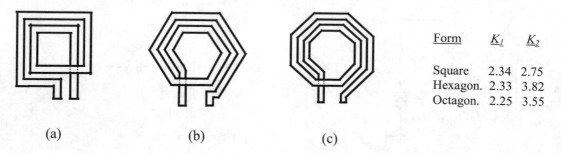

Form	K_1	K_2
Square	2.34	2.75
Hexagon.	2.33	3.82
Octagon.	2.25	3.55

(a)   (b)   (c)

Fig. 1.37 Typical forms of on-chip spiral inductors. (**a**) square, (**b**) hexagonal, (**c**) octagonal inductors. Inset: K_1 and K_2 parameters for these forms

The on-chip inductors are mostly realized in the forms shown in Fig. 1.37. The value of such an inductor depends on the dimensions and number of turns of the spiral. Since the length of the conductor corresponding to a certain loop area is minimum for a circle, the quality factor of a polygon inductor is higher than that of a square inductor. There are several simple formulas in the literature to calculate the inductance of a planar spiral inductor. One of them, named the "modified Wheeler formula," which provides reasonable accuracy for simple first step calculations, is given in [20]:

$$L = K_1 \mu_0 \frac{n^2 d_{avg}}{1 + K_2 \rho} \tag{1.56}$$

where K_1 and K_2 are form-dependent coefficients that are given in Fig. 1.37; μ is the magnetic permeability of the material, which is equal to $\mu_0 = 1.257 \times 10^{-8}$ [henry/cm] for silicon and is equal to the permeability of vacuum; and n is the number of turns of the spiral. The average diameter, d_{avg} and the fill factor, ρ are defined as follows:

$$d_{avg} = \frac{d_{out} + d_{in}}{2}, \qquad \rho = \frac{(d_{out} - d_{in})}{(d_{out} + d_{in})} \tag{1.57}$$

where d_{out} and d_{in} are the outer and inner diameter of the inductor, respectively. The practically realizable range of the on-chip inductors is $L = 0.5 \ldots 10$ nH, and the corresponding quality factors range between $Q = 3 \ldots 15$.[17]

Example 1.4
Calculate the self-inductance of the spiral inductor shown in Fig. 1.38, and the quality factor at 10 GHz assuming that the only loss is related to the DC resistance of the aluminum strip.

The outer diameter of the coil is 100 µm. The inner diameter can be calculated as $d_{in} = 100 - (4 \times 10 + 2 \times 5) = 50$ µm. Then the average diameter and the fill factor are:

$$d_{avg} = 0.5(100 + 50) = 75 \mu m, \qquad \rho = (100 - 50)/(100 + 50) = 1/3$$

Using the values given in Fig. 1.37:

$$L = 2.34 \times \left(1.275 \times 10^{-8}\right) \frac{(2^2) \times (75 \times 10^{-4})}{1 + (2.75 \times 0.333)} = 460 \times 10^{-12} = 0.46 \text{ [nH]}$$

---

[17] The quality factor of an inductor at a certain frequency ω, can be defined as $Q = L\omega/R_s$, if all losses are represented with a series resistor R_s and the inductor is in resonance at ω with an ideal (lossless) capacitor (See Chap. 4).

Fig. 1.38 Plan view of a
typical square inductor

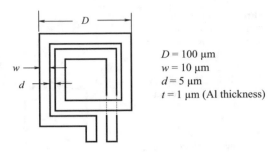

$D = 100\ \mu m$
$w = 10\ \mu m$
$d = 5\ \mu m$
$t = 1\ \mu m$ (Al thickness)

Fig. 1.39 The equivalent
circuit of an on-chip
inductor connected
between the nodes A
and A'

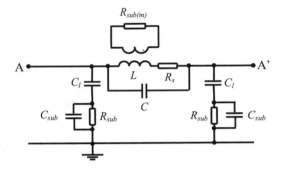

The average length of the aluminum strip is $l = 620\ \mu m$ (620×10^{-4} cm) and the cross-section $A = (10 \times 10^{-4}) \times (1 \times 10^{-4}) = 10^{-7}$ cm^2. The specific resistance of aluminum is 2.73×10^{-6} ohm. cm, now the resistance of the coil can be calculated as

$$R_s = \left(2.73 \times 10^{-6}\right)\frac{620 \times 10^{-4}}{10^{-7}} = 1.69\ \text{ohm}$$

With the assumption that the series resistance represents the only loss of the inductor, the quality factor at $f = 10$ GHz is

$$Q = \frac{L\omega}{R_s} = \frac{\left(0.46 \times 10^{-9}\right)\left(2\pi \times 10^{10}\right)}{1.69} = 17.1$$

There are several computer programs developed to calculate the inductance and quality factor of on-chip inductances with high precision [21, 22]. One of them, SPIRAL, calculates the value of the inductance shown in Fig. 1.38 as 0.42 nH, and the quality factor of the inductor at 10 GHz as 5.2. The difference between the two inductance values can be considered acceptable. But there is a very large difference between the calculated and simulated Q values. This means that there are other loss mechanisms in addition to the DC resistance of the aluminum strip. One of them is the skin effect that increases the resistance at high frequencies. Other mechanisms can be explained by the equivalent circuit shown in Fig. 1.39.

In Fig. 1.39, R_s represents the resistance of the metal strip that increases with frequency due to the skin effect, as previously mentioned. C is the parasitic capacitance between the terminals of the inductor. The metal strip forming the inductor is separated from the silicon substrate with a thick silicon dioxide layer. Consequently, there is a capacitance between the metal strip and the substrate. This distributed capacitance is represented with two equal parasitic capacitances (C_l) lumped at the

Fig. 1.40 (**a**) A center-tapped (symmetrical) inductor. Note that $L_{AB} = 2L(1+k)$, where k is smaller than unity and depends on the geometry. (**b**) Layout of the center-tapped inductor. (**c**) The equivalent circuit of the center-tapped, symmetrical inductor (k is the magnetic coupling coefficient of the two halves of the inductor)

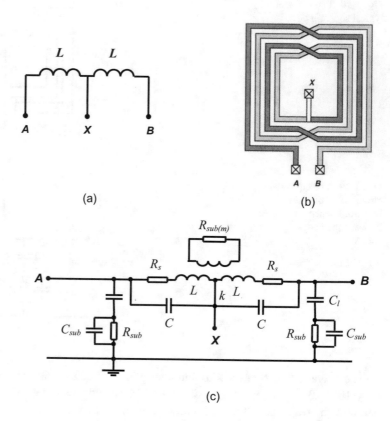

(a)   (b)

(c)

terminals of the inductor. R_{sub} is the series resistance of C_1 and depends on the doping concentrations of the regions below the structure. Capacitors parallel to R_{sub} are incorporated in the equivalent circuit to achieve a good fit to the measured data [23]. $R_{sub(m)}$ represents the loss due to the magnetically induced current in the substrate. Another effect of this "image current" that is magnetically coupled to the inductor is to reduce the value of the effective inductance.

In several applications, for example differential LNAs, differential oscillators, etc., center-tapped symmetrical inductors are needed, as shown in Fig. 1.40a. To ensure the symmetry of the parasitics of the two half-sections, a careful layout design has prime importance. A square-shaped symmetrical inductance layout is shown in Fig. 1.40b. For hexagonal and octagonal inductors, symmetrical layouts can be generated similarly.

It must not be overlooked that the equivalent inductance of two serially connected inductors (L_1 and L_2) is equal to

$$L_{eq} = L_1 + L_2 \mp 2k\sqrt{L_1 L_2}$$

where k is the magnetic coupling coefficient; it is always smaller than one (unity) and the sign is positive if the two halves of the inductor are "wound" in the same direction.

The losses due to R_{sub} can be reduced (a) by decreasing the current flowing through C_1, and (b) by decreasing the value of the resistance itself. Decreasing the capacitor current requires a thick oxide layer between the top metal layer and the silicon surface. The thickness of this oxide layer is determined by the number of layers and the thickness of the inter-layer oxides (Fig. 1.41a). To decrease the resistance on the current path, a grounded, high conductivity metal or poly-silicon shield

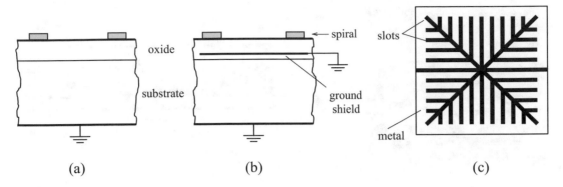

Fig. 1.41 (**a**) Cross-section of a spiral on-chip inductor. (**b**) Spiral inductor with ground shield. (**c**) Plan view of a patterned ground shield

(ground shield) can be placed between the inductor and the substrate (Fig. 1.41b). This modification effectively reduces the losses associated with the capacitive substrate current. But due to the image current induced on this layer (that is equivalent to the current of a short-circuited secondary, magnetically coupled to the inductor), the effective inductance decreases strongly. To overcome this adverse effect, a simple but efficient solution is to pattern the ground shield in such a way as to prevent the image currents on the shield (Fig. 1.41c) [24]. But it must be kept in mind that this shield does not prevent the induced currents in the silicon substrate and cannot reduce the losses represented with $R_{sub(m)}$.

The parasitic capacitances shown in Fig. 1.39 determine the "self-resonance frequency" of the inductor (See Chap. 4). Note that for frequencies higher than this frequency, the "inductor" exhibits a "capacitive" impedance.

To increase the quality factor of an on-chip inductor one solution is to etch away the silicon dioxide layer under the spiral and thus to reduce the parasitic capacitive current. But this "suspended inductor" approach is not compatible with standard processes. The bonding wires can also be used to realize low-value inductors, with considerably high Q values.[18] However, due to high tolerances and bad repeatability, they cannot be considered as standard components of ICs.

1.2.4   Pads and Packages

A chip must be protected from environmental (mechanical and chemical) hazards and must have connections to the rest of the circuit. Figure 1.42a and b show the photomicrograph of an RF chip and the cross-section of a typical RF package, respectively. As shown in Fig. 1.42a, bonding pads are placed along the edges of the chip that provide landing areas to make ultrasonic or thermocompression bonding for bonding wires to connect input, output, DC supply, and ground nodes of the IC to the corresponding legs of the package.[19] The dimensions of the metal (Al or Cu) bonding pads are in the range of 80 to 100 μm and the diameter of typical (Al or Au) bonding wires is 25 μm.

---

[18] The self-inductance of a 25-micron diameter bonding wire is approximately 1nH/mm. The Q value of a bonding wire is about 60 at 2 GHz.

[19] Another possibility is not to use any package, to fix the chip directly on to the printed circuit board (PCB) and make the bondings from bonding pads on the chip to the corresponding pads on the PCB. In this case, the chip and the bonding wires must be protected with an appropriate sealant.

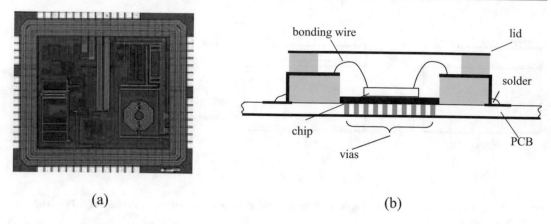

Fig. 1.42 (a) The photomicrograph of an RF chip. (b) The schematic cross-section of a typical RF package

Fig. 1.43 The circuit
diagram of a typical input
ESD protection circuit

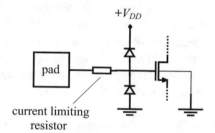

In addition to serving as bonding areas, the bonding pads must have features to protect the IC from over-voltages that may break down the gate oxides or the p-n junctions of the MOS transistors connected to this pad. Figure 1.43 shows a simple electrostatic discharge (ESD) protection circuit for an input pad. It can be easily seen that this circuit provides a lower limit equal to one diode voltage (approximately 0.7 V) smaller than the ground potential and an upper limit one diode voltage higher than V_{DD} for the voltage reaching the circuit input node, during normal operation. In the event of an external ESD pulse with very high voltage levels, the protection diodes are designed to go into reverse breakdown mode in order to drain a large current to V_{DD} or to ground, instantaneously. In order to prevent permanent damage, the reverse breakdown voltage of the diodes must be smaller than the gate oxide breakdown voltage of the input transistor. Also note that a relatively small-valued current-limiting series resistor (also called the ballast resistor) is needed to prevent overheating.

ESD protection circuits for digital and analog ICs are different in nature. The essential requirements for analog circuits are minimal parasitics that may affect the performance of the circuit, especially at the higher end of the operating frequency range, with minimal additional noise and low nonlinearity. These requirements unfortunately conflict with the physical constraints that are dictated by the protection devices.

The ESD protection diodes in particular must be designed with a relatively large effective junction area (using inter-digitated structures) in order to support the large current density that may flow in the event of an ESD pulse. The diode must also be surrounded by a single or double guard ring to prevent stray currents. The large junction area results in large parasitic capacitances that appear in parallel to

Fig. 1.44 The typical
human body model

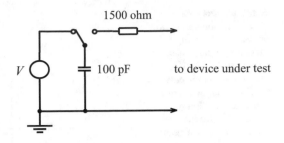

the input (or output) node, in addition to the pad metal-to-substrate capacitance. Similarly, the current limiting resistors must be realized as diffusion resistors in the substrate (to allow heat dispersion), which introduces relatively large parasitic capacitances with respect to the substrate. Hence, the RC parasitics that are due to ESD protection circuitry are virtually unavoidable in the signal path.

In addition to input and output protection diodes, the power supply networks must also contain so-called power clamp devices between V_{DD} and ground that are capable of conducting large currents without sustaining permanent damage, in the event of the supply network itself being subjected to an ESD pulse. In chips with multiple power supply domains, all independent supply networks must be interconnected using a bridge device (usually consisting of two cross-coupled junction diodes) that is inactive under normal conditions, but is capable of passing a large instantaneous current in case of an ESD event.

To simulate what happens if somebody touches one of the legs of the IC, it is necessary to model the average electrostatic capacity of the human body and the average charge on the body: One of the commonly used human body models is shown in Fig. 1.44. V is typically 1000 V D.C. The 100 pF capacitor is charged with this voltage and then discharged to one of the pads of the IC. The circuit must withstand this instantaneous current pulse without harm.

Packages have several functions. One of them is to protect the chip and the bonding wires, as mentioned before. The second function is to connect the internal nodes of the chip to the rest of the circuit via the bonding pads, the bonding wires, and the legs of the package that are soldered to the appropriate copper strips on the PCB. Finally, the metal floor of the package onto which the chip is eutectic-bonded or adhered provides a relatively low thermal resistance from the rear of the chip to the PCB. As shown in 1.42a, forming a large number of metal-filled vias between the upper copper area (the metal floor of the package) and the lower copper plate of the PCB reduces the thermal resistance from the chip to ambient and effectively improves the cooling efficiency.

The most significant parasitics of the packages are the self-inductances of the bonding wires that are connected in series to the inputs, outputs, DC supply, and ground lines. The self-inductance of a 25 μm diameter bonding wire is approximately 1 nH per millimeter. Depending on the dimensions of the package and the distance from a bonding pad to the corresponding leg, the parasitic inductance of a bonding wire is usually in the range of several nanohenries. These parasitic inductances influence the frequency characteristics of the gain and the input and output impedances, and must be included in the final simulations.

In Fig. 1.45, a chip is shown with its bonding wire inductances. For the sake of simplicity, it is assumed that there are only two different internal blocks in the chip, which are marked (1) and (2). These blocks may be different stages of an analog circuit or, if the chip is a mixed-mode IC, one block may be an analog circuit and the other, digital.

The input bonding wire inductance is in series with the input capacitance of (1) and forms a series resonance circuit. This resonance effect, which will be investigated in Chap. 4, can be useful or harmful depending on the nature of the circuit. The effects of this inductance on the frequency

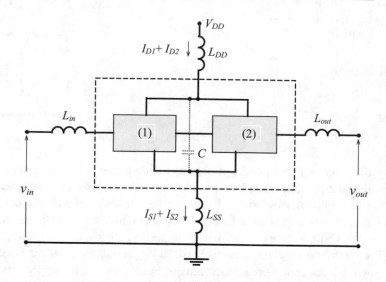

Fig. 1.45 Schematic of an IC with the bonding wire parasitics. C is a high capacity on-chip capacitor to bypass the high-frequency components of the DC supply current and hence to maintain the internal V_{DD} to internal ground voltage constant

characteristic of gain and on the input impedance must be simulated in order to find solutions to improve the unwanted results. The effects of the output bonding wire are similar.

The effects of the ground connection bonding wire are important for fully analog as well as for mixed-mode chips. In the case of an analog chip, the ground return currents of (1) and (2) (and the currents of others if there are more than 2 stages) flow over L_{SS}. The signal voltage drop on this impedance is in the input loop of (1) and the resulting frequency-dependent feedback can severely affect the performance of the circuit.

The voltage drop on L_{DD} due to the signal components of the supply currents of (1) and (2) results in a non-constant internal V_{DD}, which also causes a – relatively weak – feedback. This effect can be minimized with an on-chip bypass capacitor connected between the internal V_{DD} node and the internal ground, as shown in Fig. 1.45. This high value capacitor is usually realized using all the available area on the chip surface.

In the case of a mixed-mode circuit: if, for example, (1) is an analog block and (2) is digital, the pulsating supply and ground currents of (2) injects an unwanted noise to the analog block. To prevent (or at least to minimize) this effect, the supply and ground lines of the analog and digital blocks must be separated and must be connected to the legs of the package through different bonding wires.

These explanations show that the bonding wire inductances can cause significant problems and it is necessary to reduce them. One of the most common solutions is to use multiple bonding pads and bonding wires in parallel. Another solution is to use bonding ribbons (that exhibit smaller self-inductance per micron) instead of round bonding wires. The more effective solution is to completely eliminate the bonding wires and use flip-chip (or similar) bonding techniques for the circuits operating in the microwave frequencies.

Another very effective method to minimize the adverse effects of the DC supply and ground bonding wires is to design the circuit on the chip with differential building blocks. Since the input signal of an analog (or digital) circuit controls the sharing ratio of currents flowing through the two halves of the circuit and the sum of these currents is constant, there is essentially no signal voltage drop on L_{SS} and L_{DD}, and consequently, any feedback or cross noise injection risks are eliminated.

References

1. S. Wolf, *Silicon Processing for the VLSI Era, Vol. 3 – The Submicron MOSFET* (Lattice Press, Sunset Beach, 1995)
2. N. Shigyo, T. Shimane, M. Suda, T. Enda, S. Fukuda, Verification of saturation velocity lowering in MOSFET's inversion layer. IEEE Trans. Electron. Devices **45**(2), 460–464
3. K. Chen, J. Duster, H.C. Wann, P.K. Ko, C. Hu, MOSFET carrier mobility model based on gate oxide thickness, threshold and gate voltages. J. Solid State Electron. **39**, 1515–1518 (1996)
4. K. Chen, C. Hu, Performance and Vdd scaling in deep submicrometer CMOS. IEEE J. Solid State Circuits **33**, 1586–1589 (1998)
5. R.S. Muller, T.I. Kamins, M. Chan, *Device Electronics for integrated Circuits*, 3rd edn. (Wiley, New York, 2003)
6. S. Wolf, *Silicon Processing for the VLSI Era, Vol. 3 – The Submicron MOSFET* (Lattice Press, Sunset Beach, 1995)
7. Y.P. Tsividis, *Operation and Modeling of the MOS Transistor*, 2nd edn. (McGraw-Hill, New York, 1999)
8. U. Cilingiroglu, *Systematic Analysis of Bipolar and MOS Transistors* (Artech House, Boston, 1993)
9. Ng, Lynch, "Analysis of the gate-voltage dependent series resistance of MOSFETS" IEEE Trans. Elecr. Dev., Vol. ED-33, July 86, p 965.
10. Ng, Lynch, "The impact of the intrinsic series resistance on MOSFET scaling" IEEE Trans. Elecr. Dev., Vol. ED-34, March 87, p 503.
11. E. Gondro, *"An improved bias dependent series resistance description for MOS models"*, Compact Model Council Meeting, Santa-Clara, USA, May 1998
12. D. Foty, *MOSFET Modeling with SPICE, Principles and Practice* (Prentice-Hall, Upper Saddle River, 1997)
13. Y. Taur, T.H. Ning, *Fundamentals of Modern VLSI Devices* (Cambridge University Press, Cambridge, 1998)
14. A.B. Glaser, G.E. Subak-Sharpe, *Integrated circuit Engineering; Design, Fabrication and Applications* (Addison-Wesley, Reading, 1979)
15. Calculation of a step input into an infinite RC transmission line, Lecture notes of Prof. E. Cheever, Swarthmore College. http://www.swarthmore.edu/NatSci/echeeve1/Ref/trans/stepinf.html
16. H. Samavati, A. Hajimiri, A.R. Shahani, G.N. Nasserbakht, T.H. Lee, Fractal capacitors. IEEE J. Solid State Circuits **33**(12), 2035–2041 (1998)
17. O.E. Akcasu, High capacitance structure in a semiconductor device, U.S. Patent No. 5,208,725, May 4, 1993
18. R.L. Bunch, S. Raman, Large signal analysis of MOS varactors in CMOS-Gm LC VCOs. IEEE J. Solid State Circuits **38**(8), 1325–1332 (2003)
19. P. Andreani, S. Mattisson, On the use of MOS varactors in RF VCOs. IEEE J. Solid State Circuits **35**(6), 905–910 (2000)
20. S.S. Mohan, M.M. Hershenson, S.P. Boyd, and T.H. Lee, "Simple accurate expressions for planar spiral inductances", IEEE Journal of Solid State Circuits, vol, 34, no. 10, pp. 1419–1424, Oct. 1999.
21. "ASITIC": Analysis and Simulation of Spiral Inductors and Transformers for ICs. EECE, University of California, Berkeley
22. "SPIRAL", OEA International, Inc., Morgan Hill, CA, USA
23. K.B. Ashby, I.A. Koullias, W.C. Finley, J.J. Bustek, S. Moinian, High Q inductors for wireless applications in complementary silicon bipolar process. IEEE J. Solid State Circuits **31**(1), 4–9 (1996)
24. C.P. Yue, S.S. Wong, On-chip spiral inductors with patterned ground shields for Si based RFICs. IEEE J. Solid State Circuits **33**(5), 743–752 (1998)

Basic MOS Amplifiers: DC and Low-Frequency Behavior

2

Basic MOS amplifiers are the main building blocks of a vast array of analog signal processing systems as well as other analog electronic circuits. The overall performance of a complex circuit strongly depends on the performances of its basic building blocks. In this section, the main properties of these basic circuits will be investigated.

2.1 Common-Source (Grounded-Source) Amplifier

The basic structure of a common-source amplifier is shown in Fig. 2.1a. The gate of the NMOS transistor, M1, is biased with a DC voltage source V_{GS}, to conduct the appropriate DC (quiescent) current. A signal source (v_i) is connected in series with the bias voltage to control the drain current. The load resistor R_D helps to convert the drain current variations into output voltage variations (output signal). Since the output of a MOS amplifier is usually connected to the gate of another MOS amplifier, the DC and low-frequency load coming from the subsequent stage is negligible[1]; therefore, the only load is R_D . In Fig. 2.1b, this "DC load line" with a slope equal to $(-1/R_D)$ is drawn on the output curves of M1. For a given value of the gate bias voltage (for example V_{GS1}), the drain current is I_{D1}, which corresponds to the intersection of the load-line and the output curve corresponding to V_{GS1}. This intersection point is called the "operating point" or the "quiescent point" and denoted with Q.

- From its upper limit of V_{DD} down to the lower limit of the saturation region ($V_{DS(sat)}$), the output voltage is proportional with the drain current, which has a quadratic relation to the input voltage, manifesting itself as a severe nonlinear distortion on the output voltage.[2]
- To obtain the maximum output voltage swing without excessive distortion or clipping, the quiescent drain-source voltage must be set in the middle of the "dynamic range" of the output signal, as shown in Fig. 2.1.
- It is obvious that for a given input signal amplitude, higher R_D values result in higher output signal amplitudes, or in other words, higher voltage gains.

---

[1] The effects of the input capacitance and the input conductance that appears due to the Miller effect at high frequencies will be investigated later on.

[2] In a velocity-saturated transistor, the drain current is almost linearly related to the gate voltage and consequently the nonlinear distortion is smaller.

© The Author(s), under exclusive license to Springer Nature Switzerland AG 2021
D. Leblebici, Y. Leblebici, *Fundamentals of High Frequency CMOS Analog Integrated Circuits*,
https://doi.org/10.1007/978-3-030-63658-6_2

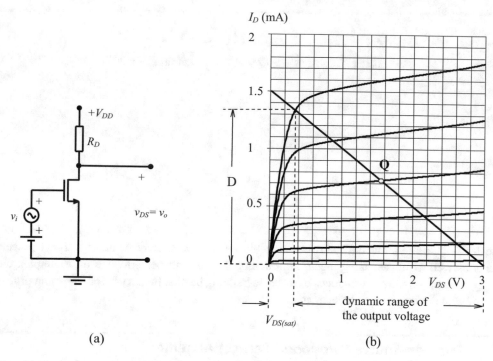

(a)   (b)

Fig. 2.1 (a) The basic resistor loaded MOS amplifier. (b) The output characteristic curves, the DC load line, and the operating point (Q). The drain current dynamic range is marked with D. The NMOS transistor (AMS-035) has W/L ratio of (35 μm/0.35 μm). $R = 2$ kΩ and $V_{GS} = 0.9$ V

- When the instantaneous value of the input voltage increases in the positive direction, the total gate-source voltage increases, resulting in an increase on the drain current and an increase on the voltage drop on R_D. This means that the output voltage decreases, or changes in the opposite direction with the input voltage, i.e., for a sinusoidal input signal the output signal is 180° out-of-phase with the input.

Resistors are not preferred components for CMOS integrated circuits, because they consume a large amount of area on the chip and are not directly compatible with standard CMOS processes. To eliminate the resistor, a MOS transistor can be used instead. There are several alternatives of this solution, but the most convenient approach is to use a PMOS transistor biased with an appropriate gate-source voltage (V_{GS2}), as the load of the NMOS input transistor (Fig. 2.2a). In this circuit, the PMOS transistor acts as a nonlinear resistor controlled by its gate-source DC bias voltage (Fig. 2.2b). From another point of view, this load can be considered as a DC current source, provided that M2 is in the saturation region, i.e., $|V_{DS2}| > |(V_{GS2}-V_{TH2})|$. The current of this source is equal to the quiescent drain current of M1, and its internal resistance is the output resistance of M2, for a given gate bias (Fig. 2.2c). Since M2 has no signal control mission and acts only as a passive load, this type of amplifier will be called "passive MOS loaded amplifier."

Apart from the elimination of the resistor, this circuit has several advantages compared to the resistor-loaded circuit. In Fig. 2.2d, the nonlinear load curve is superimposed over the output curves of M1 in such a way that $V_{DS1} + |V_{DS2}| \cong V_{DD}$. Since the drain currents of M1 and M2 have equal magnitudes, the operating point corresponds to the intersection of the characteristic curve of the load transistor (M2) and the curve of the input transistor corresponding to its bias voltage, V_{GS1}. To provide a symmetrical variation for the output voltage around the quiescent value, this point must be

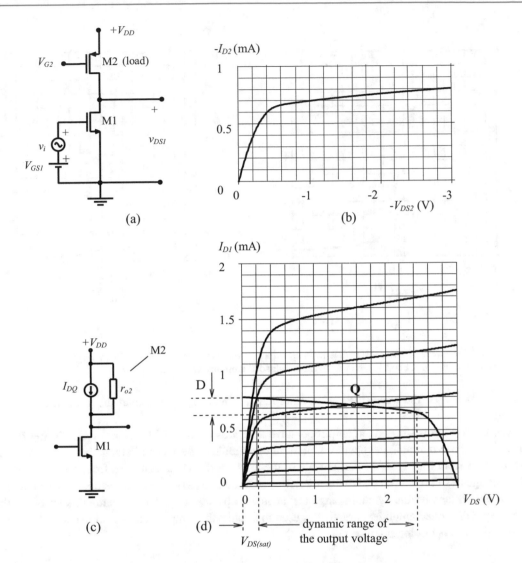

Fig. 2.2 (**a**) Passive PMOS transistor loaded common-source NMOS amplifier. (**b**) The nonlinear load line (the output characteristic curve of the load transistor for the specified V_{GS2} value), (**c**) The input transistor (M1) and the equivalent of the load transistor (M2). (**d**) The output characteristic curves of M1 together with its load. The drain current dynamic range is marked with D. M1 and M2 are AMS-035 transistors with aspect ratios of 35 μm/0.35 μm and 60 μm/0.6 μm, respectively. Gate biases are $V_{GS1} = 0.9$ V and $V_{GS2} = 1.3$ V ($V_{G2} = 1.7$ V)

set in the middle of the dynamic range, which is the region where both M1 and M2 are in saturation mode. This operating range has a width equal to $V_{DD} - (V_{DS1(sat)} + |V_{DS2(sat)}|)$.

The amount of change (i.e., the dynamic range) in the drain current corresponding to the dynamic range of the output voltage is quite small. This means that the full swing of the output voltage can be obtained with a comparatively small swing of the input signal voltage. Consequently: (a) this circuit produces a higher voltage gain, (b) since this circuit operates in a small portion of the nonlinear I_D - V_{GS} curve, the linearity is better, in other words, the nonlinear distortion is smaller.

In Fig. 2.3, the voltage transfer curves of (a) a resistor-loaded and (b) a PMOS transistor-loaded common-source amplifier, operating under identical conditions, are given. The comparison of these curves shows that the small-signal voltage gain (that corresponds to the slope of the curve at the

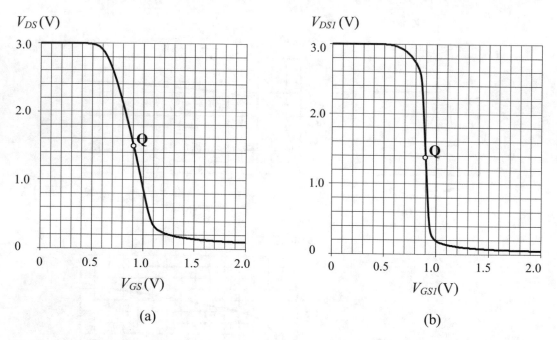

Fig. 2.3 PSpice simulation results showing the input voltage to output voltage transfer curves of (**a**) the resistor-loaded amplifier shown in Fig. 2.1 and (**b**) the passive PMOS-loaded amplifier shown in Fig. 2.2

operating point) of the transistor-loaded amplifier is obviously higher than the gain of a resistor-loaded amplifier.

The passive transistor-loaded amplifiers have a serious problem: The appropriate biasing of the load transistor to maintain the operating point in the middle of the dynamic range is very critical. As can be seen from Fig. 2.2d, a small change of the load curve (which can arise from the value of the bias voltage of M2 or can be due to the tolerances of M2 and/or M1) can shift the operating point, and the output voltage usable dynamic range can decrease considerably. To overcome this problem, the passive MOS-loaded amplifiers are always used in circuits that employ an overall negative feedback to stabilize the operating points.

2.1.1 Biasing

The DC gate bias voltage of a common-source amplifier is usually supplied by the output of the previous stage, together with the input signal. This means that the previous stage must be designed in such a way that the quiescent DC voltage on the output node has the appropriate value to bias the common-source stage at the pre-defined operating point.[3] The type of the transistor of the common-source stage must be chosen depending on the level of the DC voltage on the output of the previous stage. Usually, an NMOS transistor is preferable for DC voltages close to ground (0 V), and a PMOS transistor is preferable for DC voltages close to the positive supply (Fig. 2.4).

---

[3] Another solution is to connect the input of the second stage to the input of the first stage over a "coupling capacitor" and bias the second stage independently. It is obvious that in this case the amplifier has a cut-off frequency for low frequencies that depends on the value of the coupling capacitor and the total resistance in the loop.

Fig. 2.4 (**a**) NMOS input common-source amplifier. (**b**) PMOS input common-source amplifier. A indicates the output node of the previous stage and B indicates the DC source to bias the load transistor in each case

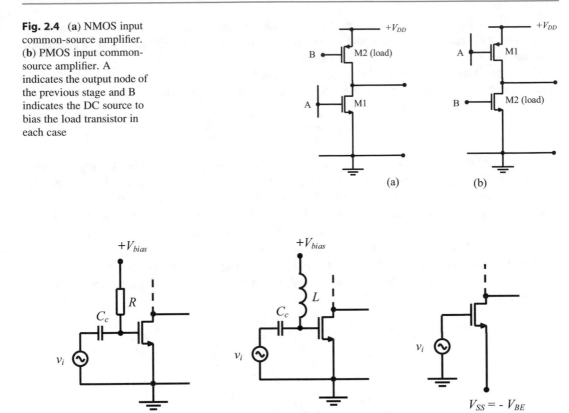

Fig. 2.5 Biasing of the input transistor. (**a**) and (**b**) AC coupling, (**c**) DC coupling of the signal to the input of the amplifier

For common-source input stages, the gate has to be independently biased with a DC voltage source. If capacitive coupling of the signal to the gate of the input transistor is permissible, the gate can be connected to the bias voltage source through a high-value resistor or an RF "choking coil" (in radio frequency applications) as shown in Fig. 2.5a and b.

The bias voltage can be obtained from the main DC voltage source (for example V_{DD}) with a resistive or MOS voltage divider. If the input signal source has to be directly coupled to the gate of the transistor (the quiescent DC voltage on the input node is zero), the source must be connected to an appropriate negative voltage source, to maintain the appropriate gate-source DC bias (Fig. 2.5c).

For passive MOS transistor loaded amplifiers, the fixed DC bias voltage of the load transistor can be obtained from V_{DD} using a suitable voltage divider.

2.1.2 The Small-Signal Equivalent Circuit

For an amplifier biased at an appropriate operating point, the characteristic curve representing the voltage-current relation of the device can be approximated with the tangent of this curve, as long as the magnitude of the signal applied to the input is small enough compared to the bias voltage at this point. This corresponds to the "linearization" of the device characteristics that permits us to use the simple and well-known network analysis procedures, instead of the complicated nonlinear solution methods that are practically impossible for hand calculations.

For a non-velocity saturated NMOS transistor biased in the normal saturation region, the drain current can be expressed in terms of the gate-source and drain-source voltages as

$$I_D = \frac{1}{2}\beta(V_{GS} - V_{TH})^2 \left(\frac{1 + \lambda V_{DS}}{1 + \lambda(V_{GS} - V_{TH})}\right) \tag{2.1}$$

where $\beta = KP(W/L)$ and $KP = \mu C_{ox}$. It was mentioned in Chap. 1 that, especially for small geometries (for very thin gate oxides) the mobility strongly decreases with the transversal electric field in the gate region. Therefore, in (2.1), the mobility value corresponding to the quiescent gate-source voltage must be used. Additionally, if there is a substrate bias, the threshold voltage corresponding to this condition must be used. With these precautions, the quiescent drain current can be calculated in terms of V_{DS} and V_{GS}.

If we expand (2.1) into a Taylor series around the operating point and assume that variations of voltages and currents are sufficiently small compared to the quiescent values, we can neglect all higher-order terms and express the drain current as

$$I_D + \Delta I_D \cong I_D + \left.\frac{\partial I_D}{\partial V_{GS}}\right|_{\Delta V_{DS}=0} \times \Delta V_{GS} + \left.\frac{\partial I_D}{\partial V_{DS}}\right|_{\Delta V_{GS}=0} \times \Delta V_{DS}$$

or

$$\Delta I_D \cong \left.\frac{\partial I_D}{\partial V_{GS}}\right|_{\Delta V_{DS}=0} \times \Delta V_{GS} + \left.\frac{\partial I_D}{\partial V_{DS}}\right|_{\Delta V_{GS}=0} \times \Delta V_{DS} \tag{2.2}$$

(2.2) Indicates that the change of the drain current is (approximately) a linear function of the variations of the gate and drain voltages. The coefficients of this linear expression are

$$g_m = \left.\frac{\partial I_D}{\partial V_{GS}}\right|_{\Delta V_{DS}=0} \quad \text{and} \quad g_o = \left.\frac{\partial I_D}{\partial V_{DS}}\right|_{\Delta V_{GS}=0} \tag{2.3}$$

which can be calculated as the partial first derivatives of (2.1) at the operating point. They are the previously defined "transconductance" and "output conductance" parameters of the transistor[4] for this operating point. Now (2.2) can be written as

$$\Delta I_D = g_m \cdot \Delta V_{GS} + g_o \cdot \Delta V_{DS} \tag{2.4}$$

Each of these two small-signal parameters of the MOS transistor for any (V_{GS}, V_{DS}, I_D) operating point can be calculated from (2.1) and (2.3). For the transconductance parameter, we use (cf. Chap. 1):

$$g_m = \beta(V_{GS} - V_{TH})(1 + \lambda V_{DS})\frac{2 + \lambda(V_{GS} - V_{TH})}{2\left[1 + \lambda(V_{GS} - V_{TH})^2\right]} \tag{2.5}$$

This expression can be simplified for small λ values as

---

[4] Throughout this book, the output conductance and the output resistance of a transistor will be represented interchangeably by g_o or g_{ds} and r_o or r_{ds}, respectively.

$$g_m \cong \beta(V_{GS} - V_{TH}) \qquad (2.5a)$$

Although it is an approximate formula, especially for short-channel transistors (where the λ parameter is not negligibly small), (2.5a) is very convenient for hand calculations. Another simple and useful expression that gives the transconductance in terms of the drain DC current is

$$g_m \cong \sqrt{2\beta I_D} \qquad (2.5b)$$

The corresponding expressions for PMOS transistors are

$$|I_D| = \frac{1}{2}\beta(V_{GS} - V_{TH})^2 \left(\frac{1 + \lambda|V_{DS}|}{1 + \lambda|V_{GS} - V_{TH}|} \right)$$

$$g_m \cong \beta|(V_{GS} - V_{TH})|$$

$$g_m \cong \sqrt{2\beta|I_D|}$$

In case of velocity saturation (cf. Chap. 1), the drain current can be written as

$$|I_D| = kWC_{ox}(V_{GS} - V_{TH})v_{sat} = g_m(V_{GS} - V_{TH}) \qquad (2.6)$$

and the transconductance is found as:

$$g_m = \frac{dI_D}{dV_{GS}} = kWC_{ox}v_{sat} \qquad (2.7)$$

which is valid for both NMOS and PMOS transistors. (2.7) indicates that for a short-channel MOS transistor operating in the velocity saturation regime, the transconductance parameter does not depend on the operating point and is proportional to the gate-width.

The output conductance parameter of an NMOS transistor calculated from (2.1) and (2.3) is

$$g_o = I_D \frac{\lambda}{(1 + \lambda V_{DS})} \qquad (2.8)$$

and can be simplified for small λ values as

$$g_o \cong I_D\lambda \qquad (2.8a)$$

For PMOS transistors, due to the negative polarity of the source-drain voltage and the drain current, the output conductance parameter must be expressed as

$$g_o = |I_D| \frac{\lambda}{(1 + \lambda|V_{DS}|)} \cong |I_D|\lambda \qquad (2.8b)$$

From (2.4) we can also derive a different set of definitions for g_m and g_o that is useful for obtaining the values of these parameters by direct measurements, or from the characteristic curves of the transistor:

$$g_m = \frac{\Delta I_D}{\Delta V_{GS}}\bigg|_{\Delta V_{DS}=0} \qquad \text{and} \qquad g_o = \frac{\Delta I_D}{\Delta V_{DS}}\bigg|_{\Delta V_{GS}=0} \qquad (2.9)$$

Fig. 2.6 The small-signal equivalent circuit of a MOS transistor biased at a certain operating point in the saturation region

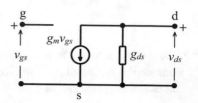

The transconductance is the most important parameter of the transistor and expresses the control capability of the gate-source voltage over the drain current. The definition of g_m given in (2.3) indicates that its value is equal to the slope of the tangent at the operating point of the $I_D = f(V_{GS})$ curve (drawn for the quiescent value of V_{DS}). Similarly, g_o is the slope of the tangent at the operating point of the $I_D = f(V_{DS})$ output characteristic curve, corresponding to the quiescent value of V_{GS}.

In (2.4), the Δ variations of the voltages and the current can be replaced with the instantaneous values of the small-signal components on the quiescent DC values and the expression can be written as

$$i_d = g_m v_{gs} + g_{ds} v_{ds} \tag{2.10}$$

which can be represented with a circuit as shown in Fig. 2.6. This "small-signal equivalent circuit" is very useful for solving the relations of signal currents and voltages in a circuit containing MOS transistors. For high frequencies, it is necessary to include other physical and parasitic components into this equivalent circuit.

The basic rules of deriving the small-signal equivalent of a circuit can be summarized as follows:

- Calculate (determine) the DC quiescent voltages and currents of each transistor
- Calculate the small-signal parameters (g_m and g_{ds}) corresponding to these operating points
- Replace the transistors in the circuit with their small-signal equivalents
- Include all passive components of the circuit
- Short-circuit all ideal DC voltage sources (supply voltages, bias voltages, etc.)
- Open-circuit all ideal DC current sources
- If the DC voltage and/or current sources are not ideal (i.e., if they have serial or parallel internal impedances), include them in the small-signal equivalent circuit
- Connect the signal source(s), and solve the circuit using conventional linear circuit analysis methods
- To obtain reasonable results from hand calculations, apply sufficiently small magnitudes to the driving signal sources such that all output signals remain in the dynamic ranges corresponding to these outputs[5]

In Fig. 2.7a and b, the small-signal equivalent circuits corresponding to the amplifiers shown in Figs. 2.1 and 2.2 are given. From Fig. 2.7a, the output voltage (v_o) can be easily calculated in terms of the input signal voltage (v_i) and then the small-signal voltage gain of the circuit as

---

[5] Computer aided (for example SPICE) analyses have the same problem. Once the software forms the small signal equivalent circuit, it assumes that the circuit is linear for all amplitudes of the input signal(s). It is the user's responsibility to apply reasonable input signals in order to keep the magnitudes of the output signals within the dynamic ranges. In case of doubt, it is advisable to run a "transient" simulation to be sure that there is no excessive nonlinear distortion or clipping on the output signals.

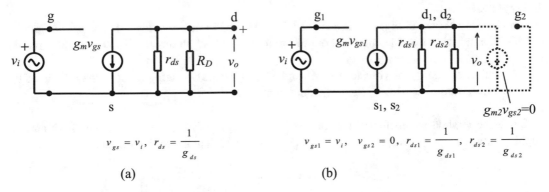

Fig. 2.7 The small-signal equivalent circuits of (**a**) a resistor-loaded common-source amplifier, (**b**) a passive MOS transistor-loaded common-source amplifier

$$A_v = \frac{v_o}{v_i} = -g_m(r_{ds}//R_D) = -g_m \frac{r_{ds}R_D}{r_{ds} + R_D} \tag{2.11}$$

which reduces to

$$A_v = -g_m R_D \tag{2.11a}$$

for

$$r_{ds} \gg R_D.$$

The small-signal voltage gain of a passive loaded common-source amplifier can be calculated from Fig. 2.7b. Since there is no signal on the gate of the load transistor, it acts only as a resistor having a value equal to its output resistance at its operating point:

$$A_v = \frac{v_o}{v_i} = -g_{m1}(r_{ds1}//r_{ds2}) = -\frac{g_{m1}}{(g_{ds1} + g_{ds2})} \tag{2.12}$$

If we insert the values of g_{m1}, g_{ds1}, and g_{ds2} in terms of the DC drain current, we obtain

$$A_v \cong -\sqrt{2\beta_1 I_{D1}} \frac{1}{I_{D1}\lambda_1 + |I_{D2}|\lambda_2} = -\sqrt{\frac{2\beta_1}{I_{D1}}} \frac{1}{(\lambda_1 + \lambda_2)} \tag{2.12a}$$

This expression indicates that:

- The voltage gain of a common-source amplifier is higher for lower DC quiescent currents (in other words, for lower power consumption)
- The voltage gain can be increased with load and/or input transistors having smaller λ parameters (larger channel lengths), keeping the aspect ratios constant

However, it must be kept in mind that the performance of the amplifier deteriorates at high frequencies due to the increased output internal resistance and increased parasitic drain junction capacitances, as will be explained in Chap. 3.

Example 2.1

An amplifier as shown in Fig. 2.2a is designed for AMS 0.35 μm technology. The supply voltage is 3.2 V. Channel widths and lengths for both of the transistors are 35 μm and 0.35 μm, respectively.

(a) Calculate the gate bias voltages for M1 and M2, to fix the operating point in the middle of the operating range, i.e., 1.6 V, for 1 mA quiescent current.

Since the lambda parameters of the transistors (especially of M2) are considerably high and the supply voltage is low, it is appropriate to use (1.14a) to calculate the drain currents, which is

$$|I_{D2}| \cong \frac{1}{2}\beta_2\left(V_{GS2} - V_{THp}\right)^2 \frac{1 + \lambda_2|V_{DS2}|}{1 + \lambda_2\left|\left(V_{GS2} - V_{THp}\right)\right|}$$

for M2.

The parameters of M2 for AMS 0.35 μm technology are given as: $\mu_{n0} = 137$ [cm^2/V.s], $C_{ox} = 4.56 \times 10^{-7}$ [F/cm^2], $V_{THp} = -0.7$ [V], and $\lambda_2 \approx 0.2$ [V^{-1}]. $(W/L)_2$ is given as 100 and V_{D2} must be in the middle of the output voltage dynamic range, i.e., 1.6 V. β_2 can be calculated as

$$\beta_2 = \mu_p C_{ox}(W/L)_2 = 137 \times \left(4.56 \times 10^{-7}\right) \times 100 = 6.25 \times 10^{-3} \ [\text{A/V}^2]$$

Inserting $|(V_{GS2} - V_{THp})| = V_{Go2}$ and $(1 + \lambda_2 |V_{DS2}|) = (1 + 0.2 \times 1.6) = 1.32 = a$, to simplify the expressions, V_{Go2} can be solved as

$$V_{Go2} = \frac{1}{2\beta_2 a}\left[2\lambda_2|I_{D2}| \mp \sqrt{\left(2\lambda_1|I_{D2}|\right)^2 + 8|I_{D2}|\beta_2 a}\right]$$

$$V_{Go2} = \frac{1}{2\left(6.25 \times 10^{-3}\right)1.32}\left[2 \times 0.2 \times 10^{-3} \mp \sqrt{\left(2 \times 0.2 \times 10^{-3}\right)^2 + 8 \times 10^{-3}\left(6.25 \times 10^{-3}\right) \times 1.32}\right]$$

$$= 0.516\,\text{V}$$

Since M2 is a PMOS transistor, V_{GS2} and V_{THp} are negative:

$$V_{GS2} = -V_{Go2} + V_{Tp} = -0.516 + (-0.7) = -1.216\text{V}$$

Similarly, for the NMOS M1 transistor with $\mu_{n0} = 475.8$ [cm^2/V.s], $V_{THn} = 0.5$ [V], $\lambda_2 \approx 0.073$ [V^{-1}] and $(W/L)_1 = 100$, the gate-source bias voltage of M1 can be calculated as $V_{GS1} = 0.79$ V.

(b) Calculate the voltage gain.

From (2.12a):

$$A_v = -\sqrt{2 \times \left[475.8 \times \left(4.56 \times 10^{-7}\right) \times 100\right]\frac{1}{10^{-3}}\left(\frac{1}{(0.073 + 0.2)}\right)} = -24.13 \ (27.6 \text{ dB})$$

(c) Check the results with PSpice simulations.

It is common practice to use the calculated bias voltage values as inputs to SPICE. But due to the approximations in the expressions, the hand-calculated bias voltage values do not provide the targeted drain currents in most cases and fine-tuning becomes necessary. It is useful to develop a tuning strategy to prevent tedious iterations and to converge on the appropriate operating conditions. The

DC-sweep feature of SPICE is a useful option. For this example, we will apply the following procedure:

- Fix the drain voltage of the load transistor to the targeted value (connect a 1.6 V DC supply between the drain node of M2 and the ground)
- DC-sweep the gate bias voltage of M2 in a reasonable interval. Note the value of the gate voltage (V_{G2}) corresponding to the targeted drain current (-1 mA in this case)
- Disconnect the 1.6 V supply
- Apply V_{G2} to the gate of M2
- DC-sweep the gate bias voltage of M1 (V_{GS1}) and observe the drain node voltage of M1 as a function of V_{GS1}. This is the voltage transfer curve of the amplifier and the V_{GS1} value corresponding to $V_{DS1} = 1.6$ V is the appropriate bias voltage.

The obtained PSpice results are:

$$V_{GS1} = 0.9658V, \quad V_{G2} = 1.775V(V_{GS2} = -1.425V) \quad \text{and} \quad A_V = -18.57(25.37dB)$$

The discrepancies between the hand calculation and simulation results arise from several secondary effects that are not included in the expressions. For example, the mobility degradation due to the transversal field, and the series source resistance, are the most important effects that result in the lowering of the β parameter of the transistor. Certainly, it is possible to take this effect into account as explained in Chap. 1, but since a fine-tuning is unavoidable, it is wiser not to complicate the calculations further.

Another important fact to note is that the gate-source voltages of M1 and M2 are given with high precision, since the positions of the operating point and the quiescent current strongly depend on the bias voltages. Therefore, as already mentioned, this type of circuit must be used in an appropriate negative feedback loop, to keep the operating point in the middle of the voltage transfer curve.

(d) Repeat the simulation for $I_D = 1$ mA and $L_2 = 1$ μm, $W_2 = 100$ μm.

The PSpice simulation results obtained after a fine-tuning of the operating point (which is necessary to bring the operating point to the middle of the voltage transfer curve) are as follows:

$$V_{GS1} = 0.987V, \quad V_{G2} = 1.775V \quad (V_{GS2} = -1.425V) \quad \text{and} \quad A_v = -33.2(30.42dB)$$

The voltage gain is considerably (more than 5 dB) increased as expected, at the expense of higher area consumption and higher parasitic capacitance on the output node. Certainly, it is possible to increase the dimensions of M1 as well, but the lambda parameter of 0.35 μm NMOS transistors are already small and to increase the gate length does not provide a significant increase in the voltage gain. (The increase in the C_{gs} and C_{dg} parameters is another drawback affecting the high-frequency performance of the circuit, as will be explained in Chap. 3).

A simple but important modification on a common-source amplifier is shown in Fig. 2.8a, where a resistor is connected between the source terminal and the ground. From the small-signal equivalent circuit (Fig. 2.8b), the voltage gain of this circuit can be calculated as

$$A_v = -g_m \frac{R_D}{(1 + g_m R_S) + \frac{(R_D + R_S)}{r_o}} \tag{2.13}$$

which can be simplified for $r_o \gg (R_D + R_S)$ as

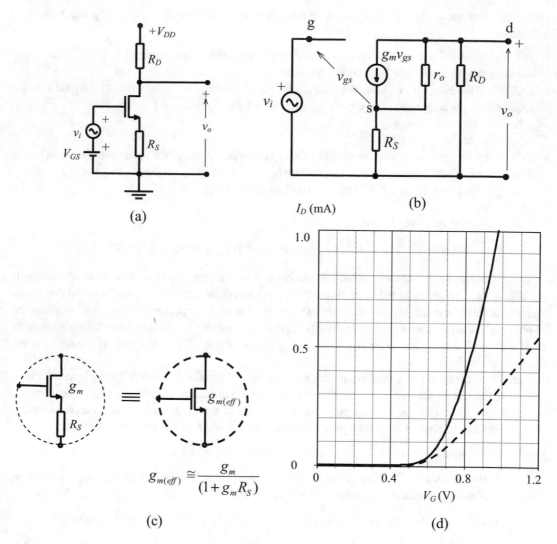

Fig. 2.8 (a) Resistor-loaded common-source amplifier with a series source resistor. (b) The small-signal equivalent circuit. (c) The definition of the effective transconductance. (d) Transfer characteristic of a MOS transistor without R_S (solid line) and with $R_S = 500\ \Omega$ (dashed line)

$$A_v \cong -\frac{g_m}{(1 + g_m R_S)} R_D \qquad (2.13a)$$

If we compare (2.13a) with (2.11a), we conclude that a resistor connected in series to the source terminal reduces the transconductance of the transistor to a smaller value, which we will call the "effective transconductance":

$$g_{m(eff)} = \frac{g_m}{1 + g_m R_S} \qquad (2.13b)$$

Another effect of R_S is the linearization of the input voltage-output current transfer characteristic and the consequent decrease in the nonlinear distortion. To visualize this effect, the transfer characteristics of a MOS transistor-alone and the transfer characteristics of the transistor-resistor combination are shown in Fig. 2.8d. Note the decrease in the slope of the curve (which is the transconductance) and the improvement of the linearity. This technique is extensively used to improve the linearity (or decrease the non-linearity related distortions) of MOS amplifiers at the expense of reducing the magnitude of the gain. It is obvious that the gate bias voltage of such a transistor must be

$$V_G = V_{GS} + I_D R_S$$

where V_{GS} is the gate-source voltage that corresponds to I_D .

Example 2.2

Let us calculate the effects of the parasitic source resistance of a 35 μm/0.35 μm NMOS transistor. The *KP* parameter of this technology is given as 170 μA/V². Assume that the dominant part of the parasitic series resistance is the intrinsic component, and that the extrinsic source region resistance and the contact resistances are negligible. The total series resistances for NMOS transistors is given as *RDSW* = 345 ohm per micron width.

(a) Calculate the parasitic series source resistance of the 35 μm/0.35 μm transistor.

$$R_S + R_D = \frac{RDSW}{W} = \frac{345}{35} = 9.86 \cong 10 \quad \Rightarrow \quad R_S \cong 5 \text{ ohm}$$

(b) For $I_D = 1$ mA, calculate the decrease in the transconductance due to this series resistance and the necessary increase in the gate bias voltage to compensate the DC voltage drop on this resistance.

From (2.13b), the reduction of the transconductance is found as

$$\frac{g_{m(eff)}}{g_m} = \frac{1}{1 + g_m R_S}$$

$$g_m R_S = \sqrt{2\beta I_D} = \sqrt{2KP\frac{W}{L}I_D} \times R_S$$

$$g_m R_S = \sqrt{2 \times \left(170 \times 10^{-6}\right) \times 100 \times 10^{-3}} \times 5 = 0.029$$

$$\frac{g_{m(eff)}}{g_m} = \frac{1}{1 + 0.029} = 0.97$$

The DC voltage drop on R_S is $5 \times 10^{-3} = 5$ mV.

(c) Repeat (b) for $I_D = 10$ mA.

A factor of 10 increase of I_D increases $g_m R_S$ by a factor of $\sqrt{10} = 3.16$, to $0.029 \times 3.16 = 0.09$ Hence,

$$\frac{g_{m(eff)}}{g_m} = \frac{1}{1 + 0.09} = 0.92$$

The DC voltage drop on R_S is $R_S I_D = 5 \times (10 \times 10^{-3}) = 50$ mV.

Problem 2.1

Derive an expression for a velocity-saturated transistor to calculate the reduction of the transconductance as a function of the source region series parasitic resistance.

2.2   Active Transistor Loaded MOS Amplifier (CMOS Inverter as Analog Amplifier)

The passive PMOS load of the NMOS input transistor in the amplifier shown in Fig. 2.2a can be "activated" by connecting its gate to the input node, instead of a constant bias voltage (Fig. 2.9a). This circuit is nothing but a CMOS logic inverter, which is investigated in depth in all books on digital CMOS circuits [1, 2]. The input voltage-output voltage transfer curve of such an inverter is given in Fig. 2.9b, where the output logic (1) and logic (0) regions are marked with (A) and (C), respectively. In the transition region (B), the output voltage is – strongly and rather linearly – controlled by the input voltage. This means that if we bias the input node in such a way that the DC quiescent voltage of the output node is in the middle of this linear region (i.e., equal to $V_{DD}/2$), this circuit can be used as an analog amplifier. An even more advantageous solution is to design the circuit in such a way that the input DC bias voltage also is equal to zero, which permits direct coupling to single-ended signal sources and loads.

The voltage transfer curve of an inverter amplifier shown in Fig. 2.9b shows us that the output signal can change from zero to V_{DD} (rail to rail). But for analog amplification with low nonlinear distortion, it is convenient to exclude the upper and lower portions of the transfer curve and restrict the output signal dynamic range into the region where both M1 and M2 are in saturation mode.

In Fig. 2.9a, assuming that neither M1 nor M2 are in the velocity saturation regime, for an input voltage V_i and the corresponding output voltage V_o, the drain currents are

$$I_{D1} = \frac{1}{2}\mu_N C_{ox}\left(\frac{W_1}{L_1}\right)(V_i - V_{THn})^2(1 + \lambda_N.V_o)$$

$$I_{D2} = -\frac{1}{2}\mu_P C_{ox}\left(\frac{W_2}{L_2}\right)(V_i - V_{DD} - V_{THp})^2[1 + \lambda_P(V_{DD} - V_o)]$$

Fig. 2.9 The CMOS inverter as a linear amplifier (**a**) and its voltage transfer curve (**b**)

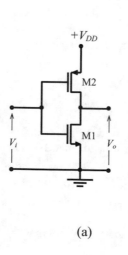

(a)

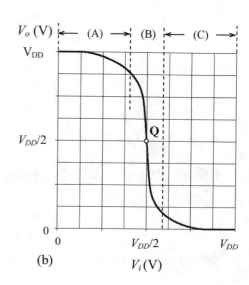

(b)

Fig. 2.10 (a) A CMOS inverter amplifier with R_F feedback resistor. (b) PSpice simulation results with no feedback resistor (A), with $R_F = 10$ k ohm (B) and with $R_F = 5$ k ohm (C). (Transistors: $L = 0.6$ μm, $W_1 = 6$ μm, $W_2 = 15$ μm)

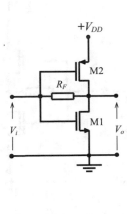

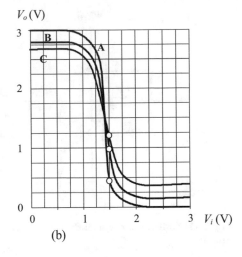

(a) (b)

and they are equal in magnitude ($I_{D1} = -I_{D2}$). This yields an expression which is useful for calculating the necessary conditions to set the operating point in the middle of the transfer curve (i.e., $V_i = V_o$, where both are equal to $V_{DD}/2$).

$$\frac{W_2}{W_1} = \frac{L_2}{L_1}\frac{\mu_N}{\mu_P}\left(\frac{V_i - V_{THn}}{V_i - V_{DD} - V_{THp}}\right)^2 \frac{1 + \lambda_N \cdot V_o}{1 + \lambda_P(V_{DD} - V_o)} \qquad (2.14)$$

Since the input voltage-output voltage transfer curve is very steep around the operating point as shown in Fig. 2.9b, the sensitivity of the output voltage with respect to the tolerances of the bias voltage and the transistor parameters is very high. Therefore, it is quite difficult to set the operating point at the desired position. The dimensions (ratios) calculated according to (2.14) usually do not give an operating point with sufficient accuracy in the middle of the linear operating range.[6]

A simple solution to improve the stability of the operating point is to connect a resistor between the input and output nodes as shown in Fig. 2.10a, thereby forcing the output voltage to approach the input DC voltage, which provides a current feedback to the input node. In Fig. 2.10b, simulation results are shown for the transfer curves of a CMOS inverter amplifier – which is intentionally not optimized – with and without a feedback resistor. It can be seen that the operating point indeed moves toward the middle of the output range as expected, however, the price of this improvement is a reduction of the slope (the small-signal voltage gain of the amplifier) and of the dynamic range.

Another possibility to extend the usability of this circuit is to use symmetrical twin power supplies as shown in Fig. 2.11. This solution provides two important advantages: (a) since the input DC bias voltage is zero, the AC (or pulse) signal source can be directly connected to the input node, (b) the load of the amplifier – which can be resistive or reactive – can be directly connected between the output node and the ground. It must be kept in mind that transistors must be able to deliver the necessary current to swing the voltage of the output node up or down to its maximum or minimum value. In case of a capacitive load, since the transistors are operating in the saturation region and hence acting as current sources, the current supplied by M1 or M2 discharges or charges the load

---

[6] Especially for small geometries, for which NMOS transistors usually operate in the velocity saturation region (but not the PMOS transistors, as explained in Chap. 1), (2.1) is not valid anymore. Therefore a fine-tuning with SPICE simulation becomes compulsory.

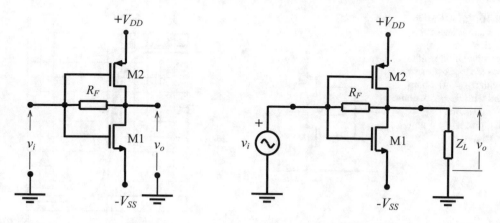

Fig. 2.11 CMOS inverter amplifier biased by two DC power supplies to force the input and output quiescent voltages to zero

capacitor with a constant slope. Consequently, the output voltage can reach its minimum or maximum value within a certain time. This "slew rate" effect will be investigated in Chap. 3.

The small-signal equivalent of this circuit and rearranged forms of it to facilitate the solution are given in Fig. 2.12a, b and c, respectively. From 2.12c, the voltage gain can be easily obtained as

$$A_v = -\frac{\overline{g}_m - G_F}{\overline{G}_L + G_F} \tag{2.15}$$

which reduces to

$$A_v = -\frac{\overline{g}_m}{\overline{G}_L} \tag{2.16}$$

for $G_F = 0$. Comparing (2.15) and (2.16), one can conclude that there are two separate signal paths from input to output: one amplifier path and a passive divider path. To prevent the domination of the passive path the condition of $\overline{g}_m > G_F$ must be fulfilled.

Due to the R_F feedback resistor, the input conductance of the amplifier is not zero anymore. It can be calculated from the small-signal equivalent circuit given in Fig. 2.12b as:

$$g_i = \frac{\overline{G}_L + \overline{g}_m}{\overline{G}_L + G_F} G_F \tag{2.17}$$

For $G_F \gg \overline{G}_L$ the input conductance is approximately equal to $\overline{G}_L + \overline{g}_m \cong \overline{g}_m$. It means that this is a low input resistance circuit; therefore, this circuit can also be considered as a "transresistance amplifier," suitable to use with high internal impedance sources.

The value of the transresistance can be calculated as

$$R_m = \frac{v_o}{i_i} = \frac{v_o}{v_i} \frac{v_i}{i_i} = A_v . r_i = -\frac{\overline{g}_m - G_F}{\overline{g}_m + \overline{G}_L} \frac{1}{G_F} \tag{2.18}$$

and reduces to

$$R_m \simeq -R_F$$

for $G_F, \overline{G}_L \ll \overline{g}_m$.

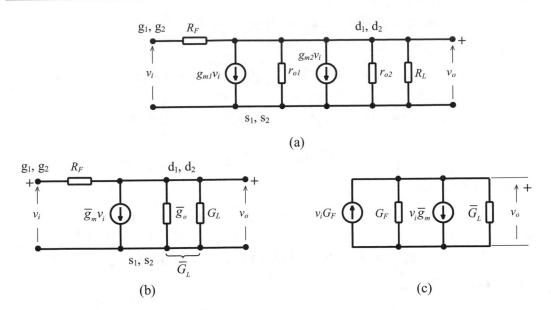

Fig. 2.12 The small-signal equivalent circuit of the inverter amplifier given in Fig. 2.10

Example 2.3

A CMOS inverter amplifier will be designed. The DC supply is a $\pm$ 1.5 V twin voltage source. The quiescent DC current must be 0.5 mA.[7] The design will be made for AMS 0.35 μm technology. The channel lengths of M1 and M2 will be chosen as 0.35 μm .

(a)   Calculate the channel widths of M1 and M2.

From the drain current expression of M1:

$$\left(\frac{W_1}{L}\right) = \frac{2I_{D1}}{(KP)_N(V_{GS1} - V_{THn})^2(1 + \lambda_N V_{DS1})}$$

$$\left(\frac{W_1}{L}\right) = \frac{2 \times \left(0.5 \times 10^{-3}\right)}{\left(170 \times 10^{-6}\right)(1.5 - 0.5)^2(1 + 0.073 \times 1.5)} = 5.3 \rightarrow W_1 \cong 1.86\mu m$$

The channel width of M2 can be calculated similarly, or from (2.14) as $W_2 \cong 6.4\mu m$.

(b) Calculate the voltage gain.

To calculate the voltage gain from (2.16), the transconductances and the output conductances of M1 and M2 must be found.

The β parameters of M1 and M2 are

$$\beta_1 = (KP)_N(W_1/L) = \left(170 \times 10^{-6}\right) \times 5.5 = 0.9 \times 10^{-3}\left[A/V^2\right]$$

---

[7] The value of the DC quiescent current is related to the slew-rate of the amplifier and must be high for a high slew-rate.

$$\beta_2 = (KP)_P(W_2/L) = (58 \times 10^{-6}) \times 18.9 = 1.096 \times 10^3 \ [A/V^2]$$

Transconductances

$$g_{m1} = \sqrt{2 \times (0.9 \times 10^{-3}) \times (0.5 \times 10^{-3})} = 0.95 \, \text{mS}$$

$$g_{m2} = \sqrt{2 \times (1.1 \times 10^{-3}) \times (0.5 \times 10^{-3})} = 1.05 \, \text{mS}$$

$$\overline{g}_m = 0.95 + 1.05 = 2 \, \text{mS}$$

Output conductances:

$$g_{o1} \cong \lambda_N I_{D1} = 0.073 \times (0.5 \times 10^{-3}) = 36.5 \times 10^{-6} \, \text{S}$$

$$g_{o2} \cong \lambda_P |I_{D2}| = 0.2 \times (0.5 \times 10^{-3}) = 100 \times 10^{-6} \, \text{S}$$

Since there is no external resistive load,

$$\overline{G}_L = (36.5 + 100) \times 10^{-6} = 0.1365 \, \text{mS}$$

Now the voltage gain:

$$A_v = -\frac{2 \times 10^{-3}}{0.1365 \times 10^{-3}} = -14.7 \, (23.4 \, \text{dB})$$

(c) Compare these results with PSpice simulation results.

With the calculated channel widths, PSpice gives a zero offset voltage transfer curve, but the quiescent current of the circuit is 0.2 mA, which is considerably lower than the targeted value. To increase the DC current to 0.5 mA, a procedure as explained in Example 2.1 must be applied. At the end of this fine-tuning, the channel widths are found as 2.5 μm and 15.5 μm, respectively, and the voltage gain as 18.4 (25.3 dB).

2.3 Common-Gate (Grounded-Gate) Amplifier

We have seen that the basic principle of a common source amplifier is to control the drain current with the gate-source voltage. This principle is also valid for the "common-gate" or the "grounded-gate" amplifier given in Fig. 2.13a. V_{SG} biases the transistor in inversion mode; a channel exists as long as $V_{GS} > V_{TH}$, and a current (I_D) flows through this channel. The DC voltage of the output node is $V_{DG} = V_{DD} - I_D R_D$. We know that the condition to keep the transistor in the saturation region is $V_{DS} \geq (V_{GS} - V_{TH})$. This condition can be expressed in terms of the output node quiescent voltage as

$$V_{DG} \geq -V_{TH} \tag{2.19}$$

This means that for a common-gate amplifier the DC voltage can decrease down to zero, or can even become negative. Consequently, the output voltage dynamic range of this circuit is larger than

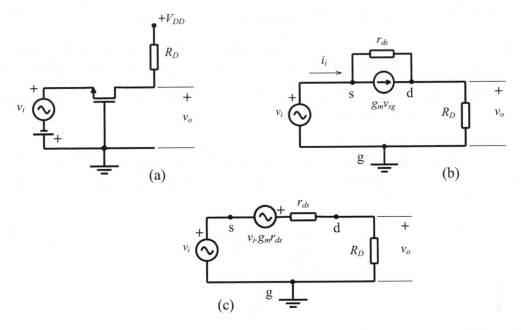

Fig. 2.13 (a) The resistor-loaded common-gate amplifier, (b) the small-signal equivalent circuit, and (c) the modified equivalent circuit following a Thevenin transformation

that of the common-source amplifier. The appropriate output quiescent voltage is the mid-point of the output dynamic range.

The circuit has an input current exactly equal to the output current, in other words, the current gain is equal to unity. If the input current is kept constant, the output of a common-gate circuit acts as an almost ideal current source. The utilization of this property for an important class of amplifiers, namely the "cascode" circuits, will be explained later.

The small-signal equivalent circuit of a common-gate amplifier (Fig. 2.13b) can be drawn with a rearrangement of the equivalent circuit given in Fig. 2.7, while Fig. 2.13c can be obtained with a Norton-Thevenin transformation. From Fig. 2.13c, the input current can be solved as

$$i_i = v_i \frac{1 + g_m . r_o}{r_o + R_D} \tag{2.20}$$

Since $v_o = i_i R_D$, the voltage gain can be calculated as:

$$A_v = \frac{v_o}{v_i} = \frac{1 + g_m r_o}{(r_o + R_D)} R_D \tag{2.21}$$

which can be reduced to

$$A_v \cong +g_m R_D \tag{2.21a}$$

for $g_m \gg (1/r_o)$ and $r_o \gg R_D$.

The comparison of (2.11a) and (2.21a) indicates that:

- The magnitude of the small-signal voltage gain of a common-gate amplifier is (at least, approximately) equal to the gain of a common-source amplifier using the same transistor dimensions, operating point, and load.
- Unlike the common-source amplifier, the sign of the gain of a common-gate amplifier is positive. This means that the output signal is in phase with the input signal.

Another important difference of the common-gate amplifier is its input resistance. It can be intuitively understood that, due to the high input current equal to the output current, the common-gate amplifier has a low input resistance. From (2.20), the small-signal input resistance can be calculated as

$$r_i = \frac{r_o + R_D}{1 + g_m \cdot r_o} \qquad (2.22)$$

and reduces to

$$r_i \cong \frac{1}{g_m} \qquad (2.22a)$$

for $g_m \gg (1/r_o)$ and $r_o \gg R_D$. This low input resistance circuit can also be considered as a transresistance amplifier for high-internal resistance signal sources, with:

$$R_m = R_D \qquad (2.23)$$

2.4   Common-Drain Amplifier (Source Follower)

The third basic single-transistor amplifier configuration is the common-drain amplifier, generally called the "source follower." The circuit diagram of a source follower is shown in Fig. 2.14a. The drain is directly connected to V_{DD} and the load resistance is connected between the source terminal and the ground. The DC current flowing through R_S is equal to the drain current in magnitude. The input DC bias voltage has to be equal to the sum of V_{GS} and the voltage drop on R_S due to the corresponding drain current.

The initial small-signal equivalent circuit and a rearranged form of it are given in Fig. 2.14b and c, respectively. From Fig. 2.14c, the small-signal voltage gain can be found as

$$A_v = \frac{g_m \cdot \overline{R}}{1 + g_m \cdot \overline{R}} \qquad (2.24)$$

where $\overline{R}$ is the parallel equivalent of the load resistor (R_S) and the output internal resistance of the transistor, corresponding to the operating point. (2.24) can be reduced to

$$A_v \cong +1 \qquad (2.24a)$$

for $g_m \cdot \overline{R} \gg 1$, which is valid for many practical cases. If we interpret (2.24a) together with the fact that there is a constant DC voltage difference equal to V_{GS} between the input terminal and the output terminal, we understand why this circuit is called a "voltage follower" (Fig. 2.14).

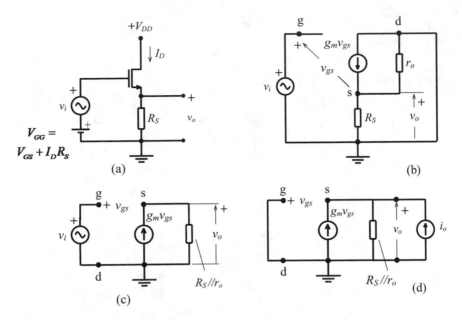

Fig. 2.14 (a) The basic source follower. (b) The small-signal equivalent circuit and (c) its rearranged form. (d) The small-signal equivalent circuit to calculate the output internal resistance

To understand better why the source follower has extensive use in electronic circuits in spite of unity voltage gain, the output internal resistance of the circuit must be calculated. In Fig. 2.14d, the equivalent circuit is arranged to calculate the output internal resistance according to the definition. From this circuit, the small-signal output internal resistance can be easily found as

$$r_o = \frac{v}{i_o} = \frac{\overline{R}}{1 + g_m \cdot \overline{R}} \tag{2.25}$$

which can be reduced to

$$r_o \cong \frac{1}{g_m} \tag{2.25a}$$

for $g_m \cdot \overline{R} \gg 1$. We know that the transconductance of a MOS transistor operating in the non-velocity saturated regime and velocity-saturated regime are

$$g_m \cong \sqrt{2\beta I_D} = \sqrt{2\mu C_{ox} \frac{W}{L} I_D}$$

$$g_m = k \cdot C_{ox} W \cdot v_{sat}$$

respectively. This means that it is possible to obtain small output internal resistance values using appropriate W and I_D values for a non-velocity saturated transistor. For a velocity saturated transistor, the solution is even simpler as the output internal resistance of the circuit is inversely proportional with the gate width. As already explained in principle, a "high-input impedance-low internal output impedance" circuit is useful as a buffer, for efficient signal transfer from a high-internal resistance (non-ideal) voltage source to a low-impedance load.

Fig. 2.15 (a) The source follower with a current source load. (b) Using twin power supplies for zero output DC voltage. (c) Using a MOS transistor as current source.

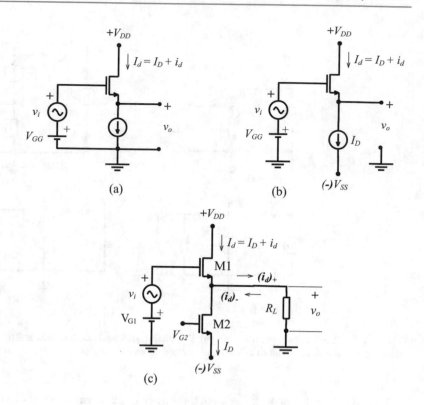

(a)   (b)

(c)

A different configuration of the source follower uses a current source instead of the resistor as shown in Fig. 2.15a. If there is a resistive load to be driven by this amplifier, it is usually more convenient to use twin DC power supplies to bring the quiescent voltage of the output node to zero and hence to prevent any DC current flowing through the load resistor (Fig. 2.15b). The DC current source connected to the source of the transistor is usually a MOS transistor, as shown in Fig. 2.15c. This transistor (M2) must be biased to conduct a current equal to the drain DC current of M1, when the voltage of the output node is equal to zero. Certainly, both of these transistors must remain in the saturation region for the whole output voltage dynamic range.

For a negative input signal, the drain (and source) current of M1 decreases by $(i_d)_-$. To maintain a constant current through the current source, a current equal to $(i_d)_-$ has to flow through R_L. Due to the direction of this current, the output voltage decreases. Depending on the value of the signal amplitude, I_{D1} can decrease down to zero. For this extreme case, $(i_d)_-$ becomes equal to I_S. To maintain M2 in the saturation region,

$$V_{SS} - I_{D2}.R_L = V_{DS2} \geq (V_{GS2} - V_{TH}) \tag{2.26}$$

must be fulfilled for a given value of R_L. The output voltage corresponding to this case (the negative peak value of the output voltage) is $\breve{v}_o = -I_S.R_L$.

For a positive signal voltage the drain current of M1 increases by $(i_d)_+$. This difference current flows through R_L, and the output voltage increases. For $v_{o(max)} = |v_{o(min)}|$, the maximum value of $(i_d)_+$ must be equal to I_S. For this case, the drain current of M1 becomes equal to $2.I_S$. To maintain M1 in saturation region,

$$\left(V_{DD} - v_{o(max)}\right) \geq \left(\hat{V}_{GS1} - V_{TH}\right) \tag{2.27}$$

must be satisfied, where $\hat{V}_{GS1}$ is the gate-source voltage of M1 corresponding to $I_{D1} = 2.I_S$.

Example 2.4

Design a source follower as shown in Fig. 2.16a for AMS 0.35 micron technology. The design goals are zero output quiescent voltage and minimum ±1 V output swing. The channel lengths are 0.35 micron. Calculate (a) channel widths and gate bias voltages, (b) the voltage gain, (c) output internal resistance of the amplifier, (d) with the calculated channel width values perform a PSpice simulation. Fine-tune the bias voltages to obtain the calculated currents necessary for the design goal.

(a)   Since the output quiescent voltage is zero and the output swing is ±1 V, the minimum value of the output voltage is $\breve{v}_o = -1$ V. To keep the power consumption minimum, at this extreme point of operation M1 must be driven to cut-off and the full current of the current source (M2) must flow from R_L. Therefore, $\breve{v}_o = -I_S R_L$ and $I_S = I_{D2} = 1/600 = 1.67$mA. Under these conditions, M2 must fulfill the condition given with (2.26), which is necessary to maintain M2 in the saturation region:

$$(V_{GS2} - V_{TH}) \leq 1.6 - \left[(1.67 \times 10^{-3}) \times 600\right] = 0.6V$$

With a 0.1 V safety margin, we will prefer to use $(V_{GS2} - V_{TH}) = 0.5$ V, since $V_{TH} = 0.5$ V, $V_{GS2} = 1$ V and $V_{G2} = -0.6$ V. This means that M2 must conduct 1.67 mA for $V_{GS2} = 1$ V. From the drain current expression of M2 the aspect ratio can be calculated as

$$\left(\frac{W}{L}\right)_2 = \frac{2 \times I_S}{(KP)_N (V_{GS2} - V_{TH})^2} = \frac{2 \times (1.67 \times 10^{-3})}{(170 \times 10^{-6}) \times (0.5)^2} \cong 78.5$$

which corresponds to $W_2 = 22.5$ μm.

At the maximum of the output voltage ($\hat{v}_o = +1$ V), a current of 1.67 mA must flow in the opposite direction through the load, R_L. For this extreme case, the current of M1 must be equal to the sum of this current and the current of M2, i.e., $\hat{I}_{D1} = 3.34$mA, and M1 must remain in the saturation region according to (2.27):

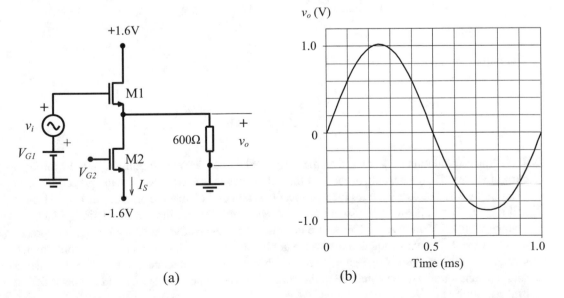

(a)   (b)

Fig. 2.16 (a) The schematic diagram of the source follower to be designed. (b) The PSpice.TRAN output for a 1.5 V amplitude, 1 kHz sinusoidal input signal

$$\left(\widehat{V}_{GS1} - V_{TH}\right) \leq 1.6 - 1 = 0.6\text{V}$$

where $\widehat{V}_{GS1}$ is the voltage corresponding to $\widehat{I}_{D1}$. With a 0.1 V safety margin we will use 0.5 V. Hence,

$$\left(\frac{W}{L}\right)_1 = \frac{2 \times \widehat{I}_{D1}}{(KP)_N \left(\widehat{V}_{GS1} - V_{TH}\right)^2} = \frac{2 \times \left(3.34 \times 10^{-3}\right)}{(170 \times 10^{-6}) \times (0.5)^2} \cong 157$$

which corresponds to $W_1 = 55$ μm.

Now the bias voltage of M1 under the quiescent condition must be calculated, corresponding to a drain current that is equal to I_S, in order to keep the load current equal to zero and to fulfill the zero output quiescent voltage condition:

$$(V_{GS1} - V_{TH})^2 = \frac{2 \times \left(1.67 \times 10^{-3}\right)}{(170 \times 10^{-6}) \times 157} = 0.125$$

which gives the bias voltage of M1 as $V_{G1} = V_{GS1} = 0.85$ V.

(b) The voltage gain can be calculated from (2.24). $\overline{R}$ in this expression is the parallel equivalent of the output resistances of M1 and M2, and the load resistance, which is approximately equal to R_L for this case. Therefore,

$$A_v \cong \frac{g_{m1}R_L}{1 + g_{m1}R_L}$$

The transconductance of M1 can be calculated as

$$g_{m1} = \sqrt{2(KP)_N(W/L)_1 I_{D1}} = \sqrt{2 \times \left(170 \times 10^{-6}\right) \times 157 \times \left(1.67 \times 10^{-3}\right)} = 9.4\text{mS}$$

Then the voltage gain can be found as:

$$A_v \cong \frac{\left(9.4 \times 10^{-3}\right) \times 600}{1 + \left[\left(9.4 \times 10^{-3}\right) \times 600\right]} = 0.85$$

(c) The output internal resistance from (2.25);

$$r_o \cong \frac{600}{1 + \left(9.4 \times 10^{-3}\right) \times 600} \cong 90\,\text{ohm}$$

(d) The PSpice simulation performed with the calculated values does not initially match the targeted current values. Therefore, a fine-tuning of the bias voltages is necessary, with the procedure used in Example 2.1. The bias voltages for a quiescent current of 1.67 mA are obtained as $V_{G1} = 1.34$ V and $V_{G2} = -0.3$ V. Figure 2.16b shows the transient simulation result for a 1 kHz sinusoidal input voltage with 1.5 V amplitude. This figure indicates that the positive and negative half periods of the output voltage are not equal, indicating a nonlinearity, which is normal for maximum voltage swing. In addition, the voltage gain is smaller than the hand-calculated value. This indicates a smaller transconductance value, due to the parasitic series source resistance and the degradation of the mobility.

Problem 2.2

A source follower as shown in Fig. 2.16a will be used to drive a transmission line having a characteristic impedance of 600 ohm. To prevent signal reflections, the transmission line must be terminated with resistors equal to its characteristic impedance, at both ends. The input impedance of the transmission line properly terminated at the output end is equal to the characteristic impedance, which is purely resistive. (a) Design the circuit. (b) Calculate the value of the maximum output swing. (c) Calculate the voltage gain. Improve the design with PSpice.

2.5 The "Long-Tailed Pair"

Probably the most important and most frequently used basic amplifier circuit is the "long-tailed" pair. Figure 2.17 shows several types of the long-tailed pair, along with a short evolution history. All of these circuits permit zero DC quiescent voltages for both inputs, provided that twin DC supplies (V_{DD} and V_{SS}) are used. In the circuits shown in Fig. 2.17, the input transistors are NMOS transistors. It is also possible (and in some cases advantageous) to use PMOS transistors as input transistors. In this case, it is obviously necessary to replace the PMOS load transistors with NMOS ones.

The basic structure of the long-tailed pair is shown in Fig. 2.17a. The circuit is symmetrical in nature; M1 and M11 are identical transistors and the load resistors are equal. Under quiescent conditions, M1 and M11 equally share the tail current, I_T, which is supplied by a passive MOS current source, M3.

Instead of the drain resistors, passive PMOS transistors (M2, M12) biased in their saturation regions can be used (Fig. 2.17b). Similar to the passive transistor loaded common-source amplifier, the biasing of the load transistors is very critical. To overcome this serious problem, and to fix the DC quiescent voltages of the output nodes, the – so-called – "common-mode feedback, CMFB" is an effective and extensively used technique, and will be investigated later.

In the configuration shown in Fig. 2.17c, the input transistors are loaded with diode-connected (low impedance) PMOS transistors (M2, M12) and their currents are mirrored to the load resistors via M4 and M14, to reduce the Miller effect. It is obvious that the mirroring coefficient (current gain) can be higher than unity, which serves to increase the voltage gain. Note that in this circuit both the input and the output quiescent voltages can be set to zero.

In the circuit shown in Fig. 2.17d, the load resistors are replaced with passive NMOS current sources (M5 and M15). The currents of these sources must be equal to $B \cdot (I_T/2)$, where B is the current gain of the PMOS current mirrors loading the input transistors. It is advantageous to bias M5 and M15 from the same DC voltage source that biases M3. In this case, the gate widths of M5 and M15 – in principle – must be equal to $(B/2) \cdot W_3$. Since the λ parameters of NMOS and PMOS transistors are (mostly) not equal, to bring the output quiescent voltages to the desired value (to zero in the case of a twin DC power supply), usually a fine-tuning on the channel widths of M5 and M15 becomes necessary. It is also important to mention that the sensitivity of the output DC quiescent voltage of this configuration is considerably lower than that of the circuit given in Fig. 2.17b. Therefore, the need for CMFB is not as severe.

If a symmetrical or differential output is not needed, there is an appropriate solution to obtain a single-ended output (Fig. 2.17e). This "active loaded long-tailed pair" configuration is extensively used as the input stage of operational amplifiers.

All these configurations (except the circuit given in Fig. 2.17e) are fully symmetrical in nature, and can be used in different operating modes:

- When a signal is applied to one of its input terminals and the other input is grounded, the circuit operates as a single-ended input amplifier providing two symmetrical output signals, v_o and v'_o

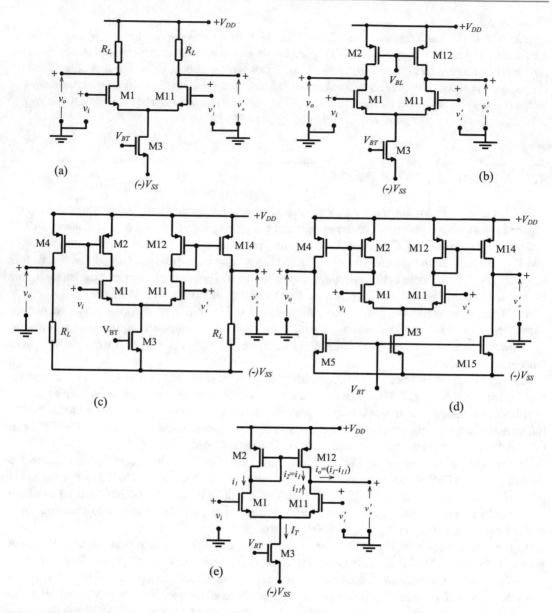

Fig. 2.17 Evolution of the basic differential amplifier: (**a**) Resistance loaded differential-input differential-output long-tailed pair. (**b**) The load resistors replaced with passive PMOS loads. (**c**) Current-mirror loads to reduce the Miller effect. (**d**) Replacement of the load resistors shown in (c) with passive NMOS loads. (**e**) The differential input – single-ended output amplifier

simultaneously, with opposite phase and equal amplitudes, each with respect to ground as shown in Fig. 2.18a.

- It is possible to use the difference of these two output signals as a floating differential output voltage: $v_{od} = (v_o - v'_o)$ (see Fig. 2.18b).
- The long-tailed pair can be used as a "difference amplifier" providing a differential (or twin single-ended) output voltage proportional with the difference of two single-ended, independent input signals, $(v_i - v'_i)$ (see Fig. 2.18c).

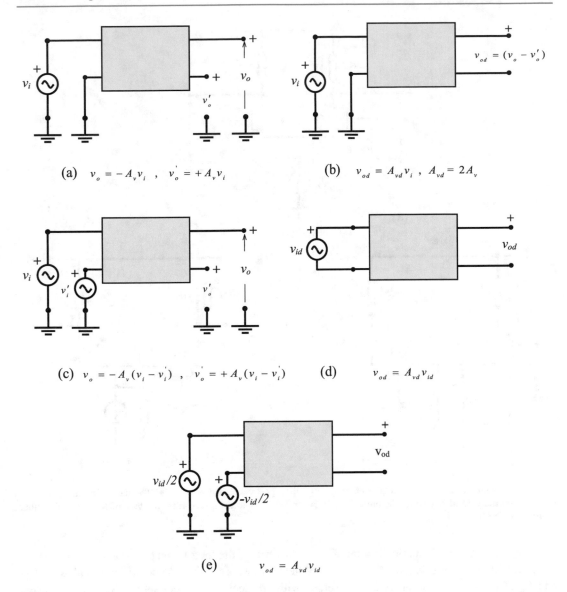

Fig. 2.18 (**a**) Single-ended input, twin single-ended outputs. (**b**) Single-ended input, differential output. (**c**) Difference amplification for two independent input signals. (**d**) Floating differential input. (**e**) Differential inputs with respect to ground

- A floating input signal (Fig. 2.18d) or an input signal differential with respect to the ground (Fig. 2.18e) can be applied between two inputs. In the case of a floating input signal, the gates of M1 and M11 must be connected to ground with two high-value resistors in order to provide DC bias. Since the output signal is also differential, the circuit is called "fully differential" or a "differential input-differential output" amplifier.
- In case the output signals are defined as currents, the circuit – by definition – is called a "transconductance amplifier," or "operational transconductance amplifier, (OTA)."

To find the gain expressions of the configurations given in Fig. 2.17, it is possible to draw the small-signal equivalent circuits and then calculate the gain. Here we will prefer a different approach, using the knowledge acquired in previous sections.

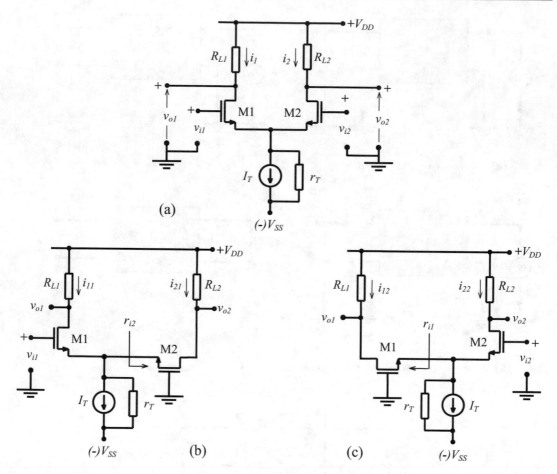

Fig. 2.19 Calculation of the signal currents in a long-tailed pair via superposition: (**a**) The basic circuit. (**b**) Circuit driven from the gate of M1 while the other input is grounded. (**c**) Circuit driven from the gate of M2 while the other input is grounded

In Fig. 2.19a, the schematic diagram of the basic form of the long-tailed pair is given. To enhance the generality of the results, the load resistors of the transistors are chosen with different values as R_{L1} and R_{L2}. The tail current source is represented with an ideal DC current source I_T, and its internal resistance (r_T). Under quiescent conditions, the drain currents of M1 and M2 are equal to $I_T/2$, and consequently, their transconductances are equal to $g_m = \sqrt{\beta_N I_T}$.

For the case of a signal applied to the gate of M1 where the other input is grounded, the circuit can be re-drawn as shown in Fig. 2.19b, in order to examine it from a different viewpoint. We can consider M1 as a transistor with a resistor (r_{i2}) in series to its source terminal. Here, r_{i2} is the small signal input resistance of M2 operating in the grounded gate configuration,[8] which according to (2.22a) is equal to $1/g_m$. The effective transconductance of a MOS transistor with a resistor in series to its source terminal (see (2.13b)) was found as

---

[8] The internal resistance of the tail current source is always much higher than the input resistance of the grounded gate circuit, therefore it can be neglected.

$$g_{m(eff)} = \frac{g_m}{1 + g_m R_S}$$

In this case, since the source resistance is equal to $1/g_m$, the effective transconductance of M1 is $g_{m(eff)} = g_m/2$. Hence the signal current of M1 that is the current flowing from R_{L1} becomes

$$i_{11} = v_{i1}\frac{g_m}{2} \quad \text{for } v_{i1} \neq 0 \text{ and } v_{i2} = 0 \tag{2.28}$$

Since the drain current of a grounded gate circuit is equal to its source current, the current flowing from R_{L2} is

$$i_{21} = -i_{11} = -v_{i1}\frac{g_m}{2} \tag{2.28a}$$

Similarly, in the case of $v_{i2} \neq 0$ and $v_{i1} = 0$ (Fig. 2.19c), the currents flowing from R_{L1} and R_{L2} are

$$i_{22} = v_{i2}\frac{g_m}{2} \tag{2.29}$$

and

$$i_{12} = -i_{22} = -v_{i2}\frac{g_m}{2} \tag{2.29a}$$

Under small signal conditions the circuit can be considered as linear, therefore, the superposition principle is applicable. Hence, the drain signal current components of M1 and M2 become

$$i_1 = i_{11} + i_{12} = \frac{g_m}{2}(v_{i1} - v_{i2}) \tag{2.30}$$

and

$$i_2 = i_{21} + i_{22} = -\frac{g_m}{2}(v_{i1} - v_{i2}) \tag{2.30a}$$

These results can be applied to the circuits given in Fig. 2.17, to calculate the voltage gain or the transconductance.

For the resistance loaded circuit shown in Fig. 2.17a, the output voltages are

$$v_o = -\overline{R}_L i_1 = -\overline{R}_L \frac{g_m}{2}(v_i - v_i') \tag{2.31}$$

where $\overline{R}_L$ is the parallel equivalent of R_L and the output resistance of M1.[9]

Similarly,

$$v_o' = -\overline{R}_L i_1' = \overline{R}_L \frac{g_m}{2}(v_i - v_i') \tag{2.31a}$$

and the differential output voltage,

$$\left(v_o - v_o'\right) = -\overline{R}_L g_m\left(v_i - v_i'\right) \tag{2.31b}$$

Therefore, the single-ended voltage gains,

---

[9] The load of such an amplifier is usually the gate input of the following stage, which is usually purely capacitive. If there is a resistive component of the load, certainly it must be taken into account as an additional parallel component to $\overline{R}_L$.

$$A_v = \frac{v_o}{(v_i - v_i')} = -\frac{1}{2}g_m\overline{R}_L \tag{2.32}$$

$$A_v' = \frac{v_o'}{(v_i - v_i')} = \frac{1}{2}g_m\overline{R}_L \tag{2.32a}$$

and the differential voltage gain,

$$A_{vd} = \frac{(v_o - v_o')}{(v_i - v_i')} = -g_m\overline{R}_L \tag{2.32b}$$

These expressions can be easily applied to the configurations given in Fig. 2.18 and interpreted accordingly.

For the passive transistor-loaded long-tailed pair shown in Fig. 2.17b, the gain expressions can be readily obtained by replacing R_L with the output resistance of M2 and M12.

In the circuit shown in Fig. 2.17c, the currents of M1 and M11 are mirrored to the load resistors via M2-M4 and M12-M14 current mirrors. If the mirroring coefficients of these mirrors are B, where B can be unity or larger, the gain expressions become

$$A_v = \frac{v_o}{(v_i - v_i')} = -\frac{1}{2}g_mB\overline{R}_L \tag{2.33}$$

$$A_v' = \frac{v_o}{(v_i - v_i')} = \frac{1}{2}g_mB\overline{R}_L \tag{2.33a}$$

$$A_{vd} = \frac{(v_o - v_o')}{(v_i - v_i')} = -g_mB\overline{R}_L \tag{2.33b}$$

where $\overline{R}_L$ is the parallel equivalent of R_L and the output resistance of M4 or M14.

For the voltage gain expressions of the circuit shown in Fig. 2.17d, $\overline{R}_L$ is the parallel equivalent of the mirror transistors M4 or M14 and the output resistances of the passive load transistors M5 or M15. This circuit is more suitable to use as an OTA. Since under quiescent conditions the currents of M4 and M5 (similarly M14 and M15) are equal, the output signal currents i_o and i_o' are zero. For v_i and v_i' input drives, the output currents are

$$i_o = -Bi_1 = \frac{1}{2}g_mB(v_i - v_i') \tag{2.34}$$

and

$$i_o' = -Bi_1' = -\frac{1}{2}g_mB(v_i - v_i') \tag{2.34a}$$

Therefore, the differential transconductance is

$$G_{md} = \frac{(i_o - i_o')}{(v_i - v_i')} = g_mB \tag{2.35}$$

For the single-ended circuit given in Fig. 2.17e, the signal current of M1 is mirrored to the output of M2. Since the current of M2 is equal in magnitude, but opposite in phase, the total signal current entering the drain node of M2 is equal to $2\,i_1$. Hence, the output voltage and the voltage gain become

$$v'_o = 2i_1 R'_L = g_m R'_L (v_i - v'_i) \tag{2.36}$$

and

$$A_v = \frac{v'_o}{(v_i - v'_i)} = g_m R'_L \tag{2.36a}$$

Example 2.5

A fully symmetrical OTA as shown in Fig. 2.17d will be designed for AMS 035 micron technology. The DC supply voltages are ±1.6V. The target value of the transconductance is 3 mS. The channel lengths of the active transistors, the quiescent current of the circuit and the B factor are chosen as 0.35 μm, 4 mA, and 1, respectively. The design will be made with hand calculations, and then fine-tuned with PSpice simulations.

Since $B = 1$, the drain DC current of the input transistors is equal to 1 mA, and the transconductance of the input transistors must be 3 mS. Then, the β coefficient and the aspect ratio of M1 and M11 can be calculated from $g_m = \sqrt{2\beta I_D}$ as

$$\beta_1 = \frac{g_{m1}^2}{2I_{D1}} = \frac{(3 \times 10^{-3})^2}{2 \times (10^{-3})} = 4.5 \times 10^{-3} \; [A/V^2]$$

and

$$\left(\frac{W}{L}\right)_1 = \frac{\beta_1}{(KP)_N} = \frac{4.5 \times 10^{-3}}{170 \times 10^{-6}} = 26.5 \rightarrow W_1 = 9.3\,\mu m$$

From the drain current expression,

$$(V_{GS1} - V_{THN})^2 \cong \frac{2I_{D1}}{(KP)_N (W/L)_1}$$
$$= \frac{2 \times (10^{-3})}{(170 \times 10^{-6}) \times 26.5} = 0.44 \rightarrow (V_{GS1} - V_{THN}) = 0.67\,V$$

The gate-source bias voltage of M1 (and M11) is found as $V_{GS1} = 1.17$ V. Since the quiescent voltage of both inputs are equal to zero, the DC voltage of the source of M1 and M11 is $V_{S1} = -1.17$ V.

We know that all transistors in this circuit must operate in the saturation region, even under worst-case conditions. The worst-case condition for M1 and M2 occurs when the full tail current (2 mA in our case) flows over this branch, under a high positive drive on the input of M1. For $\hat{I}_{D1} = I_T = 2$ mA, the gate drive can be found as $\left(\hat{V}_{GS1} - V_{THN}\right) = 0.94\,V$. Therefore, to maintain the saturation condition of M1, the condition of $V_{DS1} \geq \left(\hat{V}_{GS1} - V_{THN}\right)$ must be fulfilled. With a safety margin, we choose $V_{DS1} = 1$ V under maximum drive of M1. The voltage of the drain nodes of M1 and M2 can be found as $V_{D1} = V_{DS1} - V_{S1} = -0.17$ V. Then the gate-source voltage is -1.77 V and the drain current of M2 with this gate drive must be 2 mA. From these considerations, the aspect ratio and the gate width of M2 (and M12) can be calculated as

$$\left(\frac{W}{L}\right)_2 \cong \frac{2|I_{D2}|}{(KP)_P (V_{GS2} - V_{THP})^2}$$
$$= \frac{2 \times (2 \times 10^{-3})}{(58 \times 10^{-6}) \times [-1.77 - (-0.7)]} = 60.5 \rightarrow W_2 = 21.2\,\mu m$$

To obtain a high internal resistance, the channel length of the tail current source (M3) will be chosen as 1 μm. The aspect ratio and the channel width of M3 can be calculated such that the transistor is in the saturation region for $I_{D3} = I_T = 2$ mA. The drain-source voltage of M3 is $V_{DS3} = V_{S1} - V_{SS} = -1.17 - (-1.6) = 0.43$ V. Therefore, $V_{DS3} \geq (V_{GS3} - V_{THN})$ must be fulfilled. With a safety margin, we can choose $(V_{GS3} - V_{THN}) = 0.4$ V, therefore $V_{GS3} = 0.9$ V and $V_{G3} = -0.7$ V. Now, it is possible to find the aspect ratio of M3.

$$\left(\frac{W}{L}\right)_3 = \frac{2 \times (2 \times 10^{-3})}{(170 \times 10^{-6}) \times (0.4)^2} = 147 \rightarrow W_3 = 51.5\,\mu m$$

Since $B = 1$, the dimensions of M4 (and M14) are the same as M2.[10] Note that M5 and M15 are not on the signal path. Therefore, it is possible to use a longer channel length value, for example, 1 μm as in the tail transistor. The channel widths can be calculated for $I_{D5} = 1$ mA, $V_{DS5} = 1.6$ V, and $V_{G5} = V_{G3} = -0.7$ V as $W_5 = 73.5$ μm.

As in the previous examples, simulations performed with the calculated values do not initially match the targeted drain currents, but they provide a reasonable starting point for iterations. To maintain the saturation conditions of transistors, it is wise to keep the calculated bias voltages and fine-tune the widths of the transistors for targeted currents. The PSpice netlist obtained after the fine-tuning procedure is given below. The circuit diagram and the input voltage to output current transfer curves are shown in Fig. 2.20.

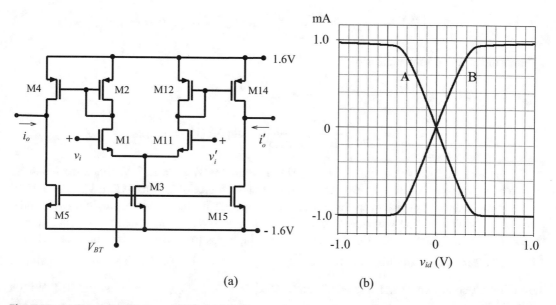

(a)   (b)

Fig. 2.20 (a) The circuit diagram. (b) The input voltage (v_{id}) to short-circuit output current transfer curve of the OTA; Curve A: I(R7), Curve B: I(17). The value of the transconductance is 3.016 mS

---

[10] If the drain-source voltages of M2 and M4 are not equal as in our example due to the channel shortening effect, the drain currents are not exactly equal. This only acts on the value of B, and consequently the final value of the transconductance. Certainly it is possible to fine-tune the width of M4 and M14 during simulations.

DESIGN:OTA-AMS 035

.lib "cmos7tm.mod"
VDD 100 0 1.6
VSS 200 0 -1.6
M1 6 1 3 200 modn L=.35U W=24U ad=20.5e-12 as=20.5e-12 Pd=48u Ps=48u
M11 5 2 3 200 modn L=.35U W=24U ad=20.5e-12 as=20.5e-12 Pd=48u Ps=48u
M2 6 6 100 100 modp L=.35U W=50U ad=42.5e-12 as=42.5e-12 PD=100U PS=100U
M12 5 5 100 100 modp L=.35U W=50U ad=42.5e-12 as=42.5e-12 PD=100U PS=100U
M3 3 4 200 200 modn L=1U W=230U ad=195.5e-12 as=195.5e-12 PD=460U PS=460U
M4 7 5 100 100 modp L=.35U W=48U ad=41e-12 as=41e-12 PD=96U PS=96U
M5 7 4 200 200 modn L=1U W=111U ad=92.4e-12 as=92.4e-12 PD=222U PS=222U
M14 17 6 100 100 modp L=.35U W=48U ad=41e-12 as=41e-12 PD=96U PS=96U
M15 17 4 200 200 modn L=1U W=111U ad=92.4e-12 as=92.4e-12 PD=222U PS=222U
R7 7 0 1
R17 17 0 1
vtb5 4 0 -0.7
VIN1 1 0 dc 0 ac 1m
e2 2 0 1 0 -1
.DC VIN1 -1 1 10M
.AC DEC 20 10MEG 10G
.PROBE
.END

Problem 2.3
It is intended to increase the transconductance of the circuit designed in Example 2.5 to 10 mS. One of the solutions is to increase the B factors of the load current mirrors. Modify the design in this way.

Problem 2.4
To improve the linearity of the circuit designed in Problem 2.3, an appropriate resistor can be connected in series to the source terminals of the input transistors, at the price of the reduction of the transconductances. Modify the design in this way to reduce the total transconductance of the circuit to 3 mS. Compare the transfer curves with that of Example 2.5.

Problem 2.5
Repeat the design given in Example 2.5 with PMOS input transistors. Discuss the advantages and disadvantages of this new design.

2.5.1   The Large Signal Behavior of the Long-Tailed Pair

From Fig. 2.20b, we see that the linear dependence between the input and the output voltages is limited to a certain region around the quiescent point. For larger differential input voltage swings, first the linearity of the output signal variation deteriorates, then the control ability of the input signal completely disappears. The calculated small signal behavior of the amplifier is valid only in the region where the output signal is a linear function of the input signal. For example, it can be seen from Fig. 2.20b that for the circuit designed above, the linear operation range is approximately ± 0.3 V

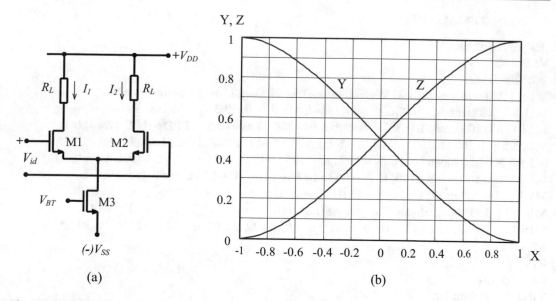

Fig. 2.21 (a) The basic long-tailed pair. (b) The normalized drain currents of M1 and M2 as a function of the differential input voltage. $\left(X = V_{id}/\sqrt{2I_T/\beta}, Y = I_1/I_T, Z = 1 - Y\right)$

around the quiescent point. For larger input signal amplitudes, nonlinear distortion and even clipping on the output signal is unavoidable.

In addition to this observation, Fig. 2.20b shows that for a sufficiently large differential input signal (approximately ±0.4 V for our example) the output current switches – almost completely – from one output to another. This is important information indicating that a long-tailed pair can be used as a switching circuit.

Since all differential circuits based on the long-tail structure exhibit the same basic properties, we will investigate the large signal behavior of the basic long-tailed pair shown in Fig. 2.21a. M1 and M2 are biased in their saturation region and equally share the tail current under quiescent condition (i.e., $v_{id} = 0$). The drain currents can be written in terms of the gate-source voltages as:

$$I_1 \cong \frac{1}{2}\beta\left(V_{gs1} - V_{TH}\right)^2 \qquad I_2 \cong \frac{1}{2}\beta\left(V_{gs2} - V_{TH}\right)^2$$

then

$$\left(V_{gs1} - V_{TH}\right) \cong \sqrt{\frac{2I_1}{\beta}} \qquad \left(V_{gs2} - V_{TH}\right) \cong \sqrt{\frac{2I_2}{\beta}}$$

The differential input voltage:

$$V_{id} = V_{g1} - V_{g2} = V_{gs1} - V_{gs2} = \sqrt{\frac{2I_1}{\beta}} - \sqrt{\frac{2I_2}{\beta}} \qquad (2.37)$$

with

$$I_2 = I_T - I_1$$

$$\frac{V_{id}}{\sqrt{2I_T/\beta}} = \sqrt{\frac{I_1}{I_T}} - \sqrt{1 - \frac{I_1}{I_T}} \tag{2.37a}$$

Using this normalized expression, the variations of I_1 and I_2 as a function of the differential input voltage v_{id} can be plotted as shown in Fig. 2.21b. As expected, for $V_{id} = 0$, $I_1 = I_2 = (I_T/2)$. I_1 monotonically increases with V_{id} and I_2 decreases, such that the sum of these currents remains equal to I_T. For

$$V_{id} = \sqrt{2I_T/\beta} \tag{2.38}$$

the tail current is completely switched to M1 and the drain current of M2 becomes zero. Similarly, for $V_{id} = -\sqrt{2I_T/\beta}$ the drain current of M2 becomes equal to I_T and I_1 reduces to zero. This means that the circuit can be used as a two-position switch, to toggle a current (the tail current) between two branches. According to (2.37), to switch a certain current, the input voltage switching interval is narrower for higher β values.

The simulation results of the single-ended and differential output voltages of a resistance-loaded long-tailed pair are given in Fig. 2.22a, as a function of the differential input voltage. The linear control regions of the transfer characteristics and the switching property of the circuit are clearly shown. The small-signal voltage gain can be derived as the slope of the transfer curve at the quiescent point.

One of the important aspects of a long-tailed pair is its behavior when the voltages applied to the inputs are identical, i.e., when the inputs are driven by a "common-mode signal." According to (2.31) and (2.31a), under common-mode input signal conditions (i.e., $v_i = v'_i$), the signal components of the output voltages must be zero. In other words, the voltages of output nodes must remain constant. But in reality, as shown by the simulation results in Fig. 2.22b, v_o and v'_o vary with the common-mode input signal, and their variations are identical. As seen from this figure, there are two different regions

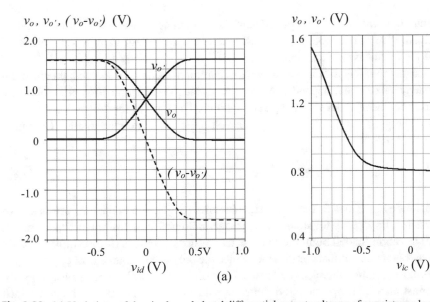

Fig. 2.22 (a) Variations of the single-ended and differential output voltages of a resistance loaded long-tailed pair for DC sweep of the differential input voltage (Input transistors: 35 µm/0.35 µm, the tail transistor: 90 µm/1 µm, the tail current: 0.8 mA, load resistors: 2 kΩ) (b) Variations of the single-ended output voltages for common-mode input voltage

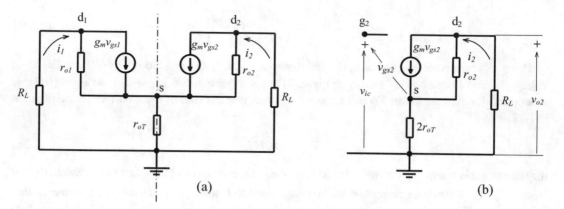

Fig. 2.23 (a) The small-signal equivalent circuit of a resistance loaded long-tailed pair (r_{oT} represents the internal resistance of the tail current source). (b) The half-circuit obtained by cutting the circuit from its symmetry axis for common-mode drive

of these curves. In the vicinity of the quiescent operating point (under small-signal conditions) the variations are small. But for large negative input voltages, the output voltages increase sharply, which is due to the transition of the tail transistor to the pre-saturation (resistive) region.

The rate of change of the single-ended output voltage with respect to the common-mode input signal under small-signal conditions is called the "common-mode gain" of the amplifier and can be calculated from the small-signal equivalent circuit shown in Fig. 2.23b, as the ratio of one of the output voltages to the common-input voltage:

$$A_{vc} = \frac{v_{o2}}{v_{ic}} = -g_m R_L \frac{1}{(1 + 2r_{oT} g_m) + \frac{2r_{oT} + R_L}{r_{o2}}} \approx -\frac{R_L}{2r_{oT}} \qquad (2.39)$$

It must be noted that for common-mode input signals, the difference between the output signals is zero and consequently, the "common-mode input to differential output voltage gain" is zero, provided that the symmetry of the circuit is perfect.

For a mixed input signal that has differential as well as common-mode components, the differential and common-mode gains can be independently calculated using the superposition principle since the circuit is assumed to be linear under small-signal conditions. The ratio of the "differential input to differential output voltage gain" to the "common-mode voltage gain" is defined as the "Common-Mode Rejection Ratio (CMRR)":

$$CMRR = \frac{A_{vd}}{A_{vc}} \approx 2g_m r_{oT} \qquad (2.40)$$

(2.40) indicates that for a high CMRR, the small-signal internal resistance of the tail current source must be as high as possible. Especially for small geometry technologies (small supply voltage values) it is not convenient to use complicated high internal resistance current sources (for example, cascode circuits) that contain more than one transistor in series. Since the tail current source is not in the signal path, it is possible to use long channel devices that provide higher internal resistance values. But since the channel widths must also be proportionally high, the parasitic capacitances are higher, which leads to lower internal impedances at high frequencies that deteriorate the CMRR at high frequencies, accordingly.

Problem 2.6

Calculate the transistor channel widths of a long-tailed pair that will be used to switch a 1 mA current, with a total input switching range of 200 mV. The parameters of this 0.18 μm technology are given as $V_{TH} = 0.4$ V, $\mu_n = 300$ cm^2/V.s and $C_{ox} = 8 \times 10^{-7}$ F/cm^2.

Problem 2.7

Derive (2.39).

2.5.2   The Common-Mode Feedback

We have seen that the output signals of a differential amplifier may have unwanted DC components that can arise from the non-proper biasing of the passive load transistors or can be the result of the non-zero common-mode gain under common-mode input signals. To eliminate, or at least to reduce the adverse effects of these unwanted DC components that occur equally on both outputs, a technique called "common-mode feedback" (CMFB) is a useful tool.[11] The common-mode feedback must be arranged in such a way that it has minimum effect on the differential gain of the amplifier, but forms an effective negative feedback to reduce the common mode gain and to control the DC components that are common for both outputs.

There are several CMFB circuits shown in the literature [1, 2]. The generally used approach for the common-mode feedback is summarized in Fig. 2.24. The output voltages have differential signal components and equal DC common-mode components, as shown in Fig. 2.24a. Assuming that the symmetry of the circuit is perfect, the difference of these signals, which is the differential output signal of the amplifier, has no common-mode component (see Fig. 2.24c). The sum of the output signals has no signal component, but a DC component, as shown in Fig. 2.24b. This sum of the common-mode components is compared with a reference voltage (which must be $2V_{oQ}$ to annul the V_{oCM}) and the resulting error signal is fed back to an appropriate node of the circuit in correct magnitude and phase, to eliminate – or reduce – the error (Fig. 2.24d).

If the output signals of the amplifier are voltages (as in a differential voltage amplifier) the feedback circuit must be designed to be effective on the common-mode voltages. But for OTAs, where the output signals of the circuit are currents by definition, the CMFB circuit must be designed to minimize the common-mode components of the output currents.

Example 2.6

As a practical example, a passive PMOS transistor-loaded long-tailed pair having a non-ideal tail current source is shown in Fig. 2.25a. The circuit was designed for 0.45 V DC level at the output nodes. The bias voltage of the load transistors was tuned as $V_{BL} = 0.674$ V to obtain this quiescent voltage. Figure 2.26a shows the very high sensitivity of the output quiescent voltages with respect to this bias voltage and it can be understood that this circuit is very sensitive to the variations of the bias voltage, as well as to the parameter tolerances of the transistors.

A simple CMFB circuit to control the output common-mode voltages is given in Fig. 2.25b. M6 and M7 in this circuit measure the voltages of the output nodes and produce related drain currents. These currents are summed on M10 and produce a gate-source voltage that is a function of the current. M8 and M9 are the same as M6 and M7, and biased with a reference voltage (V_{REF}). When

---

[11] Due to the tolerances of the devices, the symmetry of the circuit might not be perfect. For this case, the CMFB is only partially effective.

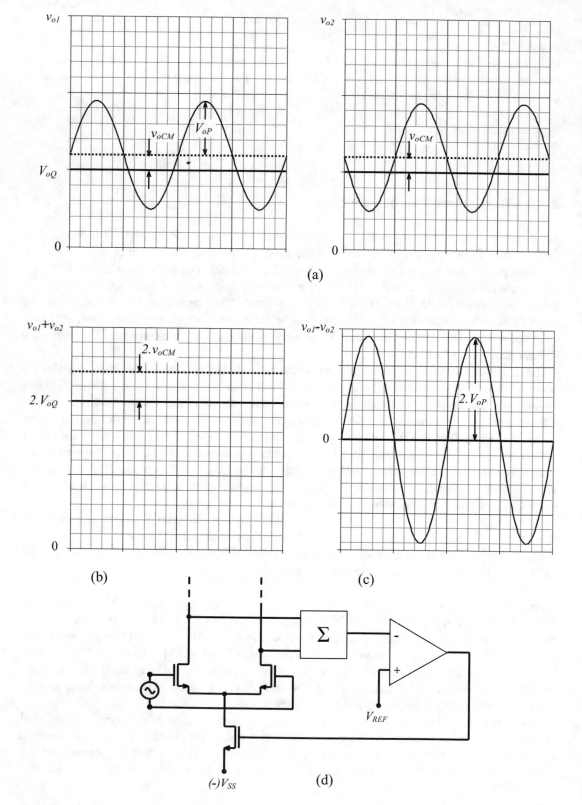

Fig. 2.24 (a) The voltages of the output nodes of a differential voltage amplifier. V_{oQ} and v_{oCM} represent the quiescent DC voltage and the CM component of the output voltage, respectively. (b) The sum and (c) the difference of the output voltages. (d) Schematic of the CMFB circuit

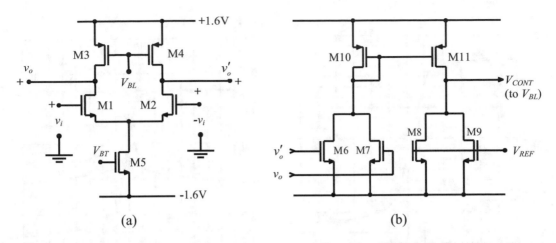

Fig. 2.25 (**a**) A passive transistor loaded long-tailed pair. Transistors are 0.35 μm AMS transistors. All channel lengths are 0.35 μm. Channel widths are 35 μm for the input transistor, 60 μm for the load transistors, and 10 μm for the tail transistor. (**b**) The simple voltage CMFB circuit. M6 to M9 are identical with L = 0.35 μm and W = 5 μm. To increase the gain of the current differencing amplifier, long channel-length devices are used as M10 and M11 (300 μm/2 μm)

the DC voltages of the output nodes are equal to V_{REF}, the drain voltage of M11 is equal to V_{G10}, which has the appropriate value to bias the load transistors. From another point of view, M10 and M11 work together as a current differencing amplifier[12] and produce the control voltage (V_{CONT}). If a CM voltage occurs on the output nodes, an error voltage component approximately proportional to the difference of the drain currents of M10 and M11 is produced and controls the load transistors to compensate the CM signal. From another point of view, this circuit compares the output common-mode voltage with the reference voltage and forces the circuit to reduce the error.

To apply CMFB to the amplifier shown in Fig. 2.25a, the outputs of the amplifier must be connected to the gates of M6 and M7, and the control voltage output of the CMFB circuit must be connected to the bias voltage terminal, V_{BL}, of the load transistors.

The single-ended and differential output voltages of this combination are shown in Fig. 2.26b under a differential input drive ($v_{i2} = - v_{i1}$). In Fig. 2.26c, the output voltages under a common-mode signal ($v_{i2} = v_{i1}$), without and with common-mode feedback are shown. It can be clearly seen that the CMFB effectively reduces the common-mode gain and clamps the output DC levels to the reference voltage. The small and relatively harmless difference between the single-ended output voltages and the reference voltage is the result of the finite loop gain of the feedback.

Example 2.7
One of the important classes of differential amplifiers is the fully differential OTA, which is extensively used in gyrator-based active filters. Since the output signals of OTAs are currents by definition, the common-mode feedback signal must be sampled from the sum of the output currents and compared to the targeted quiescent value (i.e., zero in most cases). Then, the produced error voltage must be applied to an appropriate node, in the correct phase. The output currents must be sampled without affecting the load currents. To solve this problem, in the circuit shown in Fig. 2.27a, the output transistors (M6, M7 and M16, M17) are duplicated (M26, M27 and M36, M37) and the replicated second set of output currents are used as the input signals of the feedback circuit shown in

---

[12] This circuit is inspired by the Norton Amplifier structure used in bipolar ICs in earlier years [2].

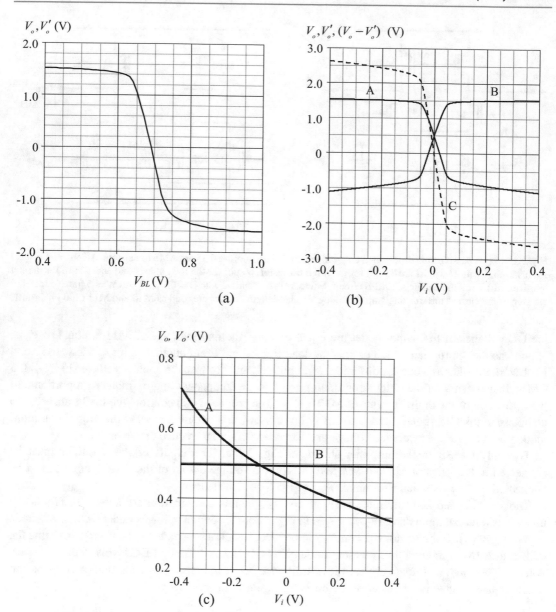

Fig. 2.26 (**a**) The variation of the single-ended output voltages of the long-tailed pair shown in Fig. 2.25a, as a function of the bias voltage of the load transistors. (**b**) The single-ended and differential output voltages of the CMFB applied amplifier for a differential drive. (**c**) The voltages of the output nodes as a function of the common-mode input signal: (A) without CMFB, (B) with CMFB

Fig. 2.27b. In this circuit, the replicated output currents are summed on a resistor (R51) and produce a feedback signal proportional to the sum of the output currents. The rest of the circuit is a simple differential amplifier, comparing the feedback signal with the target quiescent voltage and producing a suitable bias voltage for M7, M17, M27 and M37.

As an example for this current mode CMFB (CCMFB), the OTA shown in Fig. 2.27a is being designed. The tail current is chosen as 100 μA. The target transconductance value is 1 mS. In order to improve the linearity of the amplifier, series source resistances are used as shown in Fig. 2.28a. The

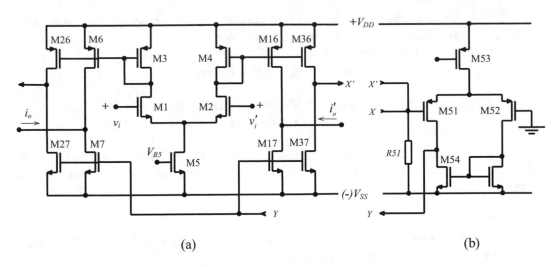

Fig. 2.27 (**a**) The schematic of the fully symmetric OTA that has separate replicated current outputs for feedback. (**b**) The simple current CMFB circuit used in this example

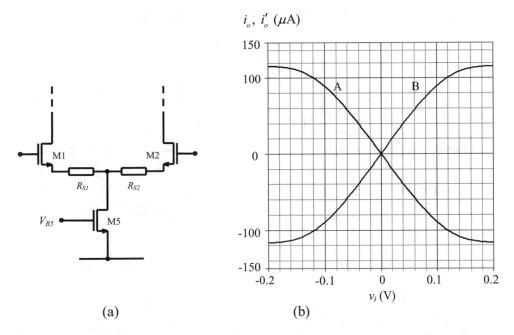

Fig. 2.28 (**a**) The source series resistances of the input transistors that decreases the effective transconductance but improves linearity as mentioned in Section 2.1. (**b**) The DC transfer curves of the OTA and the CMFB circuit given in Fig. 2.27. The input transistors have series source resistances. The dimensions of the transistor are as follows: M1, M2: 50 μm/0.35 μm, M3, M4, M6, M16, M26: 50 μm/0.35 μm, M5: 23 μm/1 μm, M7, M17, M27, M37: 13 μm/1 μm, M51, M52: 10 μm/1 μm, M54, M55: 2 μm/ 0.35 μm, M53: 40 μm/1 μm, $R_{S1} = R_{S2} = 385$ ohm, $R_{51} = 1$ kΩ. Bias voltages of M7, M17, M27, M37 are −0.7 V

core OTA was intentionally not well optimized to obtain zero output offset, such that the influence of CMFB can be observed more clearly. The results are as follows:

- The output offset currents   without   CCMFB: 11.27 µA.

                                  with   CCMFB: 0.115 µA

- The common-mode gain   without   CCMFB: 3.40 µS (CMRR = 49.4 dB)

                           with   CCMFB: 0.13 µS (CMRR = 77.7 dB)

The improvements on the values of the output offset current, the CMRR, and the shape of the simulated transfer curve given in Fig. 2.28b show the effectiveness of the common-mode feedback.

Problem 2.8

Discuss how we can improve the linearity and the CMRR of the OTA even further in the example, without (and with) increasing the current consumption.

Problem 2.9

(a) *An alternative solution to improve the linearity of a long-tailed pair is given in Fig. Prob. 2.9a. Show that this is equivalent to the classical approach given in Fig. Prob. 2.9b.*

(b) *Compare and discuss the advantages and disadvantages of these two solutions.*

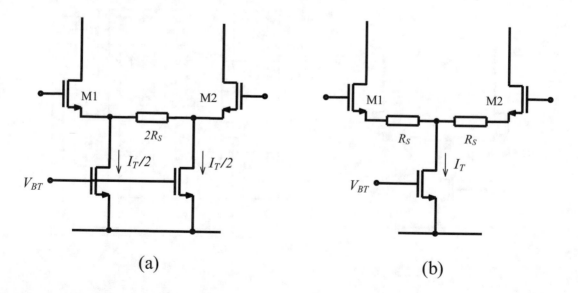

References

1. S.-M. Kang, Y. Leblebici, *CMOS Digital Integrated Circuits, Analysis and Design*, 3rd edn. (McGraw-Hill, New York, 2003)
2. A.B. Grebene, *Bipolar and MOS Analog Integrated Design* (Wiley, Hoboken, 1984)

High-Frequency Behavior of Basic Amplifiers

3

It was already discussed in earlier chapters that the amplitude and the phase responses of amplifiers change with frequency, either intentionally or unintentionally. In some applications, a specific frequency response is desired, for example, a band-pass characteristic for an LNA (here, LNA stands for "low-noise amplifier" – the low noise input stage of a receiver). To shape the frequency response according to our needs, we use reactive components such as inductors and capacitors. For other applications, a flat frequency response is required; however, gain inevitably drops at higher frequencies. The reason for this "unintentional" change of the frequency response is the "parasitic" components of the circuit: i.e., all non-avoidable reactive (usually capacitive) components related to the devices and the interconnections. To investigate the essential frequency-dependent behaviors of amplifiers, it is necessary to improve the small-signal equivalent circuit of a MOS transistor developed in Chap. 2.

The small-signal equivalent circuit of a MOS transistor biased at a certain operating point in the saturation region was given in Chap. 2, Fig. 2.6. To extend the usability of this equivalent circuit to high frequencies (in other words, to radio frequencies), it is necessary to add the parasitic capacitances and the parasitic resistances that were discussed in Chap. 1, as shown in Fig. 3.1a. In this equivalent circuit, $R_{D'}$, $R_{S,}$ and R_G are the series resistances of the corresponding regions of the device, where d', s', and g' represent the so-called internal nodes that are inaccessible from the device terminals; d, s, and g represent the external terminals. $C_{g's'}$ is the gate-source capacitance, $C_{d'g'}$ is the drain-gate overlap capacitance. C_{sb} and C_{db} are the junction capacitances of the source and drain regions and R_{sb} and R_{db} are their series resistances.

Although the equivalent circuit given in Fig. 3.1a models the small-signal behavior of a MOS transistor in the RF region with good accuracy, it is not convenient for hand calculations. To be able to obtain manageable and interpretable expressions, it is necessary to simplify this equivalent circuit. The reduced equivalent circuit shown in Fig. 3.1b is simple enough for hand calculations and provides reasonable accuracy to understand the basic behavior of the circuits, prior to the detailed SPICE simulations.

The small-signal equivalent circuits of an arbitrary electronic circuit, valid at high frequencies, can be constructed according to the basic rules given in Chap. 2, and then solved using conventional network analysis techniques. One of these techniques is the well-known method called the "Miller

The original version of this chapter was revised. The correction to this chapter is available at https://doi.org/10.1007/978-3-030-63658-6_8

D. Leblebici, Y. Leblebici, *Fundamentals of High Frequency CMOS Analog Integrated Circuits*, https://doi.org/10.1007/978-3-030-63658-6_3, Corrected Publication 2021

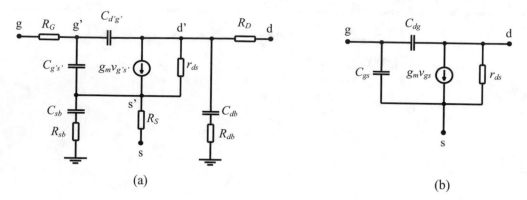

Fig. 3.1 (a) The RF small-signal equivalent circuit of a MOS transistor that contains all parasitics. (b) The simplified version for hand calculations

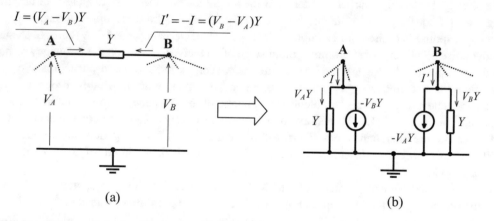

Fig. 3.2 Development of the (modified) Miller conversion

Theorem" that has been extensively used since the early days of RF electronics [1]. The method was developed to understand and calculate the unexpected effects of the anode-to-grid capacitance of a triode tube, on the input impedance of an amplifier. Before going any further, a modified form of this method, useful for straightforward calculation of the gain function as well as the input and output impedances, will be presented below:

Assume that there is an admittance Y between any two nodes (say A and B) of a circuit. It is possible to *replace* this admittance with two branches, connected between these nodes and the ground, without changing the total current balance of these nodes. The development of this conversion is given step-by-step in Fig. 3.2. The original circuit that contains an admittance Y between the nodes A and B is shown in Fig. 3.2a. The outgoing currents from these nodes due to Y are shown as I and I' where $I = -I'$. In the converted circuit, the parallel branches replacing Y have currents equal to I and I', respectively. One of the components of these currents is proportional to the node voltages ($V_A Y$ and $V_B Y$) and can be represented with two admittances equal to Y, connected between these nodes and the ground. The other component of the currents on the parallel branches is proportional to the voltages of the other node, and must be represented with voltage-controlled current sources ($-V_B Y$) and ($-V_A Y$), as shown in Fig. 3.2b.

At first sight, it seems that this transformation increases the complexity of the circuit. But it will be seen that it simplifies the calculation of gain and input and output impedances of amplifiers containing a path between the input and output nodes which is usually the case.

3.1   High-Frequency Behavior of a Common Source Amplifier

The high frequency small-signal equivalent circuit of a common source amplifier (either resistive-load as shown in Fig. 2.1a, or passive transistor loaded as shown in Fig. 2.2a) is given in Fig. 3.3a. The load of the amplifier is shown with Y_L which can be any resistive or reactive (containing inductors and/or capacitors as well) admittance. Applying the Miller transformation as explained above, the equivalent circuit can be rearranged as shown in Fig. 3.3b and c. From Fig. 3.3c, the output voltage v_{ds} can be directly calculated in terms of the input voltage, v_{gs} as

$$v_{ds} = -\frac{g_m - sC_{dg}}{Y_o} v_{gs} \tag{3.1}$$

where Y_o is the sum of the load admittance and the output internal admittance of the amplifier which is the parallel equivalent of g_{ds} and C_{dg}. As a more realistic approach, it is useful to add the parasitic

Fig. 3.3 Development of the small-signal equivalent circuit of the common source amplifier: (**a**) original equivalent circuit, (**b**) equivalent circuit after Miller transformation (**c**) equivalent circuit after simplification

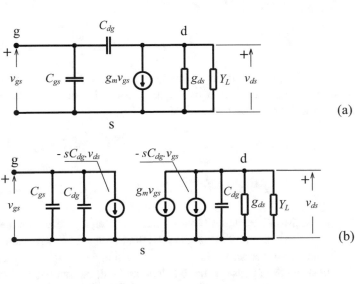

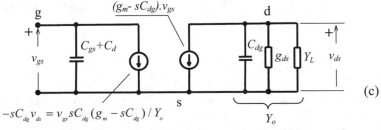

Fig. 3.4 Modification of
the equivalent circuit for
calculation of input
admittance

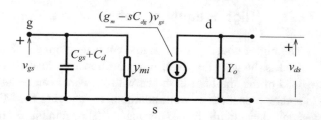

capacitance between the drain and the ground C_{op}, which is composed of the drain-bulk capacitance C_{db} and the interconnection parasitics. Now the voltage gain of the circuit can be written as

$$A_v = -\frac{g_m - sC_{gs}}{Y_o} \tag{3.2}$$

Using (3.1), the current source on the left-hand side of Fig. 3.3b can be expressed in terms of v_{gs} as shown in Fig. 3.3c:

$$i = v_{gs}\left(g_m - sC_{dg}\right)\frac{sC_{dg}}{Y_o} = -v_{gs}A_v sC_{dg}$$

In Fig. 3.4, the equivalent circuit is rearranged by replacing the current source with the "Miller admittance" y_{mi} connected between the gate and the source nodes. From Fig. 3.4, the input admittance of the amplifier can be found as

$$y_{mi} = sC_{dg}\frac{g_m - sC_{dg}}{Y_o} = sC_{dg}(-A_v) \tag{3.3}$$

$$y_i = s\left(C_{gs} + C_{dg}\right) + y_{mi} = s\left(C_{gs} + C_{dg}\right) + C_{dg}\frac{g_m - sC_{dg}}{Y_o} \tag{3.3a}$$

From (3.2) and (3.3), it is obvious that the voltage gain and the input admittance depend on the load admittance, which in many cases is the dominant part of Y_o. Therefore, it is necessary to investigate the frequency-dependent behavior of the gain and the input admittance for typical cases of the load. The most important case is the R-C load and will be investigated in the following section. The other important case, the L-C load (tuned load), will be dealt with in detail in Chap. 4.

3.1.1   The R-C Load Case

The basic small-signal equivalent circuit and the circuit after application of the Miller transformation of an R-C loaded common source amplifier are given in Fig. 3.5a and b, respectively. Note that even for the case of pure resistive load at the output node, the actual conditions correspond to the case of an R-C loaded amplifier, because of the existence of C_{dg} and the output parasitics. The total admittance at the output node ($Y_o = G_o + sC_o$) is the parallel equivalent of the external load, $Y_L = G_L + sC_L$, and $g_{ds} + s(C_{dg} + C_{op})$, where C_{op} represents the total parasitic capacitance of the output node.

Fig. 3.5 The small-signal equivalent circuit of an R-C loaded amplifier (**a**) before Miller transformation, (**b**) after Miller transformation

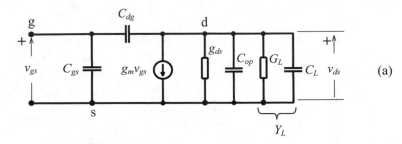

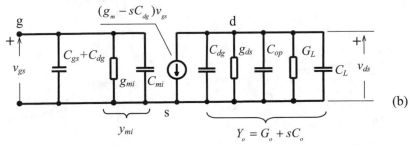

Now the voltage gain of the amplifier can be calculated from (3.2):

$$A_v = -\frac{g_m - sC_{dg}}{G_o + sC_o} = \frac{C_{dg}}{C_o}\frac{(s - s_o)}{(s - s_p)} \tag{3.4}$$

where

$$s_o = +\frac{g_m}{C_{dg}} \quad \text{and} \quad s_p = -\frac{G_o}{C_o} = -\frac{(g_{ds} + G_L)}{(C_{op} + C_{dg} + C_L)} \tag{3.4a}$$

are the zero and the pole of the voltage gain function. The low frequency ($s \to 0$) voltage gain of the amplifier can be easily obtained from (3.4) as

$$A_v(0) = -\frac{g_m}{G_o} = -\frac{g_m}{(G_L + g_{ds})} \tag{3.5}$$

The magnitude and phase characteristics of the amplifier can be obtained from the pole-zero diagram of the gain function (Fig. 3.6).

The following conclusions can be drawn from the characteristics:

- The magnitude of the gain decreases with frequency and drops to $1/\sqrt{2}$ (or 3 dB below) of its low-frequency value at ω_p corresponding to the pole
- For $\omega > \omega_p$, the magnitude characteristic of a R-C loaded amplifier decreases with a 20 dB/decade slope, up to the frequency corresponding to the "zero" of the gain function, which is usually much higher than the pole frequency (or the −3 dB frequency)

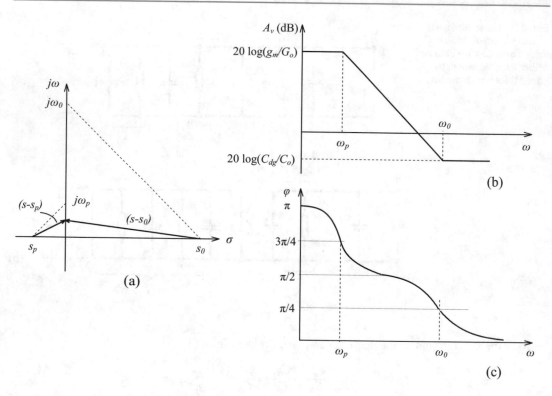

Fig. 3.6 (a) The pole-zero diagram of the R-C loaded amplifier. (b) The magnitude characteristic of the voltage gain with its asymptotes. (c) The approximate phase characteristic

- The -3 dB frequency of the amplifier is inversely proportional to the total output node capacitance. This frequency (or the bandwidth of the amplifier) can be set to a lower value by using an appropriate parallel capacitor connected to the output
- The phase of the gain is $180°$ at low frequencies, decreases to (approximately) $135°$ at the pole frequency and tends asymptotically to $90°$. But due to the existence of the right half-plane zero, it continues to decrease and reaches (approximately) $45°$ at the zero frequency and becomes asymptotic to zero at very high frequencies. This additional $90°$ phase shift, compared to the hypothetical case of $C_{dg} = 0$, can be the source of certain problems in feedback amplifiers.
- The gain-bandwidth product of the amplifier can be calculated from (3.5) as

$$GBW = |A_v| \times \frac{1}{2\pi} |s_p| \cong \frac{g_m}{2\pi C_o} \qquad (3.6)$$

The input admittance can be calculated from (3.3a), using the Miller transformation for this case. The Miller admittance can be calculated from (3.3) as

$$y_{mi} = sC_{dg} \frac{g_m - sC_{dg}}{G_o + sC_o} \qquad (3.7)$$

and in the frequency domain,

$$y_{mi}(\omega) = j\omega C_{dg}\frac{g_m - j\omega C_{dg}}{G_o + j\omega C_o} = g_{mi}(\omega) + jb(\omega) \tag{3.8}$$

The real and the imaginary parts of this admittance are

$$g_{mi}(\omega) = \frac{C_{dg}}{C_o}\left(g_m + G_o\frac{C_{dg}}{C_o}\right)\frac{1}{(\omega_p/\omega) + 1} \tag{3.9}$$

and

$$b_{mi}(\omega) = \omega\frac{C_{dg}^2}{C_o}\frac{\omega_o\omega_p - \omega^2}{\omega_p^2 + \omega^2} = \omega C_{mi} \tag{3.9a}$$

where ω_p and ω_o are the frequencies corresponding to the pole and the zero of the voltage gain function.

Equation (3.9) shows that there is an *unexpected* real part of the input admittance, which manifests itself as a frequency-dependent input conductance. The variation of this conductance as a function of ω is plotted in Fig. 3.7a. The value of this conductance is zero for $\omega = 0$, and increases with frequency up to

$$g_{mi}(\omega \to \infty) = \frac{C_{dg}}{C_o}\left(g_m + G_o\frac{C_{dg}}{C_o}\right) \tag{3.10}$$

Fig. 3.7 Variation of (**a**) the input conductance and (**b**) the input capacitance of the common source amplifier

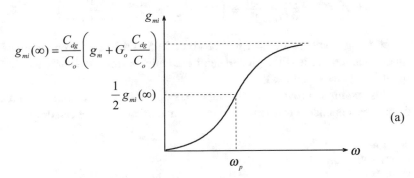

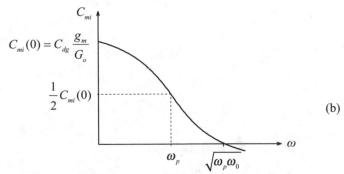

The input conductance at the pole frequency of the amplifier is

$$g_{mi}(\omega_p) = \frac{g_{mi}(\omega \to \infty)}{2} = \frac{1}{2}\frac{C_{dg}}{C_o}\left(g_m + G_o\frac{C_{dg}}{C_o}\right) \qquad (3.11)$$

Since the gate is isolated from all other parts of the device with the gate oxide, which is – almost – an ideal insulator, the input conductance of a MOS transistor is generally assumed to be zero. This is true for the DC case, and the conductance is negligibly small at low frequencies. But due to the existence of the drain-gate parasitic capacitance as seen from (3.9), this assumption is not true for the AC case in general. Especially at higher frequencies approaching the pole frequency of the amplifier, the input conductance can reach surprisingly high values.

Example 3.1
For the amplifier given in Fig. 3.8, the DC drain current (bias) of the transistor is $I_D = 150\ \mu A$, therefore, $V_{DS} = 1.5$ V. For this operating point, the small-signal parameters are $g_m = 2$ mS, $C_{dg} = 20$ fF, $C_{gs} = 90$ fF, and $g_{ds} = 30\ \mu S$. The parasitic capacitance of the output node is assumed to be 10 fF.

The low-frequency gain and the pole frequency can be calculated from (3.5) and (3.4a) as

$$A_v(0) = -\frac{2 \times 10^{-3}}{(30 \times 10^{-6} + 10^{-4})} = -15.38 \quad \Rightarrow \quad 23.74\ \text{dB}$$

$$f_p = \frac{1}{2\pi}\frac{(30 \times 10^{-6} + 10^{-4})}{(10 \times 10^{-15} + 20 \times 10^{-15} + 50 \times 10^{-15})} = 258.7\ \text{MHz}$$

The input conductance for this frequency calculated from (3.11) is

$$g_{mi}(f_p) = \frac{1}{2}\frac{20 \times 10^{-15}}{80 \times 10^{-15}}\left(2 \times 10^{-3} + 10^{-4}\frac{20 \times 10^{-15}}{80 \times 10^{-15}}\right) = 0.253 \times 10^{-3}\ \text{siemens}$$

which corresponds to an input resistance of 3.95 kΩ! Even for one-tenth of the pole frequency, the input conductance is still high and can be calculated from (3.9) as 4.95 µS, which corresponds to an input resistance of 200 kΩ.

This example proves that the input conductance of a common source MOS amplifier has a frequency-dependent resistance component and this resistance can drop to considerably low values –

Fig. 3.8 The R-C loaded
MOS amplifier investigated
in the example

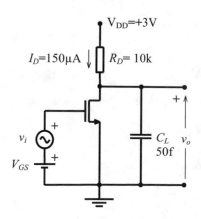

which influences the output load of the previous stage or the signal source. It is obvious that especially for high internal impedance signal sources, this frequency-dependent input conductance of the amplifier has to be taken into account.

Note that the input capacitance of the amplifier is the sum of C_{gs}, C_{dg}, and C_{mi}. In Fig. 3.7b, the variation of the Miller capacitance is plotted as a function of ω. This shows that the third component of the input capacitance has a value proportional with g_m/G_o, which is the magnitude of the Miller capacitance at low frequencies, and is usually high.

For the example given above, the value of the Miller capacitance is

$$C_{mi}(0) = C_{dg}\frac{g_m}{G_o} = 20 \times 10^{-15}\frac{2 \times 10^{-3}}{130 \times 10^{-6}} = 307.7\text{fF}$$

and the total input capacitance is

$$C_i = C_{gs} + C_{dg} + C_{mi} = 90 + 20 + 307.7 = 417.7\text{fF}$$

of which the dominant part is the Miller capacitance. C_i capacitively loads the previous stage (or the driving signal source). This high capacitive load affects the high-frequency performance of the previous stage and is the main reason for the deterioration of the high-frequency gain of multistage amplifiers.

In Fig. 3.9, the PSpice simulation results are shown for the amplifier given in Fig. 3.8. The transistor is an AMS 0.6 μm NMOS device with $L = 0.6$ μm, $W = 60$ μm. The drain current is 150 μA. The small-signal parameters of this transistor are approximately equal to the values used in the example. The frequency axes for the magnitude and phase characteristics (Figs. 3.9a and b) are intentionally drawn up to 30 GHz, to see the effects of the zero of the gain function. The variations of the input admittance and the input capacitance, which is the sum of C_{gs}, C_{dg}, and the C_{mi} Miller capacitance, are given with Fig. 3.9c and d. The simulation results reasonably agree with the values calculated using the analytical expressions.

3.2   The Source Follower Amplifier at Radio Frequencies

As explained in Chap. 2, the source follower is typically used as a buffer to couple a high internal impedance source to a low impedance load. The simplified circuit diagram of a source follower is given in Fig. 3.10a. The load can have any form: a capacitive (R-C) load, an inductive (R-L) load, or a tuned (L-C) load. In this section, we will deal with the R-C load only and leave the tuned load case to Chap. 4 (Frequency-Selective RF Circuits). The small-signal equivalent circuit of an R-C loaded source follower is given in Fig. 3.10b, and rearranged in Fig. 3.10c to ease the solution, where

$$G = G_s + g_{ds} \text{ and } C = C_s + C_{op}$$

From 3.10c, the output and input voltages can be written as

$$v_o = \frac{i_i + g_m v_{gs}}{Y}, \ v_i = v_{gs} + v_o \tag{3.12}$$

where

$$v_{gs} = i_i/sC_{gs} \text{ and } Y = G + sC \tag{3.13}$$

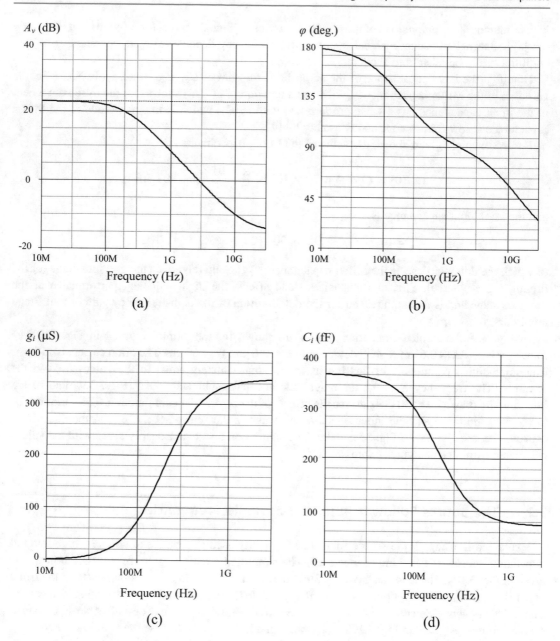

Fig. 3.9 PSpice simulation results for the circuit shown in Fig. 3.8: (**a**) The magnitude characteristic, (**b**) the phase characteristic of the voltage gain, (**c**) the variation of the input conductance, (**d**) the variation of the input capacitance as a function of frequency. The important figures of performance are $A_v(0) \cong 23$ dB, $f_p \cong 192$ MHz, $g_i(f_p) \cong 230$ μS, $C_i(0) \cong 360$ fF

From (3.12) and (3.13), the voltage gain can be solved as

$$A_v = \frac{g_m + sC_{gs}}{(g_m + G) + s(C_{gs} + C)} \tag{3.14}$$

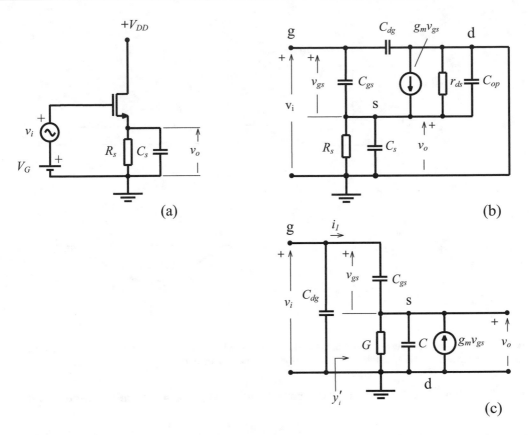

Fig. 3.10 The source follower amplifier circuit: (**a**) simplified circuit diagram, (**b**) small-signal equivalent circuit, and (**c**) rearranged equivalent circuit for calculation of gain and input impedance

which can also be expressed in terms of the pole and zero frequencies as

$$A_v = \frac{C_{gs}}{(C_{gs} + C)} \frac{(s - s_0)}{(s - s_p)} \tag{3.14a}$$

where

$$s_0 = -\frac{g_m}{C_{gs}} \quad \text{and} \quad s_p = -\frac{(g_m + G)}{(C_{gs} + C)} \tag{3.15}$$

The low-frequency gain can be seen from (3.14) as

$$A_v(0) = \frac{g_m}{(g_m + G)}$$

which – naturally – fits to (2.24). The magnitude and phase characteristics of the amplifier can be obtained from the pole-zero diagram of the gain function (Fig. 3.11).

From Fig. 3.10c, it can be seen that the input admittance of the source follower is the sum of the admittance shown with y_i' and (sC_{dg}). y_i' is i_i/v_i, and i_i can be calculated as

$$i_i = (v_i - v_o)sC_{gs} = v_i(1 - A_v)sC_{gs}$$

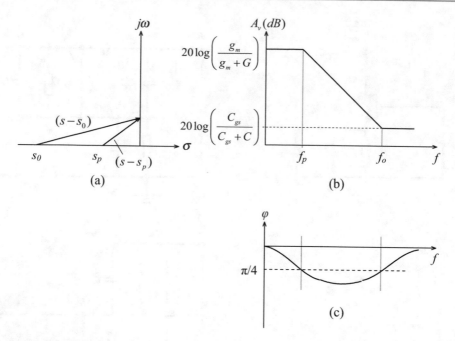

Fig. 3.11 (**a**) The voltage gain pole-zero diagram of the R-C loaded source follower. (**b**) The magnitude characteristic of the voltage gain with its asymptotes. (**c**) The approximate phase characteristic

Using the value of A_v from (3.14), the input admittance (C_{dg} excluded) in the ω domain can be calculated as

$$y_i'(\omega) = \left(1 - \frac{(g_m + j\omega C_{gs})}{(g_m + G) + j\omega(C_{gs} + C)}\right) \cdot j\omega C_{gs} = g_i + j\omega C_i \qquad (3.16)$$

The input conductance, which is the real part of this expression, can be arranged as

$$g_i(\omega) = \frac{C_{gs}}{(C_{gs} + C)^2}(GC_{gs} - g_m C)\frac{1}{1 + (\omega_p/\omega)^2} \qquad (3.17)$$

or

$$g_i(\omega) = g_i(\infty)\frac{1}{1 + (\omega_p/\omega)^2} \qquad (3.17a)$$

A careful analysis of these expressions leads to the following important conclusions:

- For an R-C loaded source follower, if $GC_{gs} > g_m C$ the input conductance is positive, and similar to that of an R-C loaded common source amplifier, it increases with frequency.
- In the case of $GC_{gs} = g_m C$, the input conductance is zero for all frequencies. This is an important feature and useful for the design of very low input conductance (high input impedance) amplifiers.
- In the case of $GC_{gs} < g_m C$, the input conductance of a capacitive loaded source follower is negative, and this negative conductance increases with frequency. This property can be used

when a negative conductance is needed.[1] If there is a parasitic inductance on the gate connection of the transistor, this negative conductance together with this inductance and the input capacitance of the transistor can result in ringing behavior on the signal, or it can even lead to oscillation. For on-chip source follower input stages, the inductance of the bonding wire can cause such ringing or oscillation. In such cases, this negative conductance must be compensated with an appropriate resistance connected in parallel or series to the gate.

The total input capacitance of a source follower can be calculated as the sum of C_{dg} and the capacitance calculated from (3.16) as

$$C_{iT} = C_{dg} + C\frac{\omega_0'\omega_p + \omega^2}{\omega_p^2 + \omega^2} \tag{3.18}$$

where ω_p and ω_0' are the pole and zero frequencies of the input admittance function. The pole frequency is equal to the pole of the gain function as given in (3.15), and the zero frequency is $\omega_0' = (G/C)$. The values of the input capacitance for low frequencies and for very high frequencies are

$$C_{iT}(0) = C_{dg} + (C_{gs} + C)\frac{G}{(g_m + G)}$$

$$C_{iT}(\infty) = C_{dg} + C$$

The frequency dependence of the input conductance and the input capacitance of a source follower for $GC_{gs} \ll g_mC$ are plotted in Fig. 3.12a and b, respectively.

In Fig. 3.13a, the PSpice simulation results are given for different capacitive loads of a source follower. It is seen that for small capacitive loads the input conductance is positive, for an appropriate load it is equal to zero for all frequencies, and negative for higher capacitive loads. Figure 3.13b shows the output waveform of this amplifier, driven by a 50-ohm input resistance square wave signal source and having a 5 nH series gate inductance (that corresponds to an approximately 5 mm long bonding wire). The observed ringing corresponds to the resonance frequency of the inductance and the input capacitance of the amplifier.

Source followers usually drive low impedance loads. In some applications, for example, if the source follower drives a transmission line, the output internal impedance must keep a predefined value in a wide frequency range in order to maintain a perfect matching for optimum power transfer and to prevent reflections. Therefore, the behavior of the output impedance (or admittance) at high frequencies has to be investigated.

The simplified schematic and the high frequency small-signal equivalent circuit of a source follower to calculate the output impedance (or admittance) are given in Fig. 3.14a and b. It is assumed that the internal impedance of the signal source driving the input is resistive and is shown with R_g. A current source (i_o) is connected to the output port to find the output internal admittance,[2] and the voltage of this port will be calculated. It can be easily seen that the parallel components r_o and C_{op} can

---

[1] See Chap. 4, Example 4.2 for a practical application of this property to increase the quality factor of an inductor.

[2] For the sake of simplicity of the calculations, the output admittance is preferred.

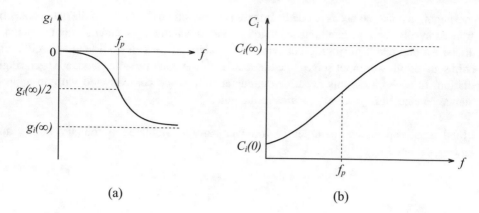

Fig. 3.12 Frequency dependence of (**a**) the input conductance and (**b**) the input capacitance of a source follower for the condition $GC_{gs} \ll g_m C$ which is usually valid

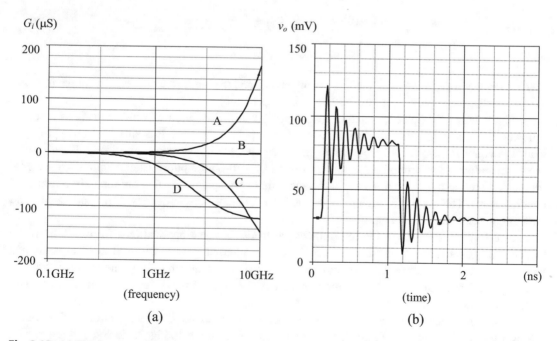

Fig. 3.13 (**a**) The frequency dependence of the input conductance of a source follower for different load capacitance values for (A) 20 fF, (B) 80 fF, (C) 200 fF, and (D) 1 pF. (**b**) The waveform of the output voltage for a square wave input signal with L = 5nH inductor in series to the gate and 200 fF source load capacitance, with the internal resistance of the pulse source being 50 Ω. (The transistor is an AMS 0.35 μm NMOS device with $W = 35$ μm, $L = 0.35$ μm. $R_L = 100$ Ω, and $I_D = 300$ μA)

be excluded to simplify the calculation (Fig. 3.14c), and then re-included.[3] The output admittance of the reduced circuit can now be calculated as

---

[3] It will be seen later that the admittance corresponding to these components is very small compared to the admittance of the reduced circuit; consequently, it is negligible.

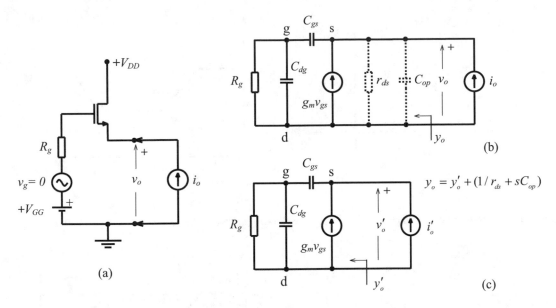

Fig. 3.14 (**a**) The conceptual arrangement to calculate the output impedance of the circuit. (**b**) The equivalent circuit of the source follower amplifier used for the calculation of output impedance (admittance). (**c**) Simplified equivalent circuit

$$y_o' = \frac{i_o'}{v_o'} = \frac{\left(G_g + sC_{dg}\right)\left(g_m + sC_{gs}\right)}{G_g + s\left(C_{dg} + C_{gs}\right)} \tag{3.19}$$

This admittance function has two zeros and one pole described as:

$$s_{01} = -\frac{G_g}{C_{dg}}, \quad s_{02} = -\frac{g_m}{C_{gs}}, \quad s_p = -\frac{G_g}{\left(C_{dg} + C_{gs}\right)} \tag{3.20}$$

and the low-frequency value of the output admittance is equal to g_m, as already shown in Chap. 2.

The frequency characteristic of y_o' depends on the relative positions of the zeros and the pole. It is obvious that $s_{01} > s_p$ and the position of s_{02} depend on the value of g_m. In Fig. 3.15, the pole-zero diagram and the frequency characteristics of $|y_o'|$ corresponding to $|s_{01}| > |s_{02}| > |s_p|$ are shown. It can be seen that under appropriate conditions, there is a possibility of pole-zero cancellation. For example, if $|s_{02}| = |s_p|$ they cancel each other out and the output conductance remains constant up to the frequency corresponding to s_{01}. In Fig. 3.16, the PSpice simulation results for a source follower are shown, without and with pole-zero cancellation.

3.3 The Common Gate Amplifier at High Frequencies

The schematic diagram and the small-signal equivalent circuit of a common gate amplifier are given in Fig. 3.17a and b. Similar to the previously investigated common source and source follower circuit, the investigation of the case of tuned load will be left to Chap. 4. For an *R-C* (or *G-C*) load, the equivalent circuit is given in Fig. 3.17c, with a Norton-Thevenin transformation to facilitate the solution, where

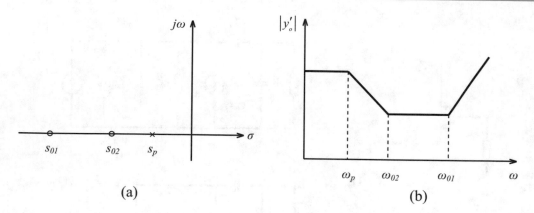

Fig. 3.15 (a) The pole-zero diagram of y_o' of a source follower. (b) The corresponding frequency characteristic

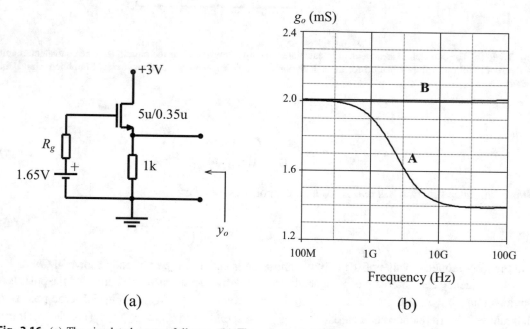

Fig. 3.16 (a) The simulated source follower. (b) The output conductance of the follower (A) without pole-zero cancellation and (B) with pole-zero cancellation. Transistor is an AMS 0.35 micron NMOS transistor with $W = 5$ μm, $L = 0.35$ μm, and $I_D = 0.7$ mA. The value of the driving signal source internal resistance is 10 k for (A) and 1 k for (B)

$$Y_L = G_L + sC_L, \quad Y_L' = Y_L + sC_{dg} = G_L + s\left(C_L + C_{dg}\right) = G_L + sC_L'$$

From Fig. 3.17c, the voltage gain of the amplifier can be solved as

$$A_v = \frac{v_o}{v_i} = \frac{g_m + g_{ds}}{C_L'} \frac{1}{(s - s_p)} \tag{3.21}$$

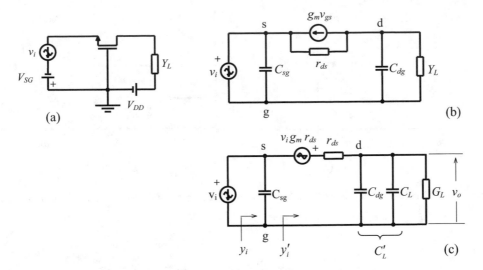

Fig. 3.17 (**a**) The circuit schematic of the common gate amplifier. (**b**) The small-signal equivalent circuit of the common-gate amplifier. (**c**) Equivalent circuit with RC load

where the pole of the gain function that corresponds to the 3 dB frequency (or bandwidth) of the amplifier is

$$s_p = -\frac{G_L + g_{ds}}{C_L'} \tag{3.22}$$

Using (3.21) and (3.22), the low-frequency gain and the gain-bandwidth product can be calculated as

$$A_v(0) = \frac{g_m + g_{ds}}{G_L + g_{ds}} \cong \frac{g_m}{G_L + g_{ds}} \tag{3.23}$$

and

$$GBW \cong \frac{1}{2\pi} \frac{g_m}{C_L'} \tag{3.24}$$

which is equal to that of the common source amplifier.

Although the magnitude of the voltage gain and the gain-bandwidth product of the common source and common gate amplifiers are equal, there are significant differences between their input and output admittances.

The input admittance of a common gate amplifier can be calculated from Fig. 3.17c, as the sum of sC_{gs} and y_i':

$$y_i = \frac{s^2 C_{gs} C_L' + s\left[C_{gs}(g_{ds} + G_L) + C_L'(g_m + g_{ds})\right] + G_L(g_m + g_{ds})}{(g_{ds} + G_L) + sC_L'} \tag{3.25}$$

that can be simplified for $C_{gs}(g_{ds} + G_L) \ll C_L'(g_m + g_{ds})$ and $g_{ds} \ll g_m$ which are usually valid:

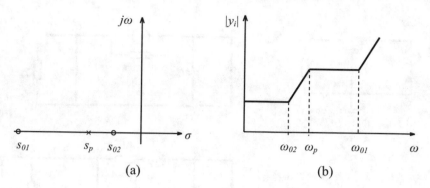

Fig. 3.18 (a) The pole-zero diagram and (b) the frequency characteristic of the input admittance of a common-gate amplifier

$$y_i = \frac{s^2 C_{gs} C'_L + s g_m C'_L + G_L g_m}{(g_{ds} + G_L) + s C'_L} = C_{gs} \frac{(s - s_{01})(s - s_{02})}{(s - s_p)} \tag{3.25a}$$

where

$$s_p = -\frac{G_L + g_{ds}}{C'_L}, \quad s_{01} \cong -\frac{g_m}{C_{gs}}, \quad s_{02} \cong -\frac{G_L}{C'_L} \tag{3.26}$$

The pole-zero diagram and the variation of the magnitude of the input admittance with frequency are shown in Fig. 3.18.

From Fig. 3.18b, we can see that:

- The magnitude of the input admittance is constant (i.e., the input admittance itself is a real quantity) up to the vicinity of $\omega_{02} = G_L/C'_L$ (which is approximately equal to the 3 dB frequency of the voltage gain).
- If $G_L \gg g_{ds}$, the magnitude of the input admittance for low frequencies (also up to the vicinity of the 3 dB frequency of the amplifier) is equal to the transconductance of the transistor. Since g_m is usually in the order of several mS, this means that the common gate circuit is a low input impedance amplifier.

The output admittance (y_o) of a common gate amplifier will be calculated from Fig. 3.19a, as the sum of $s C_{dg}$ and y'_o:

$$y'_o = \frac{i'_o}{v_o} = \frac{1}{r_{ds}} \frac{s - s'_o}{s - s'_p} \tag{3.27}$$

where

$$s'_o = -\frac{1}{R_g C_{gs}} \quad \text{and} \quad s'_p = -\frac{1 + \frac{R_g}{r_{ds}}(1 + g_m r_{ds})}{R_g C_{gs}} \tag{3.27a}$$

and obviously, the magnitude of zero is smaller than that of the pole. The magnitude of the output admittance for zero frequency (as well as for low frequencies) is:

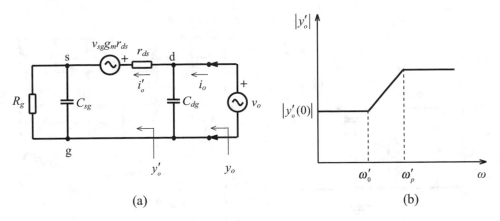

Fig. 3.19 (a) The small-signal equivalent circuit of the common-gate amplifier used for calculation of output admittance. (b) Frequency characteristic of the output admittance

$$y_o'(0) = \frac{1}{r_{ds}\left(1 + g_m R_g\right) + R_g} \tag{3.28}$$

The frequency characteristic of y_o' with its asymptotes is given in Fig. 3.19b. From (3.28) and (3.27), we can conclude that:

- The output resistance, $r_o = 1/\operatorname{Re}\left(y_o'\right) = 1/\operatorname{Re}\left(y_o\right)$ of a common gate amplifier can have very high values for high driving source internal resistance and high transconductance values. In other words, the output acts as an – almost – ideal current source.
- The frequency corresponding to $r_o C_{dg}$ dominates over the zero frequency of y_o' and determines the corner frequency of y_o.
- Since a common gate stage has a low input impedance and a high output impedance, and the output current is equal to the input current, it can be considered as a unity gain – almost – ideal current amplifier, and can be used as an intermediate stage to efficiently transfer the current of a moderate (not too high, not too low) internal impedance current source, to the load.

If we evaluate the already investigated properties of a common gate stage – namely its low input impedance, its very high output internal impedance, and its unity current gain – we can conclude that it is suitable to be used as a trans-impedance amplifier. In Fig. 3.20, the schematic diagram and the PSpice simulation results of a common gate trans-impedance amplifier are given. The current signal source and the parallel capacitance represent the input signal source, for example, a photodiode. The low-frequency value of the trans-impedance is equal to the load resistance. The 40 dB/decade slope of the frequency response indicates that there are two poles in the high-frequency region, apparently one from the input resistance, which is approximately equal to $1/g_m$, and the total capacitance parallel to the input, and the other one from the output side, and equal to $1/R_L C_{db}$. The magnitude of the trans-impedance at low frequencies, and the 3 dB frequency for a load of $R_L = 1$ kΩ and $R_L = 500$ Ω, are 1 kΩ/5.8 GHz and 500 Ω/6.72 GHz, respectively.[4]

---

[4] These figures indicate the "intrinsic" bandwidth of the amplifier (i.e., with no external load capacitance). In the case of an external load capacitance, the pole related to the output becomes $1/R_L(C_{db} + C_L)$ and dominates the frequency response.

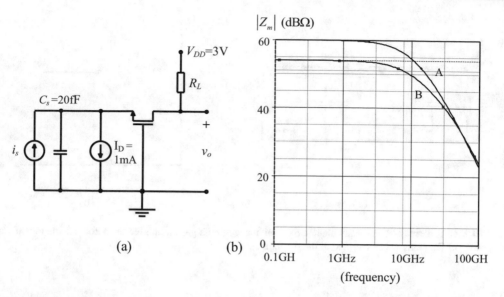

Fig. 3.20 (a) Example of a common gate trans-impedance amplifier, (b) PSpice simulation results showing the frequency characteristics for 1 kΩ (A) and 500 Ω (B) load resistance. The transistor is an AMS 0.35 μm NMOS device with L = 0.35 μm, W = 10 μm

3.4   The "Cascode" Amplifier

We have seen that two of the basic single-transistor amplifiers, namely the common source amplifier and the common gate amplifier provide high voltage gain. However, both of them have severe drawbacks. Due to the output-to-input (drain-to-gate) capacitance and the Miller effect, the input admittance of a common source amplifier deteriorates (increases) at high frequencies. The input admittance of a common gate amplifier is inherently small, therefore it is not suitable to be used as an efficient voltage amplifier, especially if the signal source has high internal impedance.

The solution is the so-called cascode amplifier[5] that provides low output-to-input feedback, high input impedance, and high voltage gain. In a cascode amplifier, a common source amplifier and a common gate amplifier are combined in such a way that the common source stage is loaded with the input of the common gate amplifier (Fig. 3.21a). The load is connected to the output of the common gate stage. The basic operating principles are as follows:

- Since the input conductance of the common gate stage is approximately equal to $1/g_{m2}$, the voltage gain of the input (common source) stage is $A_{v1} \cong -(g_{m1}/g_{m2})$ that is small in magnitude and equal to unity for $g_{m1} = g_{m2}$ (and close to unity for many cases). Consequently, the Miller admittance is small and equal to sC_{dg} up to the pole frequency of the voltage gain (see expression 3.3). This means that the high input capacitance and high input conductance problems arising from the Miller effect are effectively eliminated.

---

[5] The cascode configuration was used in early days of vacuum tube amplifiers to overcome the high plate-to-grid capacitance of triodes. The term was used for the first time in 1939 [2].

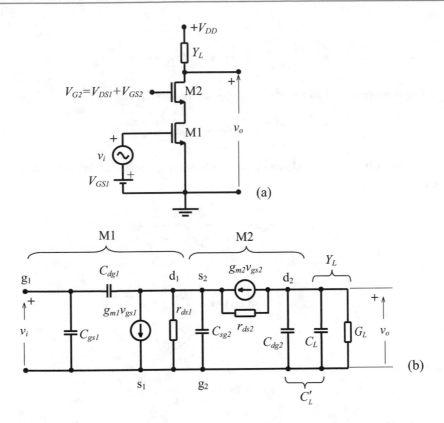

Fig. 3.21 (**a**) The circuit schematic of a cascode amplifier and (**b**) its small-signal equivalent circuit

- Consequently, the input admittance is low (the input impedance is high) in a broader frequency range.
- Since the common gate stage is driven by a high impedance signal source, namely the output impedance of the common source stage, the output internal impedance of the circuit is very high.
- Since the current gain of the common gate circuit is equal to unity, the output current of the common source stage is directly transferred to the load through the common gate stage. Therefore, the overall voltage gain of the circuit is equal to the gain of the first stage, as if the load were directly connected to the output M1.

To investigate the properties of a cascode amplifier for a capacitive load in detail, the small-signal equivalent circuit shown in Fig. 3.21b will be used.

According to (3.25a), the input admittance of M2, which is the load of M1, is

$$y_{i2} = C_{gs2} \frac{(s - s_{01})(s - s_{02})}{(s - s_p)} \tag{3.29}$$

where

$$s_p = -\frac{G_L + g_{ds2}}{C_L'}, \quad s_{01} \cong -\frac{g_{m2}}{C_{gs2}}, \quad s_{02} \cong -\frac{G_L}{C_L'} \tag{3.30}$$

From (3.3), the voltage gain of M1 loaded with y_{i2} can be written as

$$A_{v1} = -\frac{g_{m1} - sC_{dg1}}{y_{i2}} = -C_{dg1}\frac{(s - s_{011})}{y_{i2}} \tag{3.31}$$

where $s_{011} = -\frac{g_{m1}}{C_{dg1}}$. Combining (3.31) and (3.29) and renaming the poles and zeros of A_{v1} to prevent any confusion we obtain

$$A_{v1} = -\frac{C_{dg1}}{C_{gs2}}\frac{(s - s_{01}')(s - s_{02}')}{(s - s_{p1}')(s - s_{p2}')} \tag{3.32}$$

where

$$s_{01}' = s_{011} = -\frac{g_{m1}}{C_{dg1}}$$

$$s_{02}' = s_p = -\frac{G_L + g_{ds2}}{C_L'} \tag{3.32a}$$

$$s_{p1}' = s_{01} \cong -\frac{g_{m2}}{C_{gs2}}$$

$$s_{p2}' = s_{02} \cong -\frac{G_L}{C_L'}$$

According to (3.21), the voltage gain of the common gate stage, M2 is

$$A_{v2} = \frac{v_o}{v_{g2}} = \frac{g_{m2} + g_{ds2}}{C_L'}\frac{1}{(s - s_{p3}')} \cong \frac{g_{m2}}{C_L'}\frac{1}{(s - s_{p3}')} \tag{3.33}$$

where

$$s_{p3}' = -\frac{G_L + g_{ds2}}{C_L'} \tag{3.33a}$$

Now the overall voltage gain of the cascode circuit can be written as

$$A_v = A_{v1} \cdot A_{v2} = -\frac{C_{dg1}}{C_{gs2}}\frac{g_{m2}}{C_L'}\frac{(s - s_{01}')(s - s_{02}')}{(s - s_{p1}')(s - s_{p2}')(s - s_{p3}')} \tag{3.34}$$

Noting that $s_{02}' = s_{p3}'$, the gain expression (3.34) can be simplified as

$$A_v = -\frac{C_{dg1}}{C_{gs2}}\frac{g_{m2}}{C_L'}\frac{(s - s_{01}')}{(s - s_{p1}')(s - s_{p2}')} \tag{3.35}$$

This expression can be interpreted as follows:

- The low-frequency value of the gain is

$$A_v(0) = -\frac{C_{dg1}}{C_{gs2}}\frac{g_{m2}}{C_L'}\frac{(-s_{01}')}{(-s_{p1}')(-s_{p2}')} = -\frac{g_{m1}}{G_L} \tag{3.36}$$

If we compare this expression with (3.5), we can see that the low-frequency gain of a cascode circuit is greater than the gain of M1 if it were directly loaded with Y_L:

$$\frac{A_v(0)_{\text{cascode}}}{A_v(0)_{\text{c.source}}} = 1 + \frac{g_{ds1}}{G_L}$$

This is the result of the very high output internal resistance of the common gate stage, since it is driven by a high impedance source (the output of the common source stage).

- To evaluate the high-frequency performance of the circuit it is useful to compare the zero and the poles of (3.35):

$$s_{01}' = s_{011} = -\frac{g_{m1}}{C_{dg1}}$$

$$s_{p1}' = s_{01} \cong -\frac{g_{m2}}{C_{gs2}}$$

$$s_{p2}' = s_{02} \cong -\frac{G_L}{C_L'}$$

Since the zero is negative and obviously very high in magnitude compared to that of the magnitudes of s_{p1}' and s_{p2}', its effect on the magnitude and phase of the gain is usually negligible. There are two poles affecting the magnitude and phase of the gain at high frequencies. Provided that $\left|s_{p1}'\right| \gg \left|s_{p2}'\right|$, the 3 dB frequency of the cascode circuit approaches that of the simple common source amplifier having the same output load:

$$f_{3dB} \cong \frac{G_L}{2\pi C_L'} \tag{3.37}$$

and the voltage gain-bandwidth product becomes

$$GBW \cong \frac{g_{m1}}{2\pi C_L'} \tag{3.38}$$

which is equal to that of a common source amplifier.

- If the magnitudes of s_{p1}' and s_{p2}' are comparable, the high frequency roll-off occurs earlier, and at the 3 dB frequency the phase shift exceeds $\pi/4$. This means that, due to the additional pole corresponding to the input R-C component of M2, the high-frequency performance of the cascode circuit deteriorates and the voltage gain-bandwidth product decreases.

3.5   The CMOS Inverter as a Transimpedance Amplifier

In Sect. 2.2, it was mentioned that a CMOS inverter containing a feedback resistor connected between the output and input nodes is suitable to be used as a transimpedance amplifier. The schematic diagram, the small-signal equivalent circuit, and the modified equivalent circuit after the application of the Miller transformation are given in Fig. 3.22, where $\bar{g}_m$ and $\bar{g}_{ds}$ represent the sums of the corresponding parameters of M1 and M2, and C_i and C_o represent the total parallel capacitance to the input and output nodes, respectively. As explained earlier, the resistance connected between the output and input nodes helps to improve the stability of the operating point. But there is an unavoidable capacitance, namely the sum of the C_{dg} of M1 and M2, which is parallel to this resistor. Throughout the analysis, it will be assumed that there is a capacitance (C_F) parallel to the feedback resistor, equal to the sum of the total drain-gate capacitance, the parasitic capacitance, and the parallel external capacitance (if there is any).

From Fig. 3.22c the transimpedance can be calculated as

$$Z_m = -\frac{(\bar{g}_m - Y_F)}{Y_F(\bar{g}_m + \bar{y}_i) + \bar{Y}_L(\bar{y}_i + Y_F)} \tag{3.39}$$

and arranged in the s domain as

$$Z_m = \frac{sC_F - (\bar{g}_m - G_F)}{s^2(C_iC_o + C_iC_F + C_oC_F) + s[C_F(g_i + \bar{g}_{ds} + \bar{g}_m) + C_i(\bar{g}_{ds} + G_F) + C_o(g_i + G_F)]}$$
$$+[G_F(g_i + \bar{g}_{ds} + \bar{g}_m) + g_i\bar{g}_{ds}]$$

This expression can be simplified as

$$Z_m \cong \frac{sC_F - (\bar{g}_m - G_F)}{s^2C_o(C_i + C_F) + s[\bar{g}_{ds}(C_F + C_i) + G_F(C_i + C_o) + \bar{g}_mC_F)] + G_F(\bar{g}_{ds} + \bar{g}_m)} \tag{3.40}$$

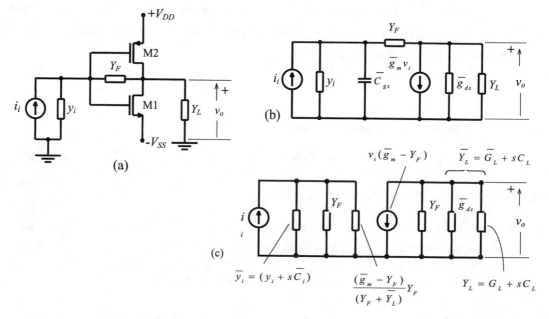

Fig. 3.22 (a) The CMOS inverter as a transresistance amplifier: the input signal is i_1 (current) and the output signal is v_2 (voltage). (b) The small-signal equivalent circuit. (c) The equivalent circuit after Miller transformation

The low-frequency value of the transimpedance (the transresistance) is

$$Z_m(0) = -\frac{(\bar{g}_m - G_F)}{G_F(\bar{g}_{ds} + \bar{g}_m)}$$

and reduces to $-(1/G_F)$ for $\bar{g}_{ds}, G_F \ll \bar{g}_m$, as already shown in Chap. 2.

The frequency characteristic of the amplifier depends on the relative positions of the zero and the two poles of this gain function. To investigate the possibilities, it is convenient to write the expression in a closed-form as

$$Z_m \cong \frac{s.D + E}{s^2.A + s.B + C}$$

The gain function has a positive-real zero:

$$s_z = -\frac{E}{D} = \frac{(\bar{g}_m - G_F)}{C_F} \tag{3.41}$$

The poles can be solved as the roots of the quadratic nominator:

$$s_{p1,p2} = \frac{B}{2A}\left(-1 \mp \sqrt{1 - \frac{4AC}{B^2}}\right) \tag{3.42}$$

These roots are:

$$\text{a) two separate negative-real poles for} \qquad \frac{4AC}{B^2} < 1$$

$$\text{b) two equal negative-real poles for} \qquad \frac{4AC}{B^2} = 1 \tag{3.43}$$

$$\text{c) one complex-conjugate pair for} \qquad \frac{4AC}{B^2} > 1$$

In Fig. 3.23, the pole-zero positioning for these cases and the corresponding shapes of the frequency characteristics are shown. The important points related to these three cases will be briefly discussed below.

(a)   Two separate negative-real poles:

In this case, provided that the two poles are sufficiently apart from each other and the right half-plane zero is far away, the 3 dB frequency is determined by the dominant pole, the slope of the magnitude curve is -20 dB/decade, and the phase shift of the output signal does not exceed $\pi/2$, up to the vicinity of the second pole. It is known from feedback theory that, if this amplifier is in a negative feedback loop, one of the poles must dominate to guarantee the stability, i.e., the first pole frequency must be sufficiently smaller in magnitude than the second one. It can be shown that if the gain of an amplifier (A_o) is reduced to (A_f) by negative feedback, in order to have a flat final frequency response,

$$\frac{\omega_{p2}}{\omega_{p1}} \geq \frac{A_0}{A_f} - 1$$

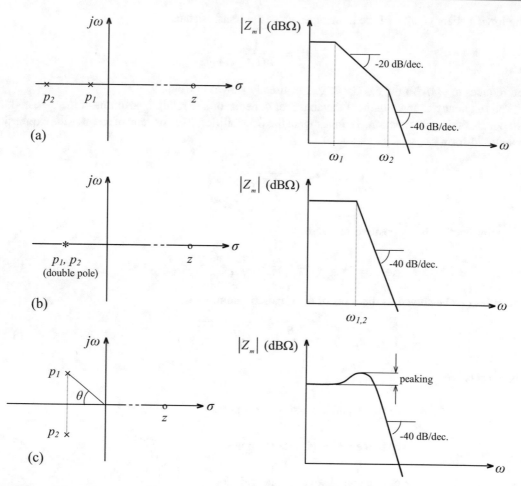

Fig. 3.23 Possible pole-zero diagrams and the corresponding frequency characteristics (the magnitude of the zero is assumed much higher than the magnitude of the poles)

must be satisfied, where ω_{p1} and ω_{p2} are the frequencies corresponding to the poles of the no-feedback (or open-loop) gain of the amplifier (Ref.: Chap. 4). It means that in case of a strong negative feedback to reduce the gain substantially, the magnitude of one of the poles must be much smaller than the other. For such a case $\frac{4AC}{B^2} \ll 1$ and (3.42) can be written as

$$s_{p1,p2} \cong \frac{B}{2A}\left[-1\mp\left(1-\frac{1}{2}\frac{4AC}{B^2}\right)\right]$$

$$s_{p1} = -\frac{C}{B} \;\text{(lower frequency "dominant" pole)}$$

$$s_{p2} = -\frac{B}{A} \;\text{(higher frequency "far" pole)}$$

The values of these poles in terms of the circuit parameters can be calculated as

$$s_{p1} = -\frac{G_F(\bar{g}_{ds} + \bar{g}_m)}{C_F(\bar{g}_{ds} + \bar{g}_m) + C_o G_F} = -\frac{1}{\dfrac{C_F}{G_F} + \dfrac{C_o}{(\bar{g}_{ds} + \bar{g}_m)}} \tag{3.44}$$

$$s_{p2} = -\frac{C_F(\bar{g}_{ds} + \bar{g}_m) + G_F C_o}{C_o C_F} = -\left(\frac{(\bar{g}_{ds} + \bar{g}_m)}{C_o} + \frac{G_F}{C_F}\right) \tag{3.45}$$

(b) Two equal negative-real poles (double-poles):

In this case, the drop of the gain is 6 dB at the frequency corresponding to the double-pole[6] and the slope of the decrease is 40 dB/decade. This configuration is apparently not suitable to be used in a feedback loop, but can be used as a flat frequency response wideband amplifier stage, suitable to give a rail-to-rail output voltage.

Inserting the circuit parameters into (3.43b), the condition to obtain such a frequency characteristic and the magnitude of the frequency corresponding to the pole can be solved as

$$C_F \cong C_o \frac{G_F}{(\bar{g}_{ds} + \bar{g}_m)} \tag{3.46}$$

$$\omega_p = \frac{B}{2A} = \frac{\bar{g}_{ds}(C_F + C_i) + G_F C_o + \bar{g}_m C_F}{2C_o(C_i + C_F)} \tag{3.47}$$

(c) Complex-conjugate poles:

The pole-zero diagram of an amplifier having two complex-conjugate poles and a positive-real zero is shown in Fig. 3.23c. From network theory, it is known that the shape of the frequency response depends on the θ angle. For $\theta > \pi/4$ (or $\xi = \cos\theta < 0.707$) the frequency characteristic exhibits a peak, and the 3 dB frequency slightly increases (Fig. 3.24a). The response of the amplifier to a step input signal correspondingly exhibits an overshoot that also increases with θ (Fig. 3.24b), and is considered harmful for many applications. The effects of the positioning of the complex-conjugate poles of an amplifier are investigated in several electronics textbooks in detail, and can be easily adapted to this case, whenever necessary.

Example 3.2

A transimpedance amplifier will be designed with a low-frequency "gain" of 2000 ohm (66 dBΩ) flat up to 1 GHz, delivering a rail-to-rail output voltage swing to a $C_L = 1$ pF external capacitive load. The internal impedance of the driving current signal source is 10 kΩ parallel to 50 fF (the output impedance of the previous stage). The design will be made using AMS 035 micron CMOS technology parameters. As one of the candidate configurations, we will design a CMOS inverter type circuit. Since any peaking on the frequency response (any overshoot on the pulse response) is not acceptable, a double-pole solution is targeted. The design will be made using the expressions derived in this section, and then fine-tuned with PSpice.

---

[6] The 3 dB frequency of a multiple-pole amplifier can be calculated as $\omega_{3dB} = \omega_p\sqrt{2^{1/n} - 1}$, where n is the number of the coincident poles. In case of $n = 2$, $\omega_{3dB} = 0.614\,\omega_p$.

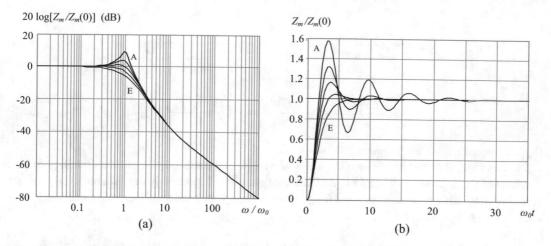

$20 \log[Z_m/Z_m(0)]$ (dB)

$Z_m/Z_m(0)$

(a)

(b)

Fig. 3.24 The normalized frequency responses and the normalized step responses for different values of θ. Curves A to E correspond to $\theta = 80, 70, 60, 45$, and 0 degrees, respectively. ($\omega_0^2 = C/A$. Curves are drawn for $\omega_z/\omega_0 = 10$. Note the decrease of the slope due to the zero)

Fig. 3.25 The circuit diagram of the transimpedance amplifier to be designed

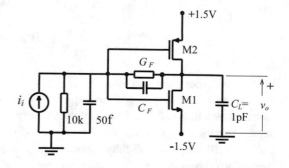

The circuit diagram is as shown in Fig. 3.25. The input capacitance (C_i) in the expressions is the sum of the signal source internal capacitance (which is given as 50 fF) and the gate-source capacitances of the transistors (which are not known yet). Similarly, the output capacitance (C_o) is the sum of the load capacitance (which is given as 1 pF) and the junction capacitances of the transistors (which are not known). The only known component is G_F, which is approximately equal to $1/Z_m(0) = 0.5$ mS. Another aspect that must be kept in mind is that the width of the PMOS transistor must be approximately (μ_{no}/μ_{po}) times bigger than that of the NMOS transistor (cf. Chap. 2). This ratio is $475.8/137 \cong 3.5$ for AMS 035 micron CMOS technology (cf. Appendix A). For maximum high-frequency performance, the gate lengths of both transistors will be chosen as 0.35 μm.

To estimate the dimensions of the transistors, expressions (3.46) and (3.47) can be used. If we insert (3.46) into (3.47) and make some simplifications assuming that $C_F \ll C_i$ and $\overline{g}_{ds} \ll \overline{g}_m$ (both being realistic) we obtain

$$\omega_p \cong \frac{\overline{g}_{ds}}{2C_o} + \frac{G_F}{C_i} \tag{3.48}$$

The bandwidth (3 dB frequency) was given as 1 GHz. Since the amplifier has a double-pole, the frequency corresponding to the pole is $f_p = 1$ GHz/0.614 = 1.628 GHz, or $\omega_p = 2\pi \times 1.628 \times 10^9 \cong 10^{10}$ rad/s. But since all parameters – except G_F – are geometry dependent, (3.48) does not lead to the solution.

Another attack point is to consider the slew-rate. The rise-time of a 1 GHz bandwidth amplifier is approximately $t_r = 0.35/f_{3dB} = 0.35/10^9 = 0.35$ ns (Chap. 2). For a rail-to-rail output step, the PMOS transistor, which acts as a current source, must be capable of charging the total load capacitance up to $V_{DD} = 1.5$ V in 0.35 ns:

$$I_P = C_o \frac{dV}{dT} = C_o \frac{1.5}{0.35 \times 10^{-9}} \tag{3.49}$$

C_o is the sum of the external load capacitance (1 pF) and the total parasitic capacitance of the output node. This parasitic capacitance can be calculated (see Appendix A) in terms of the widths of the transistors as

$$C_{jDN}(0) = 0.8W_N + 0.5(X + W_N) \cong 1.3W_N \quad [\text{fF}, W_N \text{ in } \mu\text{m}]$$

$$C_{jDP}(0) = 0.8W_P + 0.5(X + W_P) \cong 1.3W_P \quad [\text{fF}, W_N \text{ in } \mu\text{m}]$$

and the total parasitic capacitance[7]

$$C_{jDT}(0) \cong 1.3(W_N + W_P) = 1.3(1 + 3.5)W_N = 5.85W_N \quad [\text{fF}, W_N \text{ in } \mu\text{m}]$$

Now we can make an estimation for the total output capacitance and take $C_o = 1.2$ pF, which corresponds to $W_N = 34$ micron (which must be checked later on).

Now from (3.49), I_{DP} can be calculated as

$$|I_{DP}| = 1.2 \times 10^{-12} \frac{1.5}{0.35 \times 10^{-9}} = 5.14 \times 10^{-3} = 5.14\,\text{mA}$$

This is the drain current of the PMOS which flows under a gate voltage equal to $-V_{DD}$ (-1.5 V). Now the gate-width of the PMOS transistor can be calculated:

$$|I_{DP}| \cong \frac{1}{2}\mu_p C_{ox} \frac{W_P}{L}(-V_{DD} - V_{TP})^2(1 + \lambda_P V_{DD})$$

$$5.14 \times 10^3 = \frac{1}{2}137(4.56 \times 10^{-7})\frac{W_P}{0.35}(-1.5 + 0.7)^2(1 + 0.2 \times 1.5)$$

$$W_P \cong 70\,\mu\text{m}$$

The corresponding NMOS transistor channel width to conduct the same current under quiescent conditions can be calculated from (2.14) as $W_N \cong 15$ μm.

We know that the hand-calculated current values corresponding to a certain bias condition are usually higher than the currents in reality. Therefore, it is necessary to check the drain currents with DC simulations and fine-tune to the target values and adjust the quiescent voltage of the output node, as performed in Design Example 2.3. The PSpice simulation shows that the drain currents for the calculated channel widths are 1.93 mA, which is considerably smaller than the 5.14 mA target value. The previously described tuning procedure gives $W_1 = 45$ μm and $W_2 = 154$ μm. Hence, the DC

---

[7] This is the maximum (pessimistic) value of the parasitic capacitance, since it corresponds to the zero-bias voltage on the drain junctions. In reality, the junction is reverse-biased and the parasitic capacitance is correspondingly smaller.

currents become equal to the target value and the DC voltage of the output node becomes zero (i.e., no offset).

Now as a final step, the value of C_F can be calculated from (3.46). But before this step, C_o, $\overline{g}_m$ and $\overline{g}_{ds}$ must be derived.

The output parasitic capacitance, which is the sum of the junction capacitances of the drain regions, can be found for the calculated dimensions as:

$$C_{jDT}(0) \cong 5.85 W_N = 5.85 \times 45 \cong 263\,\text{fF}$$

$$C_o = C_L + C_{jDT} = 1.263\,\text{fF}$$

$\overline{g}_m$ is the sum of the transconductances and can be calculated as:

$$\overline{g}_m = g_{mN} + g_{mP} = \sqrt{2\frac{W_1}{L}\mu_n C_{ox} I_D} + \sqrt{2\frac{W_2}{L}\mu_p C_{ox} I_D} = 21.1\,\text{mS}$$

$\overline{g}_{ds}$ is the sum of the output conductances of the transistors:

$$\overline{g}_{ds} \cong I_D(\lambda_N + \lambda_P) = 5.14 \times 10^{-3}(0.073 + 0.2) = 1.4\,\text{mS}$$

Now the value of C_F can be calculated from (3.46):

$$C_F = 1.263\frac{0.5 \times 10^{-3}}{22.6 \times 10^{-3}} = 0.028\,\text{pF} = 28\,\text{fF}$$

It must be noted that the sum of the drain to source capacitances is the intrinsic component of C_F. If we calculate $(C_{dgN} + C_{dgP})$ we find that

$$(C_{dgN} + C_{dgP}) = (W_N + W_P)C_{DGO} = (45 + 154)0.12 = 23.9\,\text{fF}$$

This means that due to the intrinsic C_F, the circuit is slightly overcompensated. The PSpice simulation results with the calculated design parameters are given in Fig. 3.26. Obviously, there are certain

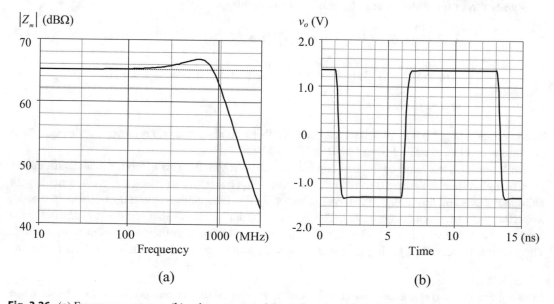

(a)   (b)

Fig. 3.26 (a) Frequency response, (b) pulse response of the designed transimpedance amplifier

discrepancies compared to the target values. Let us interpret the results and obtain hints for the further fine-tuning of the design:

From the frequency characteristic given in Fig. 3.26a, we see that:

(a)   The low-frequency gain is 65.2 dBΩ. If this 0.8 dBΩ discrepancy is not tolerable, R_F can be increased to reach the target value. But (3.48) tells us that the increase of R_F decreases the pole frequency, and consequently the 3 dB frequency.

(b)   The 3 dB frequency is 1.05 GHz, but there is a 1.7 dB peak, which may not be acceptable for some applications. To decrease the peaking, C_F must be increased (See (3.40) and (3.43)). Indeed, if we use a feedback capacitor equal to $C_F = 50$ fF, the peak disappears, the response becomes flat, but the 3 dB frequency decreases to 795 MHz. To increase the 3 dB frequency, according to (3.48) R_F must be decreased. Observations in (a) and (b) indicate that there is a trade-off problem.

(c)   If the magnitude of the gain has prime importance, the gain can be increased to the target value with $R_F = 2.2$ kΩ. The frequency response is flat and the 3 dB frequency is 764 MHz for $C_F = 45$ fF, and it has a 1 dB peak and 935 MHz bandwidth for $C_F = 12$ fF.

(d)   If a flat frequency response with 1GHz bandwidth has prime importance, this can be obtained at the price of gain decrease. For $R_F = 1.2$ kΩ and $C_F = 60$ fF, the gain is 60.6 dBΩ and the 3 dB frequency is 1.004 GHz, with 0.2 dB peak, which is acceptable for many applications.

(e)   If both the bandwidth and the gain conditions have to be fulfilled without any compromise from the original specifications, the DC current must be increased, or the circuit must be designed using an alternative technology with smaller feature sizes to decrease the parasitic capacitances.

(f)   The pulse response given in Fig. 3.26b shows that the circuit has no slew-rate limitation (the rise-time is 0.35 ns) and a reasonably nice wave shape. The PSpice netlist for the original design is given below.

```
*AMS-035u INVERTER TRANSIMPEDANCE AMPLIFIER*
vdd   10 0   1.5
vss   20 0 -1.5
cin 2 0   50f
rin 2 0 10k
*iin 0 2 pulse -1m 1m   1n 10p 10p 5n 12n
iin 0 2   ac 1u
M1 3 2 20 20 modn w=45U l=.35U, ad=38.2e-12 as=38.2e-12, pd=90e-6, ps=90e-6
M2 3 2 10 10 modp w=154U l=.35U, ad=131e-12 as=131e-12, pd=308e-6, ps=308e-6
RF2 3 2 2k
*vx 3 0 0
*cf 3 2 15f
CL 3 0 1P
.lib "cmos7tm.mod"
.ac dec 50 .01g 3G
.DC iin -1m 1m 10u
.TRAN .01N 15N
.probe
.end
```

3.6   MOS Transistor with Source Degeneration at High Frequencies

In Sect. 2.1, we have seen that a resistor in series to the source terminal of a MOS transistor decreases the transconductance to so-called effective transconductance. It is useful to generalize this effect for an impedance placed in series to the source terminal and to deal not only with the transconductance, but with all of the y parameters. In Fig. 3.27a, a MOS transistor is shown together with the impedance (the source degeneration impedance) connected in series to the source, which can be external, or the parasitic (internal) impedance of the transistor. Now we will calculate the y parameters of the "equivalent transistor" M_e shown in Fig. 3.27b, in terms of the small-signal parameters of M and its source impedance, Z_s. For the sake of the simplicity of the derivations, C_{dg} will be excluded from the y-parameter calculations, and it can be re-inserted whenever necessary.[8]

The y_{11} and y_{21} parameters, both defined under shorted output conditions, can be solved from Fig. 3.28a:

$$y_{11} \cong \frac{sC_{gs}}{1 + (g_m + sC_{gs})Z_s} \qquad (3.50)$$

Fig. 3.27 (a) A MOS transistor with source degeneration impedance. (b) The equivalent transistor

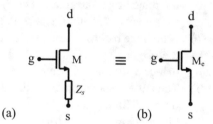

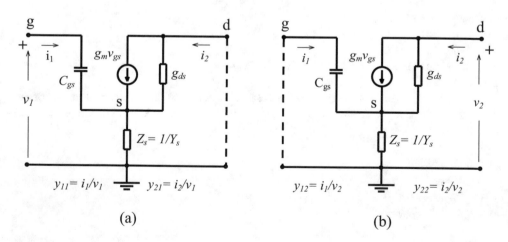

(a)

(b)

Fig. 3.28 Small-signal equivalent circuits used to derive expressions (**a**) for y_{11} and y_{21}, (**b**) for y_{12} and y_{22}

---

[8] Hint: The total equivalent y parameters of two parallel-connected two-ports correspond to the sums of the y parameters of the individual two-ports.

Fig. 3.29 (a) Inductive source degeneration. (b) Arrangement to obtain a resistive and small input impedance for a certain frequency

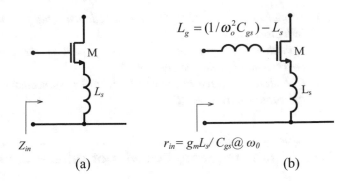

$$y_{21} \cong g_m \frac{1}{1 + g_m Z_s} \tag{3.51}$$

Similarly, the y_{12} and y_{22} parameters, both defined under shorted input conditions, can be derived from Fig. 3.28b:

$$y_{12} \cong -\frac{s C_{gs} g_{ds}}{s C_{gs} + Y_s + g_m} \tag{3.52}$$

$$y_{22} \cong \frac{g_{ds}}{1 + \frac{g_m}{(s C_{gs} + Y_s)}} \tag{3.53}$$

These expressions are valid for any type of source degeneration impedance. For an inductance (L_s) connected in series to the source, e.g., as shown in Fig. 3.29 (assuming that the effect of y_{12} is negligible), the input admittance can be written as

$$y_{in} \cong y_{11} \cong \frac{s C_{gs}}{1 + (g_m + s C_{gs}) s L_s}$$

$$z_{in} = \frac{1}{y_{in}} = \frac{1 + (g_m + s C_{gs}) s L_s}{s C_{gs}} = \frac{1}{s C_{gs}} + \frac{g_m L_s}{C_{gs}} + s L_s \tag{3.54}$$

which represents a series combination of C_{gs}, L_s, and a frequency-independent resistive component. For a specified frequency, the capacitive component can be eliminated by resonance with an appropriate inductance connected in series to the gate. This is a useful possibility for the design of resistive, low input impedance circuits like LNAs, which will be investigated in Chap. 4 in further detail.

Problem 3.1

The circuit shown in Fig. 3.29a will be used in the input stage of an amplifier operating at 2 GHz. The parameters of the transistor are given as: $\mu = 300\ cm^2/V.s$, $T_{ox} = 6\ nm$, $CDGO = 1.2 \times 10^{-10} F/m$, $L = 0.25$ and $W = 50\ \mu m$. The transistor is operating in saturation and biased for $I_D = 2\ mA$. The quality factors of inductors on the chip are $Q = 10\ @\ 2GHz$.

(a) *Calculate the transconductance of the transistor.*
(b) *Calculate the value of the input capacitance of the transistor.*

(c) *To make the real part of the input impedance equal to 100 ohms, calculate the inductance of the inductor.*

(d) *Calculate the value of the imaginary part of the input impedance and the value of the corresponding reactive element.*

(e) *Calculate the value of the reactance to be connected in series to the gate to make the input impedance purely resistive.*

3.7 High-Frequency Behavior of Differential Amplifiers

In Chap. 2, we have seen that several varieties of differential amplifiers were developed over time. In this section, we will investigate the frequency-dependent behavior of the three most important and basic differential amplifier configurations: the R-C loaded long-tailed pair, the fully differential active loaded pair, and the active current mirror loaded differential input – single-ended output long-tailed pair.

3.7.1 The R-C Loaded Long-Tailed Pair

The circuit diagram of a long-tailed pair loaded with equal impedances is given in Fig. 3.30a. The transistors M and M' are matched devices, therefore all corresponding parameters are assumed to be equal. The admittance Y_T represents the internal admittance of the tail DC current source. The circuit is differentially driven by a voltage source, v_i. Due to the complete symmetry of the circuit, the two inputs are assumed to be driven by $v_i/2$ and $-v_i/2$, respectively (cf. Chap. 2), as shown in Fig. 3.30b. The individual output voltages of M and M' are shown with v_o and v'_o. Therefore, the differential output voltage is $v_{od} = \left(v_o - v'_o\right)$.

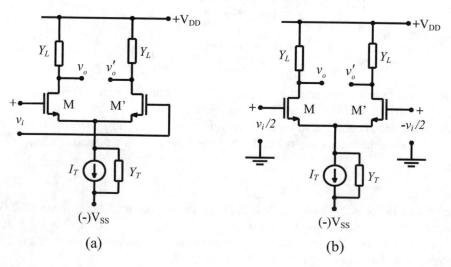

Fig. 3.30 (a) Circuit diagram of a differentially driven long-tailed pair. (b) Circuit with the equivalent individual single-ended input signals

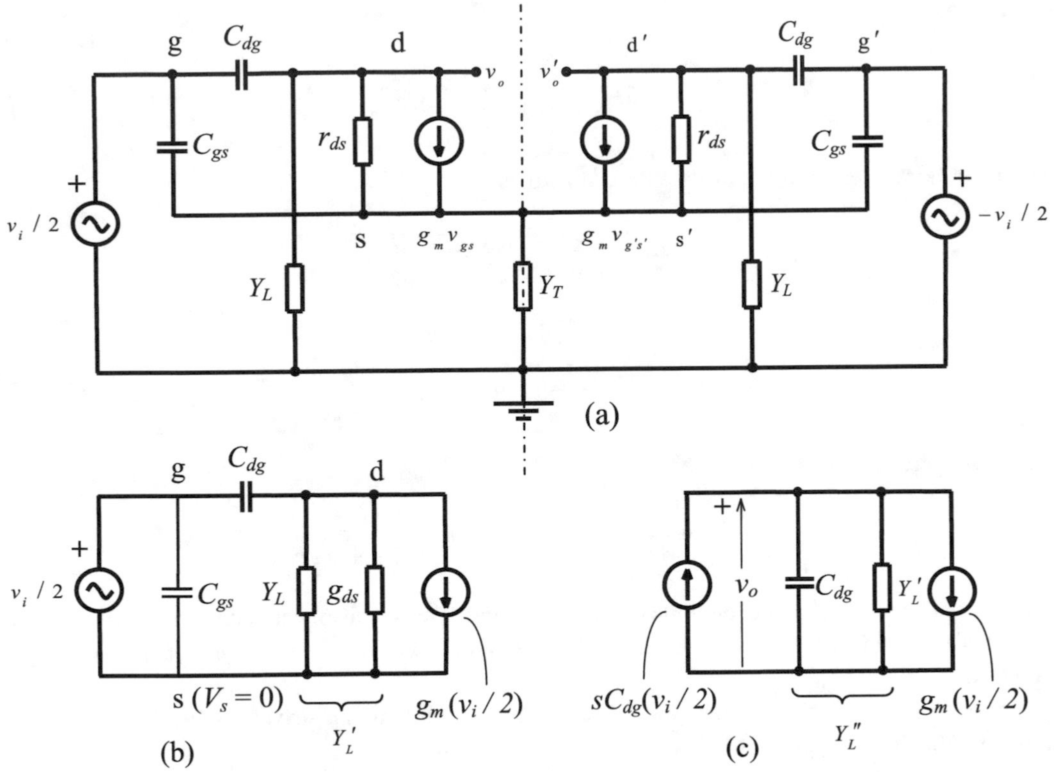

Fig. 3.31 (a) The small-signal equivalent circuit. (b) Simplification based on the symmetry of the circuit. (c) The equivalent circuit after a Norton-Thévenin conversion

The small-signal equivalent of the circuit is given in Fig. 3.31a. Since the signal voltage on the common source node is zero due to the symmetry of the circuit, it is possible to deal with each of the symmetrical halves separately, as shown in Fig. 3.31b. In Fig. 3.31c, this equivalent circuit is redrawn to facilitate the network equations.[9]

From Fig. 3.31c, the output voltage and the voltage gain from one of the inputs to the output of the corresponding transistor can be easily written as

$$v_o = \frac{v_i}{2} \left(sC_{dg} - g_m \right) \frac{1}{Y''}$$

$$A_v = \frac{v_o}{(v_i/2)} = \left(sC_{dg} - g_m \right) \frac{1}{Y''} \tag{3.55}$$

For an R-C load as $Y_L = G_L + sC_L$, Y_L'' becomes

$$Y_L' = (G_L + sC_L) + g_{ds}$$

$$Y_L'' = Y_L' + sC_{dg} = (G_L + g_{ds}) + s\left(C_L + C_{dg} \right)$$

---

[9] Note that C_{gs} is excluded from Fig. 3.31b since it is parallel to a voltage source.

and the voltage gain:

$$A_v = \frac{(sC_{dg} - g_m)}{(G_L + g_{ds}) + s(C_L + C_{dg})} \tag{3.56}$$

which can be written in terms of its pole and zero as

$$A_v = \frac{C_{dg}}{(C_L + C_{dg})} \frac{(s - s_0)}{(s - s_p)} \tag{3.57}$$

where

$$s_p = -\frac{(G_L + g_{ds})}{(C_L + C_{dg})} \quad \text{and} \quad s_0 = +\frac{g_m}{C_{dg}} \tag{3.58}$$

and the low-frequency gain becomes

$$A_v(0) = \frac{C_{dg}}{(C_L + C_{dg})} \times \frac{-s_0}{-s_p} = -\frac{g_m}{(G_L + g_{ds})}$$

as expected (See Chap. 2). It is obvious that if series (and equal) resistors are connected to the sources to improve the linearity as mentioned in Chap. 2, g_{meff} must be used instead of g_m in the gain expressions.

Expression (3.57) is the voltage gain from the input of M, which is driven by $(v_1/2)$, to its output. Similarly, the voltage gain of M' is

$$A_v' = \frac{v_o'}{-(v_i/2)} = \frac{C_{dg}}{(C_L + C_{dg})} \frac{(s - s_0)}{(s - s_p)} \tag{3.59}$$

Now the voltage gain from the differential input (v_i) to the differential output can be written as

$$A_{vdd} = \frac{(v_o - v_o')}{(v_i/2) - (-v_i/2)} = \frac{v_i}{v_{od}} = \frac{C_{dg}}{(C_L + C_{dg})} \frac{(s - s_0)}{(s - s_p)} \tag{3.60}$$

Expression (3.60) indicates that:

- An R-C loaded long-tailed pair has a pole that is determined by the capacitance and the total equivalent conductance parallel to each output node. The frequency corresponding to this pole is the 3 dB frequency of the voltage gain:

$$f_{3dB} = \frac{1}{2\pi} \frac{(G_L + g_{ds})}{(C_L + C_{dg})} \tag{3.61}$$

- It must be noted the drain junction capacitances (C_{dj}) are parallel to the external load capacitances (C_L). For small external load capacitances, C_{dj} must be taken into account and the effective load capacitance must be calculated as

$$(C_L + C_{dg} + C_{dj})$$

- The gain function has a positive zero at very high frequencies. The magnitude of this zero can be calculated as

$$s_0 = \frac{g_m}{C_{dg}} = \mu \frac{1}{L} \frac{C_{ox}}{C_{DGO}} (V_{GS} - V_T) \tag{3.62}$$

which is independent from the gate width. For example, with AMS 035 micron technology and 0.1 V gate over-drive, (3.62) gives the frequency corresponding to this pole as 82.2 GHz, which is too high to be taken into account.

As a by-product of these calculations, the signal current flowing through the total load admittance of one of the transistors, which is nothing else but the current of the voltage-controlled current source in the small-signal equivalent circuit, can be written as

$$i_s = -v_o \left[(G_L + g_{ds}) + s(C_L + C_{dg}) \right] \tag{3.63}$$

From (3.56) and (3.63) i_s can be solved and simplified as

$$i_s = -(v_i/2) C_{dg} (s - s_0) \tag{3.63a}$$

This current can be assumed constant (frequency-independent) and equal to

$$i_s \cong i_s(0) = \frac{v_i}{2} g_m \tag{3.63b}$$

since for all practical cases $\omega \ll \omega_0$. This knowledge is valuable and will be used for the investigation of current mirror loaded circuits. It is also useful to note that for arbitrary loads – for example, RLC (tuned) loads – the output voltage can be calculated simply as $v_o = -i_s Z_L$.

3.7.2 The Fully Differential, Current Mirror Loaded Amplifier

As already discussed in Chap. 2, one of the most important versions of differential amplifiers is the fully differential, current mirror loaded amplifier shown in Fig. 3.32. The circuit can be considered either as a voltage amplifier when loaded with equal impedances, or as a transconductance (transadmittance) amplifier when loaded with impedances sufficiently smaller than the output internal impedance (cf. Chap. 2).

Fig. 3.32 Circuit diagram of a fully differential, current-mirror loaded amplifier

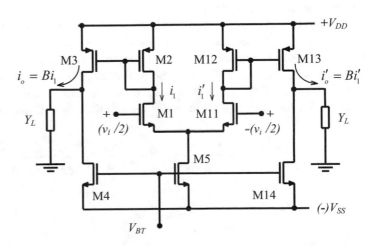

The DC operation of the circuit was investigated in Chap. 2. Now we will calculate the output currents of the circuit under a differential input voltage. According to (3.63b) the drain current is

$$i_1 = \frac{v_1}{2} g_{m1}$$

which is the input signal current of the current mirror (M2, M3). We know that the output signal current of this mirror can be expressed as (c.f. Appendix D)

$$i_3 = i_1 B = i_1 B_0 \frac{s_p}{s_0} \frac{(s - s_0)}{(s - s_p)} \tag{3.64}$$

where B_0 is the DC (or low frequency) current transfer ratio, which is equal to W_3/W_2, and s_0 and s_p are the zero and pole of B:

$$s_0 = + \frac{g_{m3}}{C_{dg3}} \text{ and } s_p = - \frac{g_{m2}}{(C_{gs2} + C_{gs3}) + C_{xp}} \tag{3.65}$$

where C_{xp} represents the extra parasitics, the sum of drain junction capacitances of M1 and M2 and the gate-to-source overlap capacitances of M2 and M3. M4 is a DC current source, biased by V_B. Since M4 is biased such that its DC current is equal to the DC current of M3, and M4 has a high small-signal output impedance, the output current (i_o) flowing through the low impedance load is equal to i_3. From (3.63b) and (3.64), the output current can be expressed in terms of the input signal voltage, and then, the transadmittance can be written as

$$y_m = \frac{i_o}{v_1} = \frac{1}{2} g_{m1} B_0 \frac{s_p}{s_0} \frac{(s - s_0)}{(s - s_p)}$$

and since $|s_0| \gg |s_p|$,

$$y_m \cong \frac{1}{2} g_{m1} B_0 s_p \frac{1}{(s - s_p)} \cong g_m s_p \frac{1}{(s - s_p)} \tag{3.66}$$

where g_m is the low-frequency value of y_m.

This is a typical single-pole gain function where the magnitude characteristic has a 3 dB frequency corresponding to the pole, and a -20 dB/decade slope. The phase characteristic reaches $\pi/4$ at the 3 dB frequency, and theoretically becomes asymptotic to $\pi/2$. But at the high-frequency end of the characteristic, due to the effects of the neglected zero or the neglected parasitics, the phase shift can exceed $\pi/2$ and can cause problems if the amplifier is in a feedback loop.

If we calculate the 3 dB frequency of the amplifier from (3.65) as

$$f_{3dB} = \frac{1}{2\pi} \frac{g_{m2}}{(1 + B_0)C_{gs2} + C_{xp}} = \frac{1}{2\pi} \frac{\mu_p C_{ox}(W_2/L)|V_{GS2} - V_{TP}|}{\frac{2}{3}(1 + B_0)W_2 L C_{ox} + C_{xp}} \tag{3.67}$$

and assume $C_{xp} \ll (1 + B_0)C_{gs2}$, we obtain:

$$f_{3dB} \approx \frac{1}{2\pi} \frac{\mu_p}{(1 + B_0)L^2} (|V_{GS2} - V_{TP}|) \tag{3.67a}$$

Although this assumption has limited validity for very small geometries, it provides a valuable design hint: Using NMOS current mirrors for M2 and M3, and consequently, using matched PMOS transistors as the input pair, results in an improvement of the 3 dB frequency by approximately (μ_n/μ_p), for comparable gate over-drive voltage values. In addition, using a PMOS input stage is usually more advantageous from the point of view of noise.

(3.67a) also indicates that the bandwidth decreases with B_0. On the other hand, as explained in Chap. 2 and seen from (3.66), the low-frequency value of the transadmittance is proportional with B_0. For the design of a high-gain, high-bandwidth transadmittance amplifier, there is a trade-off related to B_0, in terms of the total power consumption, total area consumption, the low-frequency value of the transadmittance, and the bandwidth.

The output voltages on equal output loads can be calculated as

$$v_o = \frac{i_o}{Y_L} = v_i \frac{y_m}{Y_L}, \quad v'_o = \frac{i'_o}{Y_L} = \frac{-i_o}{Y_L} = -v_1 \frac{y_m}{Y_L}$$

where the differential output voltage corresponding to a differential input is found as:

$$v_{odd} = \left(v_o - v'_o\right) = 2v_i \frac{y_m}{Y_L} = v_i \frac{1}{Y_L} g_{m1} B_0 s_p \frac{1}{\left(s - s_p\right)}$$

Thus, the differential-input-to-differential-output voltage gain can be written as

$$A_{vdd} = \frac{1}{Y_L} g_{m1} B_0 s_p \frac{1}{\left(s - s_p\right)} \tag{3.68}$$

For example, with an R-C load of $Y_L = G_L + sC_L$ the differential voltage gain becomes

$$A_{vdd} = \frac{1}{C_L} g_{m1} B_0 s_p \frac{1}{\left(s - s_p\right)\left(s - s_{pL}\right)} \tag{3.69}$$

where

$$s_{pL} = -\frac{G_L}{C_L} \tag{3.69a}$$

Consequently, the R-C loaded fully differential amplifier has two negative real poles that influence the frequency characteristics. In most cases, s_{pL} dominates and the gain becomes

$$A_{vdd} \cong \frac{1}{C_L} g_{m1} B_0 \frac{1}{\left(s - s_{pL}\right)} \tag{3.70}$$

which is safe for feedback applications. However, as mentioned in Chap. 2, it is usually necessary to apply a common mode feedback (CMFB) in such circuits to maintain the stability of the operating points. Although this feedback does not directly involve the differential signal path, additional phase shifts on the differential signal may occur due to the unavoidable parasitics, leading to unexpected stability problems (i.e., oscillations).

Example 3.3
Let us calculate the 3 dB frequency of the differential OTA (Fig. 2.20), designed in Chap. 2. The design goals were: transconductance $G_m = 3$ mS, $B_o = 1$, and the DC current of the input transistors $I_{DI} = 1$ mA (consequently the total current consumption, 4 mA). The calculated dimensions (in micrometers) for 0.35 micron AMS CMOS technology were:

M1, M11: 24/0.35
M2, M12: 50/0.35
M4, M14: 48/0.35
M5, M15: 111/1
M3: 230/1

From (3.65), the 3 dB frequency (frequency corresponding to the pole of the gain function) can be written as

$$f_{3dB} = \frac{1}{2\pi} \frac{g_{m2}}{(C_{gs2} + C_{gs3}) + C_{xp}}$$

$$g_{m2} = \sqrt{2(KP)_p (W/L)_2 |I_{D2}|}$$

$$(KP)_p = \mu_p C_{ox} = 137 \times 4.56 \times 10^{-7} = 0.624 \times 10^{-4} \left[A/V^2 \right]$$

$$g_{m2} = \sqrt{2 \times (0.624 \times 10^{-4}) \times (50/0.35) \times 1} = 4.22 \times 10^{-3} S = 4.22 \text{ mS}$$

$$(C_{gs2} + C_{gs4}) = \frac{2}{3}(W_2 + W_4)LC_{ox} + (W_2 + W_4)CDGO$$

Inserting dimensions in micrometers, specific capacitances in fF/μm² and fF/μm:

$$(C_{gs2} + C_{gs4}) = \frac{2}{3}(50 + 48) \times 0.35 \times 4.56 + (50 + 48) \times 0.12 \cong 110\,\text{fF}$$

$$C_{xp} = (W_1 + W_2)XC_j + (W_1 + W_2 + 2X)C_{jsw} + (W_1 + W_2 + W_4)CDGO$$

$$C_{xp} = (24 + 50) \times 0.85 \times 0.94 + 2 \times (24 + 50 + 1.7) \times 0.25 + (24 + 50 + 48) \times 0.12 \cong 111.6 \text{ fF}$$

$$f_{3dB} = \frac{1}{2\pi} \frac{4.3 \times 10^{-3}}{(110 + 111.6)10^{-15}} = 3.09 \text{ GHz}$$

The frequency characteristic obtained with PSpice simulation is shown in Fig. 3.33. The hand calculation and simulation results for the low-frequency value of the transconductance are in perfect agreement (3 mS and 3.02 mS, respectively). The disagreement of the 3 dB frequency (3.09 GHz vs. 2.36 GHz) is mainly due to the approximations on the analytical expressions and the neglected parasitics.

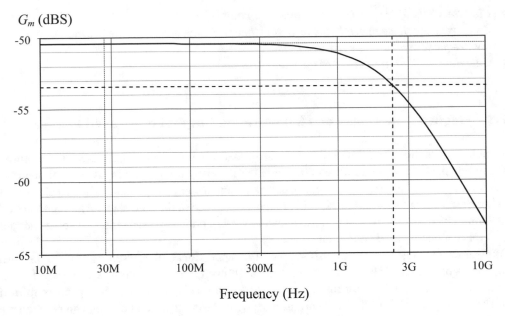

Fig. 3.33 Simulated frequency response of the amplifier. The transconductance, expressed in dBS $= 20 \log (i_{o1}/v_1)$, is -50.448 dBS which corresponds to 3.02 mS. The 3 dB frequency is 2.36 GHz and the slope is less then -20 dB/dec, up to 10 GHz

Problem 3.2

A simple, single-ended transadmitance amplifier is shown in the figure.

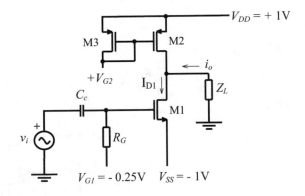

The parameters of the transistors are given as follows:

$L = 0.25 \ \mu\text{m}, T_{ox} = 5.4 \ \text{nm}$

$NMOS : V_{TH} = 0.45 \ \text{V}, CDGO = CGSO = 2.5 \times 10^{-10} \ \text{F/m}, C_j = 1.8 \times 10^{-3} \ \text{F/m}^2,$

$C_{jsw} = 4.2 \times 10^{-10} \ \text{F/m}, X = 0.5 \ \mu\text{m}.$

$PMOS : V_{TH} = -0.45 \ \text{V}, CDGO = CGSO = 4.5 \times 10^{-10} \ \text{F/m}, C_j = 1.8 \times 10^{-3} \ \text{F/m}^2,$

$C_{jsw} = 3.5 \times 10^{-10} \ \text{F/m}, X = 0.5 \ \mu\text{m}, W = 10 \ \mu\text{m}.$

The PMOS current mirror is biased for 1 mA drain current.

(a) *Find the width of M1 for $i_o = 0$ at quiescent point ($v_i = 0$).*
(b) *Z_L is very small compared to the output internal impedance of the circuit. Calculate the low-frequency value of the transadmittance.*
(c) *Calculate the 3 dB frequency of the transadmittance.*

3.7.3   Frequency Response of a Single-Ended Output Long-Tailed Pair

The circuit schematic of an NMOS differential pair loaded with a PMOS unity gain current mirror is given in Fig. 3.34. This type of differential amplifier is mostly used as the input stage of operational amplifiers, where the subsequent stage is usually a common source gain stage. Therefore, the load of the single-ended input stage is the capacitive input admittance of the gain stage. Although the single-ended long-tailed pair behaves as a transconductor, we will prefer to consider it as a voltage amplifier due to this commonly used configuration.

   We know that under a small differential AC input signal, the resulting AC drain current components of M1 and M11 (i_1 and i_{11}) are equal in magnitude and opposite in phase. The drain current of M2, which is i_1, is mirrored as $i_{12} = Bi_1$, where B is the frequency-dependent mirroring factor (current gain) of the M2–M12 current mirror. Hence the total load current can be written as

$$i_L = i_{11} + i_{12} = i_1(B + 1) \tag{3.71}$$

The drain signal current of M1 is $i_1 = g_{m1}v_{gs1}$. Since $v_{gs1} = (v_i/2)$ under small-signal conditions, the load current, the output voltage, and the small-signal voltage gain can be written as

$$i_L = \frac{1}{2}v_i g_{m1}(B + 1)$$

$$v_o = \frac{i_L}{Y_L}$$

$$A_v = \frac{v_o}{v_i} = \frac{1}{2}\frac{g_{m1}}{Y_L}(B + 1) \tag{3.72}$$

Fig. 3.34 NMOS differential amplifier with single-ended output, loaded with PMOS current mirror

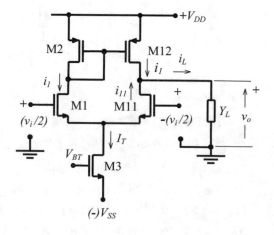

The current gain of a current mirror can be expressed as (c.f. Appendix D)

$$B = B_o \frac{s_p}{s_o} \frac{(s - s_o)}{(s - s_p)} \tag{3.73}$$

where B_o is the low-frequency current transfer ratio which is equal to unity if the transistors of the mirror are identical as in Fig. 3.34. For this case, B becomes

$$B = \frac{s_p}{s_o} \frac{(s - s_o)}{(s - s_p)} \tag{3.73a}$$

The pole and the zero of this function are

$$s_o = + \frac{g_{m12}}{C_{dg12}} = \frac{g_{m2}}{C_{dg2}} \tag{3.74}$$

$$s_p = - \frac{g_{m1}}{(C_{gs1} + C_{gs12} + C_{pT})} = \frac{g_{m1}}{(2C_{gs1} + C_{pT})} \tag{3.74a}$$

where C_{pT} represents the total parasitics of the drain node of M1, i.e., the sum of the drain junction capacitances of M1 and M2 and the gate-source overlap capacitance of M12.

Now $(B + 1)$ in (3.72) can be written as

$$(B + 1) = 1 + \frac{s_p}{s_o} \frac{(s - s_o)}{(s - s_p)} = \frac{s_o(s - s_p) + s_p(s - s_o)}{s_o(s - s_p)}$$

and simplified with $s_0 \gg |s_p|$ as

$$(B + 1) \cong \frac{(s - 2s_p)}{(s - s_p)} \tag{3.75}$$

Now the voltage gain from (3.72) becomes

$$A_v \cong \frac{1}{2} \frac{g_{m1}}{Y_L} \frac{(s - 2s_p)}{(s - s_p)} \tag{3.76}$$

The total load admittance (Y_L) is usually a parallel combination of a capacitance and a conductance:

$$y_L = g_L + sC_L$$

where the total parallel capacitance (C_L) is the sum of the external load capacitance and the junction capacitances of the drain regions of M11 and M12, and the total conductance (g_L) is the sum of the output internal admittances of M2 and M4 and the external load admittance (if there is any). Hence the gain expression becomes

$$A_v \cong \frac{1}{2} \frac{g_{m1}}{C_L} \frac{(s - 2s_p)}{(s - s_p)(s - s_L)} \tag{3.77}$$

where

$$s_L = - \frac{g_L}{C_L} \tag{3.78}$$

This gain expression can be interpreted as follows:

- The gain function has one negative real zero and two negative real poles.
- The phase shift exceeds $\pi/2$ around the second pole but approaches back to $\pi/2$ at higher frequencies.
- The gain at low frequencies ($s \to 0$) is

$$A_{vo} \cong \frac{g_{m1}}{g_L} \qquad (3.79)$$

- With no external load, the pole related to the output reaches its maximum value, which is determined by the intrinsic conductance and capacitance of the output node as $s_{L(max)} = -g_o/C_{djT}$, where g_o is the sum of the output conductances, and C_{djT} the sum of the drain region junction capacitances of M11 and M12. It can be seen that the magnitude of this pole is always smaller than the pole related to the current mirror; in other words, s_L dominates.
- With an external load, which is usually capacitive, this pole dominates more strongly and determines the bandwidth of the amplifier.
- The gain-bandwidth product of the amplifier without external load can be used as a figure of merit for optimization purposes:

$$GBW_{(max)} \cong A_{vo}f_{L(max)} = A_o\left(\frac{1}{2\pi}|s_{L(max)}|\right) \qquad (3.80)$$

Example 3.4
A differential amplifier is designed for a typical 0.18 µm CMOS technology with $L_1 = L_{11} = 0.18$ µm, $L_2 = L_{12} = 0.18$ µm, $W_1 = W_{11} = 18$ µm and $W_2 = W_{12} = 18$ µm. The drain areas are 0.5 µm. The tail current source of $I_T = 400$ µA is assumed ideal. The key device model parameters are listed as:

$T_{ox} = 4.2$ nm ($C_{ox} = 8.2 \times 10^{-7}$ F/cm^2), $V_{ToP} = -0.43$ V, $\mu_{Po} = 71.2$ cm^2/V.s, $V_{ToN} = 0.315$ V, $\mu_{No} = 326$ cm^2/V.s, $C_{GDO}=C_{GSO} = 1.58 \times 10^{-12}$ F/m, $C_{joP} = 1.14 \times 10^{-8}$ F/cm^2, $C_{swP} = 1.74 \times 10^{-10}$ F/cm, $C_{joN} = 1.19 \times 10^{-8}$ F/cm^2, $C_{sw} = 1.6 \times 10^{-10}$ F/cm.

The sum of the junction capacitances of M11 and M12 for zero bias can be calculated as $C_{jdT} = 33$ fF. Under a bias of approximately 1 V, the capacitance value is $C_{jdT} = 23$ fF. The output conductances are given as $g_{o11} = 91.5$ µS and $g_{o12} = 41.25$ µS.

The dominant pole frequency with no external load can be calculated as:

$$f_{L(max)} = \frac{1}{2\pi}\frac{g_o}{C_o} = \frac{1}{2\pi}\frac{(g_{o11} + g_{o12})}{C_{jdT}}$$

$$f_{L(max)} = \frac{1}{2\pi}\frac{(91.5 + 41.25)10^{-6}}{23 \times 10^{-15}} = 918.6\,\text{MHz}$$

It is obvious that the external load strongly affects the bandwidth. For an external load of only 50 fF and 1 pF, the calculated values of the pole frequencies are 289.4 MHz, and 20.6 MHz, respectively.

g_{m1}, which is necessary to find the low-frequency voltage gain, is calculated as

$$g_{m1} = \sqrt{2I_D\mu_n C_{ox}(W_1/L_1)} = \sqrt{2 \times (2 \times 10^{-4}) \times 245 \times 8.3 \times 10^{-7} \times 100} = 2.85\,\text{mS}$$

where $\mu_n = 245\,\text{cm}^2\,/\text{V.s}$ corresponds to a 0.5 V gate over-drive as a reasonable value (see Appendix A).

Now, A_{vo} can be calculated as:

$$A_{vo} = \frac{g_{m1}}{g_L} = \frac{2.85 \times 10^{-3}}{132.75 \times 10^{-6}} = 21.5 \ (26.6\,\text{dB})$$

These results lead to the following observations about the circuit:

The bandwidth is primarily determined by the dominant pole related to the output node, which depends on the sum of the output conductances of M11 and M12, the sum of the junction capacitances of the drain regions of M11 and M12 (both being geometry dependent) and the load capacitance. The low-frequency gain also depends on the device geometries. Therefore, it is reasonable to investigate the effects of the input and load transistor geometries on the gain-bandwidth product.

As we know, the output conductance of a MOS transistor can be approximately calculated as $g_{ds} = \lambda I_D$. The value of the channel length modulation factor (λ) is inversely proportional with the channel length [3]:

$$\lambda \cong \frac{1}{B_2.L.\sqrt{N}} = h.\frac{1}{L}$$

where h is a structural parameter. Using this expression, the output conductance can be calculated in terms of the tail current, the channel lengths of the input transistors, the ratio of the channel lengths of M11 and M12 $(l = W_{11}/W_{12})$, and the h parameter:

$$g_L = h\frac{I_T}{2}\left(\frac{1}{L_{11}} + \frac{1}{L_{12}}\right) = h\frac{I_T}{2}\frac{1}{L_{11}}(1+l) \tag{3.81}$$

The total output node capacitance can be written as

$$C_L = C_{jd11} + C_{jd12} + C_{L(ext)} = C_{jdT} + C_{L(ext)} \tag{3.82}$$

Hence the dominant pole frequency of the amplifier becomes[10]

$$f_L = \frac{1}{2\pi}\left(h\frac{I_T}{2}\frac{1}{L_1}(1+l)\right)\frac{1}{C_{jdT} + C_{L(ext)}}$$

On the other hand, the low-frequency gain can be written as

$$A_{vo} = \frac{g_{m1}}{g_L} = \frac{\sqrt{I_T\mu_n C_{ox}(W_1/L_1)}}{h\frac{I_T}{2}\frac{1}{L_1}(1+l)} = \frac{2}{h(1+l)}\sqrt{\frac{1}{I_T}\mu_n C_{ox}W_1 L_1}$$

---

[10] Since M1 and M11, M2 and M12 are identical transistors, the parameter symbols of M1 and M2 are used for both members of the pairs, from here on.

and the gain bandwidth product

$$GBW = \frac{1}{2\pi}\sqrt{I_T \mu_n C_{ox} \frac{W_1}{L_1} \frac{1}{C_{jdT} + C_{L(ext)}}} = \frac{g_{m1}}{C_{jdT}} \frac{1}{1 + \frac{C_{L(ext)}}{C_{jdT}}} \tag{3.83}$$

Expression (3.83) provides useful hints to maximize the gain-bandwidth (GBW) product:

(a) *GBW* strongly depends on the ratio of the external load capacitance to the sum of the drain junction capacitances of M11 and M12.

(b) Especially for $C_{L(ext)} \gg C_{jdT}$, increasing the tail current helps to increase the *GBW* at the expense of more power consumption.

(c) If the external load capacitance is comparable to the total output node junction capacitances, decreasing the widths of the transistors of the current mirror load increases the *GBW*. (But in this case V_{GS12} increases and affects the DC supply voltage budget. Therefore there is a lower limit for W_2, related to the DC operating conditions of the circuit).

(d) *GBW* increases with mobility (using NMOS input transistors is advantageous).

(e) *GBW* changes inversely with L_1 (using minimum channel length transistors is advantageous).

3.7.4 On the Input and Output Admittances of the Long-Tailed Pair

The two input admittances of a long-tailed pair have frequency-independent capacitive components corresponding to the physical gate-source capacitances. In addition to these components, there are frequency-dependent real and imaginary components of the input admittances resulting from the Miller effect, similar to that of a common source amplifier or a source follower. These components do not have a significant effect on the behavior of the amplifier, provided that the internal impedances of the driving signal sources are very small compared to the input impedances of the amplifier. However, in many practical cases, this condition is not valid, and the interaction of the source impedance and the input impedance of the amplifier affects the amplitude and phase characteristics of the amplifier. The input capacitance is well known and can be easily taken into account, whenever necessary. But the input *conductance* is usually ignored. Therefore we will concentrate on the real components of the input admittances of differential amplifiers in this section.

We have already seen that, although the input of a common source or a common drain amplifier is the gate terminal that is isolated from the rest of the circuit, the input admittance acquires a (positive or negative) real part at high frequencies. In this section, we will use the knowledge developed in Sects. 3.2 and 3.3 to investigate the input conductances of different types of differential amplifiers, under different driving conditions.

We know that for a differentially driven long-tailed pair, the source node is considered to be at the ground potential. Therefore, the behavior of the input admittance is the same as that of a grounded source amplifier. In Sect. 3.1, we found that the input admittance of the common source amplifier has a real part that increases with frequency, as a result of the Miller effect. For higher values of the voltage gain (i.e., for higher load resistance values), the magnitude of the input conductance increases. The simulation results showing the input conductance of a resistively loaded and differentially driven long-tailed pair (Fig. 3.35a) are in agreement with this reasoning.

If a long-tailed pair is driven from one of its inputs and the other input is grounded, the source of the input transistor is not at the ground potential anymore. Instead, the source has a load equal to the input admittance of the second transistor that is acting in this case as a passive common gate circuit.

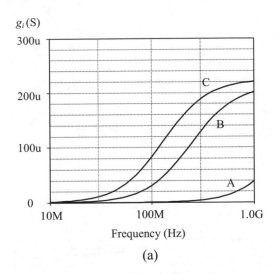

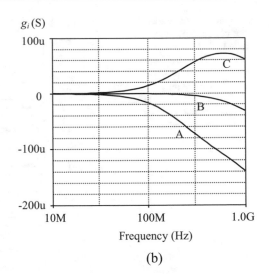

Fig. 3.35 Variation of the input conductance of a resistively loaded long-tailed pair with frequency ($L = 0.35$ μm, $W = 100$ μm, $I_T = 200$ μA). (**a**) Differential drive, (**b**) Single-ended drive. A: $R_L = 500$ ohm, B: $R_L = 5$ kohm, C: $R_L = 10$ kohm. C_L is 50 fF for all cases

We know from Sect. 2.3 that the input conductance of a common gate amplifier is equal to g_m. This conductance and its parallel capacitance, which is the sum of the source region junction capacitances of M1 and M11, and the input capacitance of M11, form an R-C load to the source of M1. From Sect. 3.2, we know that such a source load induces a *negative input conductance* component, provided that the drain is at the ground potential, or the value of the drain load impedance is low. The curve A in Fig. 3.35b corresponds to such a situation. For higher values of the drain resistance and corresponding higher gate-to-drain voltage gain, the positive Miller component of the input conductance starts to become effective, eventually compensating (curve B) or dominating (curve C) the negative component.

These results imply that the two input admittances of a single-ended-output long-tailed pair as shown in Fig. 3.34, which is driven from one of its inputs while the other input is grounded, are not equal. In case the circuit is driven from the gate of M1, which is loaded with a diode-connected (low impedance) drain load, the Miller component is negligible and the input conductance at high frequencies is negative. If the circuit is driven from the gate of M11, which has a high impedance drain load, the Miller component dominates and the input conductance at high frequencies becomes positive. The simulation results given in Fig. 3.36 illustrate this reasoning.

The output admittance of a long-tailed pair depends on the circuit configuration. For a resistively loaded or passive-transistor loaded amplifier as shown in Fig. 2.17a and b, the output admittance is the sum of the output admittance of the input transistor and the load admittance, together with the parasitic drain region junction capacitances. The direct way to increase the output resistance is to increase the channel lengths. To keep the main parameters (DC operating conditions, the transconductances of the input transistors) of the circuit, the aspect ratios must be maintained, leading to the increase of the junction capacitances. Therefore, there is a trade-off between a high output impedance and the 3 dB frequency. Another possibility is to use a cascode circuit configuration as the passive load, whenever the DC voltage budget permits.

In the OTA configuration shown in Fig. 2.17d, the current mirrors loading the input transistors have dominating roles on the 3 dB frequency of the circuit. Since the 3 dB frequency of a current mirror strongly depends on the channel length, increasing the channel lengths of the load transistors is

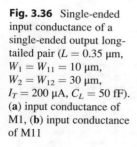

Fig. 3.36 Single-ended input conductance of a single-ended output long-tailed pair ($L = 0.35$ μm, $W_1 = W_{11} = 10$ μm, $W_2 = W_{12} = 30$ μm, $I_T = 200$ μA, $C_L = 50$ fF). (**a**) input conductance of M1, (**b**) input conductance of M11

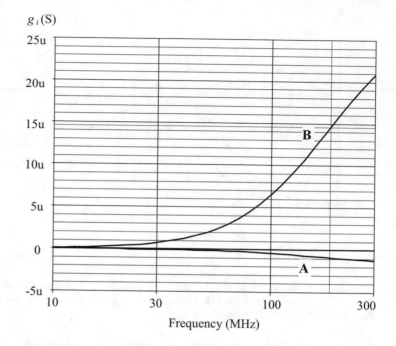

not feasible for high-frequency applications. As M5 and M15 (as well as M3) are not on the signal path, it is possible to increase the channel lengths (and maintain the aspect ratios) of these transistors. But this leads to an increase of the drain region junction capacitances that increases the internal parallel output capacitance of the circuit and affects the high-frequency performance to some extent.

We have seen that the input admittance of a differential amplifier has a frequency-dependent capacitive component and a (positive or negative) real part that can have significant values, even at moderate-to-high frequencies. The effects of these components on tuned amplifiers and gyrator based tuned circuits will be investigated in Chap. 4.

3.8　Gain Enhancement Techniques for High-Frequency Amplifiers

The gains of the basic amplifiers investigated in the previous sections may not be sufficiently high for certain applications. The gain can be increased by cascading several amplifying stages, which yields an over-all gain equal to the multiplication of the gains of the cascaded stages. In this case, it is obvious that both the gain and the bandwidth of a certain stage will be affected by the input impedance of the succeeding stage, and that the over-all cutoff frequency will be smaller than the cutoff frequencies of the individual stages. This "multiplicative" approach is extensively used for the design of high gain amplifiers and to obtain maximum cutoff frequency for a certain over-all gain. In this section, strategies on the cascading of different types of amplifiers will be discussed.

Another possibility is to increase the gain by "adding" the gains of the individual stages, which is usually called the "distributed amplification." In this case, the overall cutoff frequency becomes equal to the cutoff frequency of the identical gain stages.

3.8.1   "Additive" Approach: Distributed Amplifiers

The distributed amplifier concept was first developed in the early years of electronics engineering, initially for vacuum tubes [4], and subsequently applied to bipolar transistors, MOS transistors, and MESFETs to boost the cutoff frequency of amplifiers[11] – a feat which is not easily possible with conventional cascading techniques. The basic principle of a distributed amplifier is shown in Fig. 3.37. The circuit contains two artificial L-C transmission lines having identical signal delays τ per sector. These lines are terminated at the input and output ends with resistors equal to the characteristic impedance of the lines, Z_{01} and Z_{02}, respectively, to eliminate any signal reflections. Although Z_{01} and Z_{02} do not have to be equal, it is common practice to use $Z_{01} = Z_{02} = Z_0 = 50$ ohm.

The basic parameters of an artificial L-C transmission line can be calculated in terms of the self-inductance and parallel capacitance of a sector of the line as follows:

Characteristic impedance: $Z_0 = \sqrt{L/C}$

Cutoff frequency: $\omega_0 = \frac{2}{\sqrt{LC}}, \qquad f_0 = \frac{1}{\pi\sqrt{LC}}$

Signal delay per sector: $\tau = \frac{2}{\omega_0} = \sqrt{LC}$

The input capacitances of the identical amplifiers constitute the parallel capacitances of the input transmission line (C_1). Similarly, the output capacitances of the amplifiers (C_o) partially constitute the parallel capacitances of the output transmission line, such that the symmetry condition ($C_o + C_2) = C_1$ is satisfied. To minimize the resistive loading of the line, the input and output resistances of the

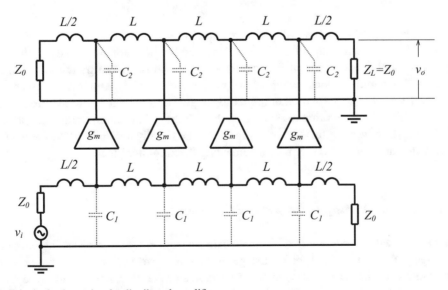

Fig. 3.37   Principal schematic of a distributed amplifier

---

[11] For example, [5–7].

amplifiers must be as high as possible with respect to the characteristic impedance of the line, which imposes the use of transadmittance amplifiers as the most appropriate type.

To understand the principle of operation of a distributed amplifier, assume that the instantaneous value of the input voltage at $t = 0$ is v_i. The input voltage propagates along the line and v_i appears at the input of the first amplifier, A_1, at $t_1 = \tau_h$,[12] at the input of the second amplifier, A_2, at $t_2 = (\tau_h + \tau)$, at the input of the third amplifier, A_3, at $t_3 = (\tau_h + 2\tau)$, and so on. The output current $i_o(v_i) = g_m v_i$ of A_1 corresponding to v_i that appears at $t_1 = \tau_h$ on node-1 of the output line propagates forward and backward along the line. The forward component of this current that is equal to $i_f(v_i) = i_o(v_i)/2$ reaches node-2 at $t = t_2$, simultaneously with the output current of A_2 corresponding to v_i. Hence the current corresponding to v_i that is propagating in the forward direction along the output line becomes the sum of the individual components, namely $2 \times i_f(v_i)$. At $t = t_3$ the forward current corresponding to v_i becomes $3 \times i_f(v_i)$, etc. Finally, the current corresponding to v_i reaching the load, Z_L, becomes $4 \times i_f(v_i)$.

From these considerations, the output voltage and the voltage gain of a distributed amplifier containing n cells can be found as

$$v_o = n \times \frac{1}{2} g_m v_i Z_o, \qquad A_v = \frac{v_o}{v_i} = \frac{1}{2} g_m Z_o n \qquad (3.84)$$

To obtain a high voltage gain, one possibility is to increase n. Another possibility is to increase the gain to a reasonable value and then to cascade the distributed amplifier blocks to reach the targeted total voltage gain,

$$A_{vT} = (A_v)^m = \left(\frac{1}{2} g_m Z_o n\right)^m \qquad (3.85)$$

where m is the number of the cascaded distributed blocks.

The bandwidth of a distributed amplifier is determined by the cutoff frequency of the individual transadmittance amplifiers and that of the transmission lines. Any type of transadmittance amplifier can be used as the amplifier block, provided that the bandwidth of the amplifier fulfills the requirement. One of the most frequently used amplifier topologies in distributed configurations is a simple MOS transistor biased in saturation region. Since the voltage gain of an individual amplifier loaded with $Z_o/2$ is low, the adverse effects of Miller feedback are usually negligible. The cascode configuration is certainly another possibility for implementing the amplifier block.

3.8.2 Cascading Strategies for Basic Gain Stages

As already mentioned, the cascading of multiple gain stages (i.e., connecting the output port of one amplifier stage to the input port of the next stage) is the most frequently used technique for building high gain amplifiers. The types and numbers of the cascaded stages depend on the type and overall gain of the targeted multistage amplifier. Here the term "gain" is used in a broader sense: it can be power gain, voltage gain, current gain, transfer admittance, or transfer impedance.

In multistage power amplifiers, the input impedance of the next stage must be equal to the complex-conjugate of the output impedance of the previous stage, in order to allow the most efficient power transfer from the output of the previous amplifier to the input of the following stage.[13] This

---

[12] τ_h is used for the signal delay of the input (or output) half-sector of the line.

[13] This is one of the basic theorems of network theory.

means that, if the output impedance of the first stage is $Z_{o1} = r_{o1} + jx_{o1}$, to satisfy the maximum power transfer condition, the input impedance of the second stage must be equal to

$$Z_{i2} = \overline{Z}_{o1} = r_{o1} - jx_{o1}.$$

Note that the real components of these two impedances are identical and the imaginary components are in series resonance. This condition dictates that the maximum power transfer condition can be fully satisfied only at a certain frequency (the resonance frequency of the two complementary reactances), and approximately satisfied in the bandwidth of the resonance circuit. The maximum power transfer condition can also be expressed in terms of the output and input admittances, i.e., $Y_{i2} = \overline{Y}_{o1} = g_{o1} - jb_{o1}$, which corresponds to the parallel resonance of the parallel reactive components, in addition to the condition $g_{i2} = g_{o1}$.

It can be seen from this brief explanation that the cascading strategy based on the maximum power transfer condition is useful only for narrowband applications, but not for wideband applications, for example, from DC to several GHz.

For wideband applications, it is necessary to *efficiently* transfer the signal (a voltage or a current) from the output port of the previous stage to the input port of the following stage.[14]

If the signal to be transferred from the output of the previous stage to the input of the following stage is a voltage, the necessary condition for efficient signal transfer is $z_{i2} \gg z_{01}$. Similarly, if the signal to be transferred from the output of the previous stage to the input of the following stage is a current, the necessary condition for efficient signal transfer is $z_{i2} \ll z_{01}$. The parallel reactive components of these impedances usually consist of unavoidable parasitic capacitances and become effective at the high end of the frequency band, in other words, they determine the limit of the usable frequency range of the amplifier.

From these considerations, we can state that for efficient and wideband signal transfer at a cascading node:

- The parallel resistance – i.e., the parallel equivalent of the output resistance of the first stage and the input resistance of the following stage – must be as small as possible. This condition increases the pole frequency related to this cascading node.
- The input resistance of the next stage must be as high as possible compared to the output resistance of the previous stage, which ensures efficient voltage transfer at this cascading node.
- Alternatively, the input resistance of the next stage must be as small as possible compared to the output resistance of the previous stage, which ensures efficient current transfer at this cascading node.

In Fig. 3.38, the four possible amplifier configurations are shown, indicating their input and output resistances. The equivalent circuit given in Fig. 3.38a is called a typical "voltage amplifier," with a high input resistance and a low internal output resistance.

The configurations shown in Fig. 3.38b–d are called, depending on the magnitude of the input and output resistances (i.e., depending on the appropriate type of the input and output signal), "transadmittance amplifier," "transimpedance amplifier" and "current amplifier," respectively.

---

[14] The following explanations are only valid provided that the input of the following stage is connected immediately next to the output of the previous stage. Otherwise, the connection must be made with a transmission line matched on both sides to prevent reflections. The allowable maximum distance without a transmission line depends on the maximum frequency content (or the rise-time) of the signal.

Fig. 3.38 The four basic amplifier configurations

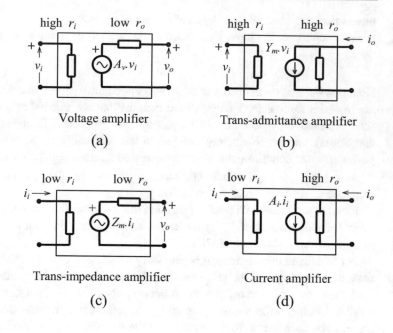

Voltage amplifier

(a)

Trans-admittance amplifier

(b)

Trans-impedance amplifier

(c)

Current amplifier

(d)

Fig. 3.39 Two appropriate configurations for a two-stage voltage amplifier. (**a**) Two cascaded voltage amplifiers, (**b**) A transadmittance amplifier cascaded with a transimpedance amplifier

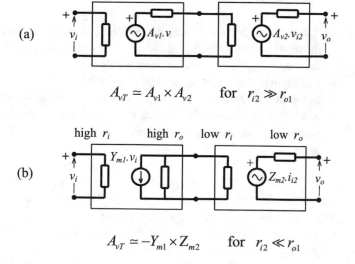

(a)

$$A_{vT} \simeq A_{v1} \times A_{v2} \qquad \text{for} \quad r_{i2} \gg r_{o1}$$

(b)

$$A_{vT} \simeq -Y_{m1} \times Z_{m2} \qquad \text{for} \quad r_{i2} \ll r_{o1}$$

To construct a two-stage cascaded wideband voltage amplifier, we can choose one of two possible solutions as shown in Fig. 3.39, by applying the conditions stated above to the cascading node for efficient signal transfer, as well as for high bandwidth: either cascading two voltage amplifiers, or cascading a transadmittance amplifier in the first stage with a transimpedance amplifier in the second stage.

Similarly, the possible configurations for two-stage transadmittance amplifiers, two-stage transimpedance amplifiers, and two-stage current amplifiers are given in Figs. 3.40, 3.41 and 3.42, respectively.

Fig. 3.40 Two appropriate configurations for a two-stage transadmittance amplifier: (**a**) A transadmittance amplifier cascaded with a current amplifier, (**b**) A voltage amplifier cascaded with a transadmittance amplifier

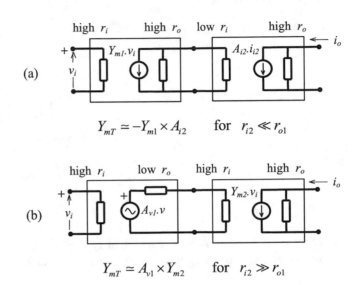

$$Y_{mT} \simeq -Y_{m1} \times A_{i2} \qquad \text{for} \quad r_{i2} \ll r_{o1}$$

$$Y_{mT} \simeq A_{v1} \times Y_{m2} \qquad \text{for} \quad r_{i2} \gg r_{o1}$$

Fig. 3.41 Two appropriate configurations for a two-stage transimpedance amplifier: (**a**) A current amplifier cascaded with a transimpedance amplifier. (**b**) A transimpedance amplifier cascaded with a voltage amplifier

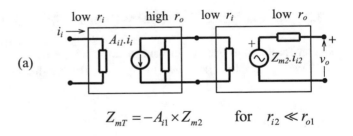

$$Z_{mT} = -A_{i1} \times Z_{m2} \qquad \text{for} \quad r_{i2} \ll r_{o1}$$

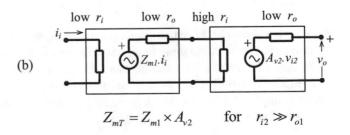

$$Z_{mT} = Z_{m1} \times A_{v2} \qquad \text{for} \quad r_{i2} \gg r_{o1}$$

Fig. 3.42 Two
appropriate configurations
for a two-stage current
amplifier: (**a**) Two
cascaded current
amplifiers, (**b**) A
transimpedance amplifier
cascaded with a
transadmittance amplifier

(a)

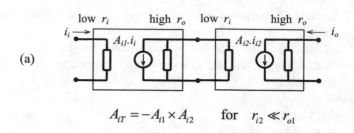

$$A_{iT} = -A_{i1} \times A_{i2} \quad \text{for} \quad r_{i2} \ll r_{o1}$$

(b)

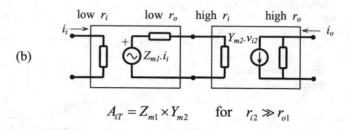

$$A_{iT} = Z_{m1} \times Y_{m2} \quad \text{for} \quad r_{i2} \gg r_{o1}$$

Appropriate configurations for a certain type of amplifier containing more than two cascaded stages can be found by using the same systematic approach.

Problem 3.3

Draw the block diagrams of possible configurations of a three-stage high gain, high bandwidth transimpedance amplifiers and compare their properties.

3.8.3   An Example: The "Cherry-Hooper" Amplifier

To construct a high-bandwidth voltage amplifier, we usually prefer the well-known basic form of a MOS voltage amplifier, as shown in Fig. 3.43. The low-frequency voltage gain of such an amplifier is

$$A_v(0) = -g_m R_L$$

and the cutoff frequency is

$$f_{(3dB)} = \frac{1}{2\pi} \frac{1}{R_L C_o} \ , \quad C_o = C_L + C_{jd}$$

where C_o is the sum of the output load capacitance C_L and of the drain parasitic capacitance C_{jd} (the junction capacitance of the drain junction). Note that both of these capacitances are controllable by the designer. In order to increase the bandwidth, R_L has to be decreased, which also decreases the voltage gain. The straightforward approach to increase the gain to the targeted value is to cascade two similar voltage amplifiers as shown in Fig. 3.39a. In this case, the pole frequency of the cascading node becomes

$$\omega_{pC} = \frac{1}{R_L C_o} \tag{3.86}$$

Fig. 3.43 The basic (simplest) form of a single-stage voltage amplifier

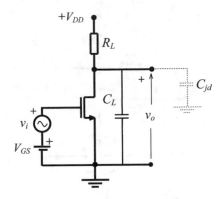

Fig. 3.44 The basic (simplest) form of a transadmittance amplifier

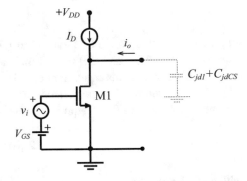

where C_o is the sum of the drain parasitic capacitance C_{jd} and the gate-source capacitance of the second stage. At the output port of the amplifier, the pole frequency is determined by R_L and the load capacitance as

$$\omega_{pL} = \frac{1}{R_L C_L} \tag{3.87}$$

Note that there is another possibility: to cascade a transadmittance amplifier and a trans-impedance amplifier as shown in Fig. 3.39b. The simplest topology of a transadmittance amplifier is shown in Fig. 3.44 where the current supplied by the DC current source is equal to the drain current of M1. The low-frequency transadmittance of the circuit is g_{m1}, corresponding to the transconductance of M1. The output parasitic capacitance is the sum of the drain parasitic capacitance C_{jd1} and the output capacitance of the DC current source C_{jdCS}, which is usually the drain bulk capacitance of a PMOS transistor and is of the same order of magnitude as C_{jd1}.

As the second stage, we need to build a low input resistance transimpedance amplifier with a low input resistance to obtain a high pole frequency at the cascading junction and a low output resistance to obtain a high pole frequency at the output node. Unfortunately, there is no basic MOS amplifier configuration fulfilling the low input impedance and low output impedance at the same time. A possibility is to use an amplifier with applied parallel voltage feedback, as shown in Fig. 3.45a.

The small-signal equivalent circuit of this amplifier is shown in Fig. 3.45b, where y_i represents the output admittance of the signal source (in our case, the output admittance of the transadmittance

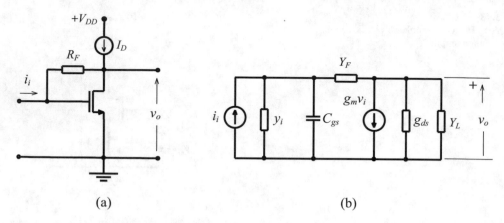

(a)   (b)

Fig. 3.45 (a) The parallel voltage feedback applied MOS amplifier as a transimpedance amplifier. (b) The small-signal equivalent circuit.

amplifier), Y_F is the parallel equivalent of (i) the feedback resistance R_F, (ii) the drain-gate capacitance of the transistor and (iii) the parallel capacitance connected to R_F (if there is any).

It can be seen that this equivalent circuit is the same as that of the CMOS inverter investigated in Sect. 3.5. Therefore, the expressions and results can be directly applied using the appropriate parameter values. The low-frequency value of the transimpedance was found as

$$Z_m(0) = \frac{(g_m - G_F)}{G_F(g_{ds} + g_m)} \tag{3.88}$$

which can be reduced to

$$Z_m(0) \cong \frac{(g_m - G_F)}{g_m G_F} = \frac{1}{G_F} - \frac{1}{g_m} \cong -\frac{1}{G_F} = -R_F \tag{3.88a}$$

for $g_{ds} \ll g_m$ and $G_F \ll g_m$. The low-frequency value of the input impedance can be calculated from the equivalent circuit as

$$Z_i(0) = \frac{R_F + r_{ds}}{1 + g_m r_{ds}} \cong \frac{1}{g_m} \tag{3.89}$$

For $r_{ds} \gg R_F$ and $g_m r_{ds} \gg 1$. Similarly, the low-frequency value of the output impedance:

$$Z_o(0) = \frac{r_{ds}}{1 + g_m r_{ds}} \cong \frac{1}{g_m} \tag{3.90}$$

According to (3.89) and (3.90), the input and output resistances of the amplifier are approximately equal to $(1/g_m)$, provided that $r_{ds} \gg R_F \gg (1/g_m)$ and $g_m r_{ds} \gg 1$. These input and output resistances can be made small enough to increase the pole frequencies of the cascading node, as well as the output node, compared to that of the two-stage cascaded voltage amplifier, given with (3.86) and (3.87).

The schematic of this two-stage voltage amplifier is given in Fig. 3.46a. The low-frequency voltage gain is

$$A_{vT}(0) = Y_{m1}(0) \times Z_{m2}(0) \cong -g_{m1} R_F$$

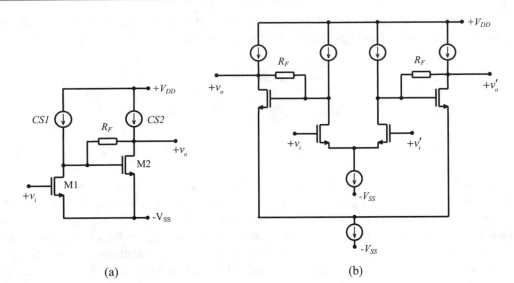

Fig. 3.46 (**a**) A voltage amplifier composed of a transadmittance stage and a transimpedance stage. (**b**) The differential version of this circuit, which is known as the Cherry-Hooper amplifier

The pole frequency corresponding to the cascading node is

$$\omega_{pC} = \frac{g_{m2}}{C_{jd1} + C_{jdCS1} + C_{gs2}}$$

where C_{jdCS1} is the drain-bulk junction capacitance of the PMOS current source of M1. Similarly, the pole frequency corresponding to the output node is

$$\omega_{pL} = \frac{g_{m2}}{C_{jd2} + C_{jdCS2} + C_L}$$

Another factor affecting the overall frequency characteristic is the parasitic capacitance parallel to the feedback resistor. As discussed in Sect. 3.5, this capacitance has an important effect on the transimpedance of the second stage and it may be possible to fine-tune the overall frequency characteristic with an additional capacitance in parallel to C_{dg2}.

The differential version of this circuit given in 3.46b corresponds to the well-known and extensively used wideband voltage amplifier topology: the "Cherry and Hooper" Amplifier [8]. This example demonstrates that in order to construct a certain type of wideband amplifier, all alternative solutions given in Figs. 3.39, 3.40, 3.41, and 3.42 must be considered in a systematic manner.

3.8.4 A Band Widening Technique: Inductive Peaking

All resistance-loaded amplifiers can be considered as "wideband amplifiers" by their nature. The low end of the band is determined by the coupling capacitors (if there are any), and the high end by the capacitance parallel to the load resistor. A wideband amplifier is characterized by its gain, which must be flat and equal to the low-frequency gain with an acceptable tolerance up to the high end of the band, and the upper cutoff frequency where the gain drops to 3 dB below the low-frequency gain.

For a resistance-loaded common source amplifier,[15] the 3 dB frequency, the low-frequency gain and the gain-bandwidth product (the figure of merit for wide band amplifiers) were previously given as

$$\omega_p = \frac{G_o}{C_L} \simeq \frac{1}{C_L R_L} \quad \text{for } r_{ds} \gg R_L$$

$$A_v(0) = -\frac{g_m}{G_o} \simeq -g_m R_L \tag{3.5}$$

$$GBW = \frac{g_m}{2\pi C_L} \tag{3.6}$$

where R_L is the load resistance of the transistor and $C_L = C_o + C_L'$ is the total capacitance of the output node, where C_o is the output capacitance of the transistor, mainly determined by the junction capacitance of the drain region, and C_L' the input capacitance of the succeeding stage (or the load). C_L' is usually given as one of the input parameters of the design problem. C_o can be initially neglected, or an estimated value depending on the technology can be given. (After calculating the dimensions of the transistor, the correct value of C_o must be found and the design must be updated, if necessary).

To reach the targeted 3 dB frequency that is equal to the pole frequency for a resistance loaded common source amplifier, the appropriate R_L value can be calculated as

$$R_L = \frac{1}{2\pi f_p C_L} \tag{3.91}$$

To obtain the targeted gain, according to (3.5) the transconductance must be

$$g_m = \frac{|A_v(0)|}{R_L} \tag{3.92}$$

The transconductance of a transistor operating in the saturation region is determined by the DC drain current and the aspect ratio of the transistor.

$$g_m = \sqrt{2\mu C_{ox} \frac{W}{L} I_D} \tag{1.33}$$

According to (1.33), any combination of (W/L) and I_D can be used to obtain the desired g_m value. Choosing lower drain current values will correspond to wider transistors, and therefore to higher parasitic capacitance values, for the same g_m value. For higher drain current values, the parasitic capacitance can be decreased but the power consumption increases. Another concern related to the drain DC current (the quiescent current) is its position on the output characteristic curves. To obtain maximum dynamic range for the output voltage and minimum nonlinear (harmonic and intermodulation) distortion for a given amplitude, the operating point must be placed at the center of the saturated operating region (as shown in Fig. 2.1), corresponding to

$$I_D = \frac{\left(V_{DD} - V_{DS(sat)}\right)}{2R_L} \tag{3.93}$$

---

[15] The following investigations can be applied to other amplifier configurations with minor modifications (whenever necessary).

which can be considered as the optimum current for the majority of applications. Using this I_D, the aspect ratio of the transistor can be calculated from (1.33):

$$\frac{W}{L} = \frac{g_m^2}{2\mu C_{ox} I_D} \tag{3.94}$$

From (3.6), it is obvious that the factor limiting the bandwidth of an amplifier is C_L, i.e., the sum of the input capacitance of the load and the output parasitic capacitance of the transistor. The only chance to reduce the value of C_L and to increase the gain-bandwidth product is to reduce the width of the transistor, which helps to reduce the parasitic junction capacitance of the drain region, at the expense of increasing the drain current and reducing the output signal dynamic range.

Another possibility to decrease the effect of the output capacitance is to compensate it with an inductance in the frequency region around the pole frequency. This technique, known as "inductive peaking", has been investigated and used from the early days of electronic engineering [9, 10] to the present day [11, 12]. The two basic approaches to compensate the load capacitance with an inductor are shown in Fig. 3.47.

The circuit shown in Fig. 3.47a is called "parallel (or shunt) compensation"; its small-signal equivalent circuit is given in Fig. 3.47b. In this approach, the output capacitance is resonated with an inductance (that increases the load impedance and hence the gain) in the vicinity of the 3 dB frequency of the basic amplifier, where the gain starts to decrease.

The circuit shown in Fig. 3.47c is called "series compensation"; its small-signal equivalent circuit is given in Fig. 3.47d. In this approach, the increase of the capacitance voltage at resonance with respect to the input voltage of the series resonance circuit is used to compensate the decrease of the output voltage (See Chap. 4, Sect. 4.1.2 for the details of main concepts).

Let us first investigate the parallel compensated amplifier. From the equivalent circuit shown in Fig. 3.47b, the voltage gain can be calculated as

$$A_v = \frac{v_o}{v_i} = -g_m \frac{R_D + sL}{s^2 L C_L + s C_L R_D + 1} \tag{3.95}$$

The gain in the frequency domain, normalized with respect to the low-frequency gain, is:

$$\overline{A} = \frac{|A_v(\omega)|}{|A_v(0)|} = \frac{1 + j\omega \frac{L}{R_D}}{(1 - \omega^2 L C_L) + j\omega C_L R_D} \tag{3.96}$$

Using the pole frequency of the basic (noncompensated) amplifier ($\omega_p = 1/R_D C_L$), the resonance frequency of the LC circuit ($\omega_o = 1/\sqrt{L C_L}$) and a parameter α that is defined as $\alpha = \omega_p/\omega_0$, (3.96) can be arranged in normalized form as

$$\overline{A} = \frac{1 + j(\omega/\omega_P)\alpha^2}{\left[1 - (\omega/\omega_P)^2 \alpha^2\right] + j(\omega/\omega_P)} \tag{3.97}$$

It must be noted that α can be written in terms of the circuit parameters and helps in the calculation of the value of L, which is the most important design parameter:

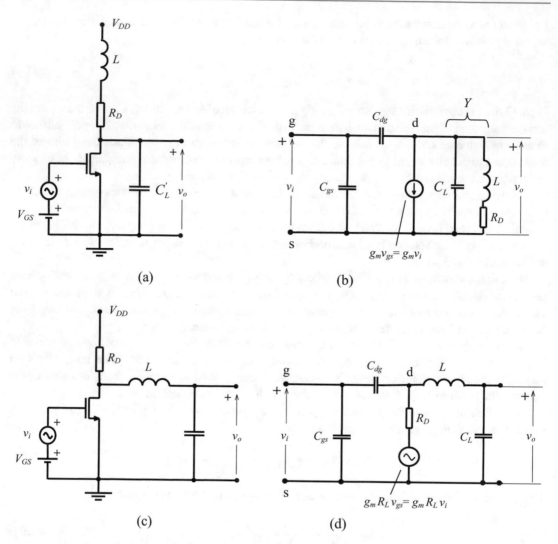

Fig. 3.47 (a) Simplified circuit diagram of a resistive loaded common source amplifier with parallel inductive peaking (b) Small-signal equivalent circuit. (c) Simplified circuit diagram of a resistive loaded common source amplifier with serial inductive peaking. (d) Small-signal equivalent circuit

$$\alpha = \frac{1}{R_D C_L} \sqrt{L C_L} \quad \rightarrow \quad L = \alpha^2 R_D^2 C_L \tag{3.98}$$

The variation of the normalized gain as a function of the normalized frequency for different values of α is given in Fig. 3.48. It can be seen from this graph that:

- Increasing α (increasing L) effectively helps to increase the bandwidth (and the gain-bandwidth product) of the amplifier. For $\alpha = 0.6$, the bandwidth is 62% higher than that of a non-compensated amplifier. For higher values of α, the increase of the bandwidth approaches 87%.
- For $\alpha = 0.7$, the frequency characteristic remains almost flat with a peaking of only 0.2 dB.
- For higher values of α, the bandwidth does not exhibit any further increase, but the peaking becomes unacceptably high for many applications.
- Therefore, it can be concluded that the optimum value of α is 0.7.

Fig. 3.48 Variation of the normalized gain as a function of the normalized frequency for $\alpha = 0$, 0.6, 0.7, 0.85 and 1

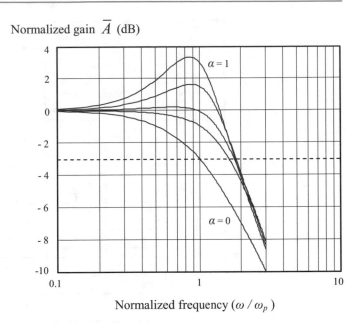

Normalized gain $\overline{A}$ (dB)

Normalized frequency (ω / ω_p)

But this conclusion has to be checked against the delay characteristic of the amplifier.

Amplifiers are typically used to amplify complex waveforms, in most cases pulses and square waves. The flatness of the gain-versus-frequency characteristic guarantees the uniform amplification of all Fourier components associated with the waveform, up to the 3 dB frequency. It must not be overlooked that to preserve the correct shape of the output waveform, the delay of all Fourier components must be the same, in other words, the delay characteristic (not the phase characteristic) must also be flat.

The relation of the phase shift and the delay for a given frequency is illustrated in Fig. 3.49. From this figure, it can be seen that the delay (τ) corresponding to a certain phase shift φ is

$$\tau = \varphi \frac{T}{2\pi} = \frac{\varphi}{2\pi f} = \frac{\varphi}{\omega} \tag{3.99}$$

The normalized delay characteristic obtained from the normalized phase characteristic is given in Fig. 3.50. From these curves, it can be seen that:

- The delay characteristic corresponding to the flat magnitude characteristic (e.g., for $\alpha = 0.7$) is not flat
- The flat delay characteristic corresponds to $\alpha = 0.6$
- Therefore, for the amplification of complex waveforms, in spite of a 13% smaller bandwidth, the appropriate value is $\alpha = 0.6$

To illustrate the effect of the phase delay, the PSpice step response simulation results for an amplifier for various values of α (corresponding to various values of L) are given in Fig. 3.51, underlining the importance of the delay characteristic for a wideband amplifier.

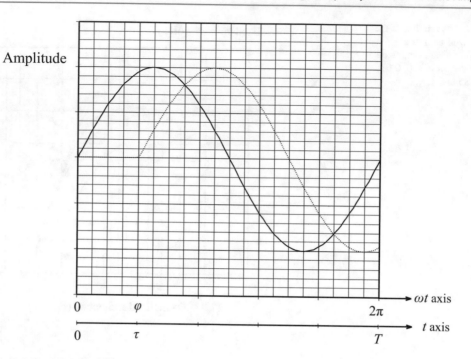

Fig. 3.49 Relation of phase shift and time delay

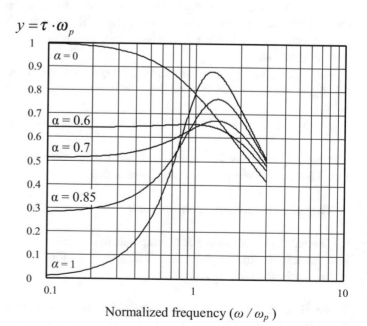

Fig. 3.50 Variation of the normalized delay as a function of the normalized frequency for $\alpha = 0$, 0.6, 0.7, 0.85 and 1

$$y = \tau \cdot \omega_p$$

Normalized frequency (ω / ω_p)

Fig. 3.51 The step response of a wideband amplifier model for +1 mV to −1 mV input step voltage. The parameters of the amplifier model are $g_m = 10$ mS, $R_L = 1$ kΩ and $C_o = 100$ fF. The responses correspond to $\alpha = 0$ ($L = 0$), $\alpha = 0.6$ ($L = 36$ nH), $\alpha = 0.7$ ($L = 49$ nH) and $\alpha = 1$ ($L = 100$ nH)

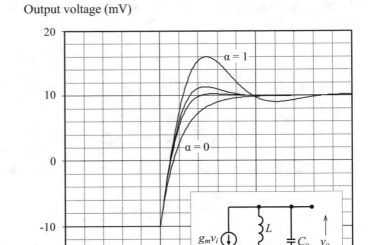

Example 3.5

Design a 3 GHz bandwidth and 14 dB voltage gain, parallel inductive peaked wideband amplifier.

> Technology: AMS035
> Supply voltage: $V_{DD} = +3$ V
> Load capacitance: $C'_L = 75$ fF
> Frequency response: Minimum delay distortion
> Signal source internal resistance: $R_s = 50$ ohm

Since the 3 dB frequency of an amplifier with minimum delay distortion (flat delay characteristic) is 62% wider than that of a nonpeaked amplifier, the starting point must be a resistance loaded amplifier with a 14 dB ($A_v = 5$) voltage gain and $f_p = 3$ GHz/1.62 = 1.85 GHz bandwidth.

The load capacitance is given as 75 fF. The other component of the total parallel output capacitance is the output capacitance of the transistor that depends on the dimensions, which has to be estimated in the beginning and must be checked later on, and updated if necessary. Let us assume the output capacitance of the transistor is also 75 fF, then the total parallel capacitance is $C_L = 150$ fF.

(3.91) gives the value of the collector load resistance:

$$R_L = \frac{1}{2\pi f_p C_L} = \frac{1}{2\pi \times \left(1.85 \times 10^9\right) \times \left(150 \times 10^{-12}\right)} = 573.5 \text{ ohm}$$

The value of the transconductance to obtain the targeted gain with this load resistance can be calculated from (3.92):

$$g_m = \frac{5}{573.5} = 8.7 \text{ mS}$$

To obtain maximum output voltage dynamic range, the drain quiescent current, according to (3.93), must be

$$I_D \cong \frac{(3-0.5)}{2 \times 573.5} = 2.18 \text{ mA}$$

The aspect ratio of the transistor can be calculated with (3.94):

$$\frac{W}{L} = \frac{\left(8.7 \times 10^{-3}\right)^2}{2 \times 300 \times \left(4.54 \times 10^{-7}\right) \times \left(2.18 \times 10^{-3}\right)} = 127.46 \quad \rightarrow \quad W \cong 45 \text{ μm}$$

Finally, the value of the peaking inductance for a flat phase delay characteristic can be found with (3.98):

$$L = (0.6)^2 \times (573.5)^2 \times \left(150 \times 10^{-15}\right) = 17.76 \text{ nH}$$

and for comparison purposes, the value of the peaking inductance for a flat gain characteristic would be:

$$L = (0.7)^2 \times (573.5)^2 \times \left(150 \times 10^{-15}\right) = 24.17 \text{ nH}$$

The PSpice simulations:

Figure 3.52 shows the circuit diagram of the amplifier with the calculated values. To obtain the calculated I_D value, a DC sweep must be applied to V_G, the gate bias voltage. The AC simulation with the calculated circuit parameters and the found gate bias voltage shows that the voltage gain is $|A_v(0)| \cong 3$, considerably smaller than the targeted value. This is due to the smaller value of the simulated transconductance as explained in Chap. 1 and, in principle, can be compensated by increasing the aspect ratio and/or the DC drain current. But the maximum output voltage dynamic range forces us to keep the calculated I_D value. Therefore, to fine-tune the gain to 5, the aspect ratio and the bias voltage must be adjusted. The obtained results are $W = 84$ μm and $V_G = 0.95$ V (which is in agreement with the pre-estimated $V_{DS(sat)} = 0.5$ V).

The PSpice netlist is given below, covering the DC sweep, AC gain, and delay characteristics, the transient response for a pulse input voltage to check and compare the delay distortion, and the transient response for a 1 GHz sinusoidal input voltage to check the output dynamic range (the L values correspond to the fine-tuned amplitude and delay characteristics):

Fig. 3.52 Schematic diagram of the amplifier

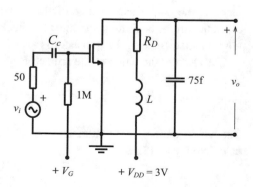

```
         *CS INDUCTIVE SHUNT PEAKED AMPLIFIER*.LIB "ams035.lib"
VDD    100   0   3
M1    1 2 0 0 modn L=.35U  W=84U ad=70e-12 as=70e-12 Pd=84u Ps=84u
CL'  1 0 75f
RL 11 1 573.5
L 100 11 16.6n
*L 100 11 24n
RG 2 200  100k
VG 200 0 .95

vin  23 0 ac 10m
*vin 23 0 pulse 10m -10m .5n 1p 1p 1n 2n
*vin 23 0 sin (0 200mV 1G)
Rs 23 22 50
cc 22 2 100p

.DC VGG .5   2   10M
.AC DEC 20 .1G 10G
*.TRAN .1n 2.5n 0 5p
.PROBE
.END
```

The simulation results for the frequency characteristics of the gain and the signal delay are shown in Fig. 3.53a and b. The pulse response corresponding to the flat gain characteristic and the flat delay characteristic is shown in Fig. 3.54. The simulation performed to check the position of the operating point is shown in Fig. 3.55.

These simulation results can be interpreted as follows:

- The delay characteristic corresponding to $L = 16.6$ nH shown in Fig. 3.53b (the solid line) is flat up to 2 GHz within 1.4% and up to 3 GHz within 5.8%.[16] The fine-tuning of the inductance value (from the calculated 17.76 nH to 16.6 nH) indicates that the assumed value of the output parasitic capacitance of the transistor was somewhat higher than the assumed value. But the difference is small and (therefore) iterative redesign is not considered to be necessary.
- The 3 dB frequency of the gain characteristic corresponding to minimum delay distortion (Fig. 3.53a, the solid line) fulfills the 3 GHz bandwidth requirement.
- The 3 dB frequency corresponding to the flat gain characteristic is even higher than expected (3.38 GHz), at the expense of excessive signal delay at high frequencies, affecting the transient response.
- Figure 3.54 shows the transient responses corresponding to the flat delay characteristic and to the flat gain characteristic. The response corresponding to the flat delay characteristic settles to the final value without overshoot, with a rise time of 110.2 ps. The response corresponding to the flat gain characteristic has 5% overshoot, which can be tolerated for many applications. The rise time for this case is 94.8 ps, which may make it preferable.
- Since the inductance values are high, the quality factors for on-chip realization will be considerably low and series resistances high. For example, the series resistance of an $L = 16.6$ nH, $Q = 5$

---

[16] Note that to eliminate the 180° basic phase shift of the amplifier, the signal delay is calculated as $(180-\varphi)/(2\pi f)$.

Fig. 3.53 PSpice
simulation results for (**a**)
the gain characteristics, (**b**)
the delay characteristics.
The solid lines correspond
to the flat delay
characteristic ($L = 16.6$
nH) and the dotted lines to
the flat gain characteristic
($L = 24$ nH)

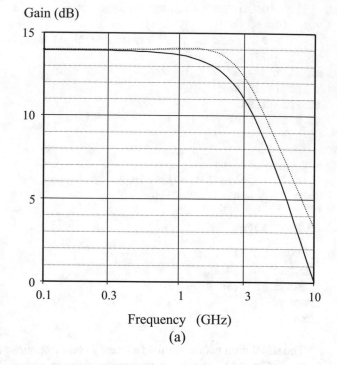

(a)

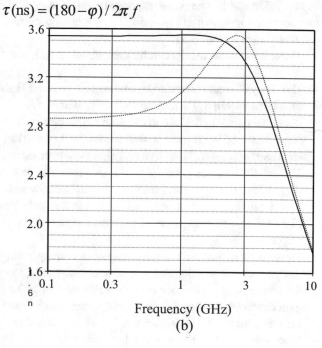

(b)

Fig. 3.54 The pulse response of the amplifier. The solid line corresponds to the flat delay characteristic ($L = 16.6$ nH) and the dotted line to $L = 24$ nH

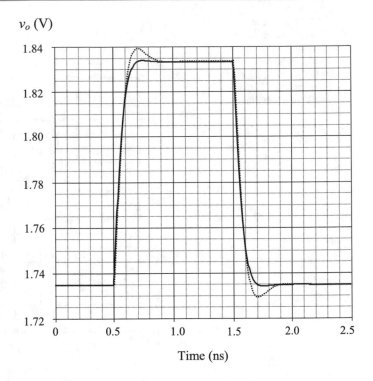

inductance at 1.85 GHz is 38.6 ohm (See Chap. 4, Sect. 4.1.1.1). Therefore, the value of R_L must be reduced to 535 ohm.

- In Fig. 3.55, the responses to 1 GHz, 500 mV amplitude, and 200 mV amplitude sinusoidal input signals are shown. The symmetrical clipping of the output waveform corresponding to 500 mV input signal indicates that the position of the operating point on the output characteristic curves (the value of the drain quiescent current) is appropriate for small nonlinear distortion.

- If the power consumption has prime importance, a drain quiescent current smaller than the value given with (3.93) can be used, at the expense of reducing the output voltage dynamic range and increasing the output capacitance due to the increase of W.

- For higher signal source resistance values, it may be necessary to take into account the effect of the $R_s C_{gs}$ low-pass section.

- The lower 3 dB frequency of the amplifier with $C_c = 10$ pF and $R_G = 100$ kohm is 160 kHz, which may be unnecessarily low for many applications. To increase the value of the lower 3 dB frequency, a solution is to reduce the value of C_c. But it must not be overlooked that C_c and C_{gs} form a capacitive, frequency-independent voltage divider and reduce the signal reaching the gate of the transistor and consequently, the voltage gain decreases in the whole band.

A final remark: After the layout and the post-layout simulations, it may be necessary to perform a second fine-tuning, due to the layout-related parasitics.

Fig. 3.55 The output
voltage of the amplifier at
1 GHz input frequency.
The solid line corresponds
to 500 mV input voltage
amplitude and the dotted
line to 200 mV amplitude

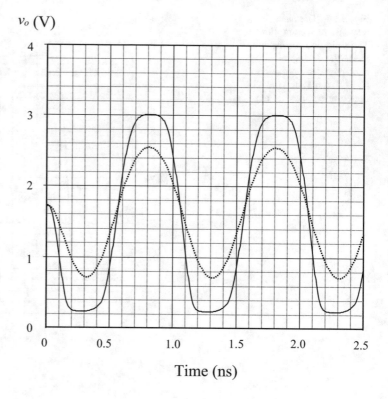

Problem 3.4
Derive the expressions for the series-peaked wideband amplifier shown in Fig. 3.47c.

Problem 3.5
(a) *Design a series-peaked amplifier fulfilling the requirements given in Example 3.4.*
(b) *Compare the results corresponding to the parallel and the series compensated amplifiers, and discuss.*

References

1. J.M. Miller, Dependence of the input impedance of a three-electrode vacuum tube upon the load in the plate circuit. Natl. Bur. Std. (U.S.) Res. Papers **15**, 367–385 (1919)
2. R.W. Hickman, F.V. Hunt, On electronic voltage stabilizers. Rev. Sci. Instr., 6–21 (1939)
3. Y.P. Tsividis, *Operation and Modeling of the MOS Transistor* (McGraw-Hill, Boston, 1987)
4. W.S. Percival, Thermionic valve circuits. British Patent Spec., No: 460,562, Granted January 1937
5. D. Leblebici, The distributed amplifier using transistors as delay elements. Ph.D. Thesis, Istanbul Technical University (1966)
6. Ren-Chih Liu, Kuo-Liang Deng, Huei Wang, A 0.6-22 GHz broadband CMOS distributed amplifier. IEEE Radio Frequency Integrated Circuit Symposium, June 2003, pp. 103–106.
7. Y. Ayasli, R.L. Mozzi, J.L. Vorhaus, L.D. Reynolds, R.A. Pucel, A monolithic GaAs 1-13-GHz traveling-wave amplifier. IEEE Trans. Microwave Theory Tech. **30**(7), 976–981 (1982)
8. E.M. Cherry, D.E. Hooper, The design of wide-band transistor feedback amplifiers. Proc. IEE, 375–389 (1963)

9. A.V. Bedford, G.L. Fredendall, Transient response of multistage video-frequency amplifiers. Proc. IRE **27**, 277–284 (1939)

10. S. Seely, *Electronic Engineering* (McGraw-Hill, New York, 1956), pp. 101–110

11. P. Staric, E. Margan, *Wideband Amplifiers* (Springer, New York, 2007)

12. S.S. Mohan, M. del Mar Hershenson, S.P. Boyd, T.H. Lee, Bandwidth extension in CMOS with optimized on-chip inductors. JSSC **35**, 346–355 (2000)

Frequency-Selective RF Circuits

4

During the early days of radio design, the tuned amplifier was one of the most important subjects of electronic engineering. It has been used as input RF amplifiers of radio receivers tunable in a certain frequency range, and as intermediate frequency (IF) amplifiers tuned to a fixed frequency. From the 1930s to the 1950s, RF amplifiers using electron tubes were one of the most important and interesting research areas of electronic engineering and were investigated in depth. TV and all other wireless systems were other application areas for tuned amplifiers.

After the emergence of the transistor, the knowledge already acquired was sufficient for transistorized tuned amplifiers, because the frequency range was still limited to several hundreds of MHz. The active filters, one of the benefits of analog ICs, replaced another class of frequency selective circuits, i.e. L-C filters. As a result of these developments, the importance of inductors and circuits containing inductors decreased and eventually these subjects disappeared from many electronic engineering curricula and textbooks.

The rapid expansion of wireless personal communication and data communication during the recent decades, and the developments in IC technology that extended the operation frequencies into the GHz range, resulted in the "re-birth" of frequency-selective circuits containing inductors. In parallel to the increase of operating frequencies reaching up to the GHz range, necessary inductance values decreased to the "nanohenry" level, which are now possible to realize as an integral part of ICs, as mentioned in Chap. 1. But the quality factor of these "on-chip inductors" is in the range of 5...15, which is considerably smaller compared to the quality factor of classical discrete "wound" inductors.[1] It must be kept in mind that the theory of tuned circuits developed in older textbooks is based on hi-Q inductors. Therefore, a re-consideration of tuned circuits for low-Q circuits is necessary.

This chapter will first feature a summary of resonance circuits; a reminder of basic definitions and behaviors with special emphasis on low-quality factor circuits will follow. Then single-tuned amplifiers, stagger tuning, and amplifiers containing coupled resonance circuits will be presented. The active and passive filters are considered outside the scope of this book, but due to the increasing importance of a class of active filters at high frequencies, gyrator-based g_m-C circuits will be investigated.

The original version of this chapter was revised. The correction to this chapter is available at https://doi.org/10.1007/978-3-030-63658-6_8

[1] The quality factors of discrete wound inductors is in the range of 100...1000, depending on the structure and material of the coil and the frequency.

4.1　Resonance Circuits

Resonance is one of the most important effects that occurs in many physical systems comprising components capable of storing energy as potential energy and kinetic energy. These systems start to swing (oscillate) when excited, i.e., when a little energy is injected into the system, for example as potential energy. This energy swings in the system back and forth between fully potential and fully kinetic phases.[2] The frequency of oscillations depends on the parameters of the system. The pendulum is the easiest way to understand oscillatory systems. An oscillatory system loses its energy in time, if there is (and there always is) a reason to consume energy in the system (for example the air friction for a pendulum), the amplitude of the oscillation eventually decreases down to zero.

In electric circuits, the capacitor is the component that is capable of storing energy in potential form. The energy stored in a capacitor is $E_p = (1/2)CV^2$ if the voltage is V. An inductor stores the energy in kinetic form. If a current I is flowing through an inductor, the energy stored in the inductor is $E_k = (1/2)LI^2$. All resistors in a circuit consume energy. The consumed (dissipated) energy in a time interval t is $E_d = I^2Rt$ in terms of the current and $E_d = (V^2/R)t$ in terms of voltage.

The components of an electrical resonance circuit are an inductor and a capacitor. Their series and/or parallel parasitic resistances, for example the resistance of the inductor material and the parallel dielectric losses of the capacitor, are the energy-consuming components of the system. There are two possibilities for forming a resonance circuit: connecting the inductor and the capacitor in series or in parallel. It will be shown that the resonance effect of a parallel resonance circuit is pronounced when it is driven by a current source. In contrast, a series resonance circuit exhibits the resonance effect when it is driven by a voltage source.

4.1.1　The Parallel Resonance Circuit

A parallel resonance circuit is shown in Fig. 4.1. The inductor is modeled with its self-inductance (L) and the series resistance (r_L) that represents all losses related to the inductor (for an on-chip inductor, this includes the resistance of the strip, the losses of the magnetically induced currents and the substrate resistance, as shown in Fig. 1.37). C is the value of the capacitance. For on-chip capacitors, since the insulator is silicon dioxide, the parallel dielectric losses are negligibly small, but there is a considerable series resistance as shown in Fig. 1.30 that is the main cause of the low Q value of an

Fig. 4.1 Parallel resonance circuit with the series resistances of the inductor and the capacitor

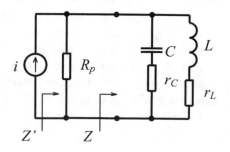

---

[2] For an excellent reading on resonance see [1].

on-chip capacitor.[3] R_p represents the internal resistance of the signal (current) source, and other parallel losses if there are any. The impedance seen by the current source (Z') is the parallel equivalent of R_p and Z (Fig. 4.1). First, we will calculate Z and then include the effect of R_p.

The impedance Z can be calculated as

$$Z = \frac{r_L + s(L + Cr_C r_L) + s^2 LC r_C}{1 + sC(r_C + r_L) + s^2 LC}$$

and in the ω domain,

$$Z(\omega) = \frac{(r_L - \omega^2 LC r_C) + j\omega(L + Cr_C r_L)}{(1 - \omega^2 LC) + j\omega C(r_C + r_L)} \tag{4.1}$$

The real and imaginary parts of (4.1) are

$$\mathrm{Re}\,\{Z\} = \frac{(r_L - \omega^2 LC r_C)(1 - \omega^2 LC) + \omega^2 C(r_L + r_C)(L + Cr_L r_C)}{(1 - \omega^2 LC)^2 + \omega^2 C^2 (r_L + r_C)^2} \tag{4.2}$$

$$\mathrm{Im}\{Z\} = \frac{\omega(L + Cr_L r_C)(1 - \omega^2 LC) - \omega C(r_L - \omega^2 LC r_C)(r_L + r_C)}{(1 - \omega^2 LC)^2 + \omega^2 C^2 (r_L + r_C)^2} \tag{4.3}$$

From these expressions, some information related to the following extreme cases can be extracted:

- For $\omega \to 0$ the imaginary part of the impedance is zero and the real part is equal to r_L, as can be intuitively seen from Fig. 4.1.
- For $\omega \to \infty$ the imaginary part of the impedance is again zero and the real part is equal to r_C.
- The imaginary part of the impedance is zero (the impedance is resistive) for

$$\omega^2_{(\mathrm{Re})} = \frac{(L - Cr_L^2)}{LC(L - Cr_C^2)} \tag{4.4}$$

In Fig. 4.2a, the root locus of Z for $r_C \ll r_L$ is shown, which corresponds to the conventional resonance circuits where the series resistance of the capacitor is negligible. As seen from the figure, the impedance is resistive and equal to r_L. The impedance is inductive up to a frequency for which the impedance again becomes resistive. The frequency corresponding to this case was found as (4.4) and can be simplified as

$$\omega^2_{(\mathrm{Re})} = \frac{1}{LC} - \frac{r_L^2}{L^2} \tag{4.5}$$

for $r_C \ll r_L$. Note that this frequency is not the same as the ω_0 natural frequency[4] that is defined as

$$\omega_0^2 = \frac{1}{LC} \tag{4.6}$$

---

[3] Note that for discrete capacitors of the classical resonance circuits, the series resistance of the capacitor is neglected during the derivation of the expressions. Therefore, these expressions do not sufficiently represent the behavior of the on-chip resonance circuits.

[4] "Natural frequency" is used for the resonance frequency of the lossless L-C combination

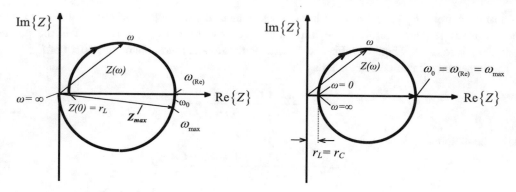

Fig. 4.2 The locus of the impedance of a parallel resonance circuit; (**a**) for $r_C \ll r_L$, (**b**) for $r_C = r_L$

In addition, it can be shown that the frequency for which the magnitude of the impedance is maximum is another important frequency and is equal to

$$\omega_{max}^2 = \frac{1}{LC}\sqrt{1 + 2r_L^2\frac{C}{L} - \frac{r_L^2}{L^2}} \tag{4.7}$$

for $r_C \ll r_L$. The differences between these three frequencies that are characteristic for a resonance circuit are small, but may have important effects on the behavior of systems containing resonance circuits. If they are assumed to be equal, this assumption may hide certain delicate properties of the circuit, as will be examined below.

For example, for an L-C oscillator, according to the Barkhausen criterion, the circuit oscillates at a frequency for which the loop gain satisfies the gain condition, and the total phase shift on the loop is equal to zero. It is obvious that, ideally, the frequency for which the magnitude of the impedance (and consequently the gain) is maximum and the frequency for which the phase shift is zero (the impedance is resistive) will coincide, with the notation we used above, $\omega_{max} = \omega_{(Re)}$. The conventional way to satisfy (at least to approach) this condition is to use a low-loss (high Q) resonance circuit as the resonator of the oscillator.

For on-chip resonance circuits, there is a possibility to equate ω_{max} and $\omega_{(Re)}$. It can be seen from (4.4) that for $r_C = r_L$, $\omega_{(Re)}$ becomes equal to ω_0 and the beginning of the locus ($\omega = 0$) coincides with its end ($\omega = \infty$) at $\text{Re}\{Z\} = r_C = r_L$. The root locus corresponding to this special case is shown in Fig. 4.2b. It can be seen that due to the symmetry of the plot, not only ω_0 and $\omega_{(Re)}$ but also ω_{max} coincide. This may be a valuable hint for the phase noise minimization of L-C oscillators.

Example 4.1
To check these results let us simulate a resonance circuit for two different cases:

(a) The inductance is $L = 2$ nH and series resistance of the inductance is $r_L = 10$ ohm. C has the appropriate value to tune the circuit to $f_0 = 5$ GHz. The series resistance of the capacitor is negligibly small.

(b) The inductance and the resonance frequencies are the same as (a). But the series resistances of L and C are the same and equal to 5 ohm.

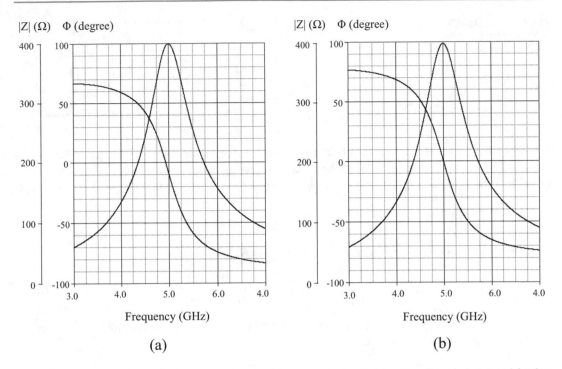

Fig. 4.3 The phase and magnitude curves of a parallel resonance circuit: (**a**) with a lossy inductor and lossless capacitor, (**b**) with a lossy inductor and lossy capacitor, if the series resistances of L and C are equal

Draw the variations of the magnitude and phase of the impedance as a function and compare.

The value of the capacitance can be calculated as 506.7 fF from (4.6). The PSpice results of the magnitude and phase are shown in Fig. 4.3a. As can be seen from these plots, the frequencies corresponding to the maximum of the magnitude of the impedance and the frequency where the impedance is resistive (i.e. where the phase angle is zero) are not same. Or from an even more important viewpoint, the phase angle corresponding to the maximum value of the impedance is not zero but $\Phi = -8.05$ degrees.

The simulation results for case (b) are shown in Fig. 4.3b. As can be seen from these plots, the frequencies corresponding to the maximum of the impedance and to the zero-phase shift are the same, or in other words, the phase shift corresponding to the maximum value of the impedance is zero, as expected.

4.1.1.1 The Quality Factor of a Resonance Circuit

Before proceeding further, let us review the basic definition of the quality factor (Q) of any oscillatory system, and apply this definition to calculate the quality factor of the circuit shown in Fig. 4.4.

The quality factor of an oscillatory system is defined as

$$Q = 2\pi \frac{\text{The total energy of the system}}{\text{The energy lost in one period}} \tag{4.8}$$

Since the total energy in an oscillatory system swings back and forth between fully potential energy and fully kinetic energy, to find the total energy of the system it is convenient to calculate the value of the potential or kinetic energy at one of these extreme conditions. For the circuit shown in Fig. 4.4, the circuit is excited with a sinusoidal voltage, $v = V \sin \omega_0 t$, whose frequency is equal to the natural

Fig. 4.4 The parallel
resonance circuit in its
most general form

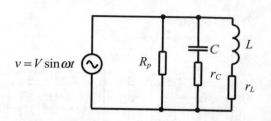

frequency ω_0 of the resonance circuit. The maximum value of the potential energy stored in the capacitor corresponds to the case when the voltage on this capacitor is maximum, i.e. the peak value of v. Then the maximum of the potential energy, which is equal to the total energy of the system, is

$$E = \frac{1}{2}CV^2 \tag{4.9}$$

The energy lost in one period (T) in the parallel resistor R_p is

$$E_p = T \times \frac{1}{2}\frac{V^2}{R_p} = \frac{1}{2f_0}\frac{V^2}{R_p} \tag{4.10}$$

To calculate the energy lost in r_L, the peak value of the sinusoidal current flowing through the inductor branch must be calculated:

$$I_L = \frac{V}{\sqrt{r_L^2 + \omega_0^2 L^2}}$$

Then the energy lost in one period in r_L is

$$E_L = \frac{1}{2f_0}V^2\frac{r_L}{r_L^2 + \omega_0^2 L^2} \tag{4.11}$$

Similarly, the energy consumed in r_C in one period can be found as

$$E_C = \frac{1}{2f_0}V^2\frac{r_C}{r_C^2 + \frac{1}{\omega_0^2 C^2}} \tag{4.12}$$

The total energy lost in one period is the sum of (4.10), (4.11), and (4.12). Hence from (4.8) and (4.9), the effective quality factor containing all of the losses of the circuit can be obtained as

$$\frac{1}{Q_{eff}} = \frac{1}{Q_p} + \frac{1}{Q_L} + \frac{1}{Q_C} \tag{4.13}$$

where

$$Q_L = \frac{r_L^2 + \omega_0^2 L^2}{\omega_0 L r_L} \cong \frac{\omega_0 L}{r_L} \tag{4.14}$$

which corresponds to the losses on r_L, in other words is the quality factor of the inductor at ω_0.

$$Q_C = \frac{\omega_0^2 C^2 r_C^2 + 1}{\omega_0 C r_C} \cong \frac{1}{\omega_0 C r_C} \tag{4.15}$$

Fig. 4.5 (a) Parallel resonance circuit with its lossy components. (b) The effective parallel resistance representing all losses of the circuit at f_0

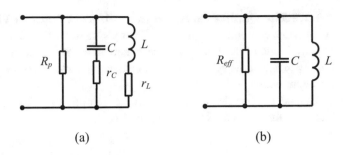

(a) (b)

which corresponds to the losses on r_C, or it is the quality factor of the capacitor at ω_0, and

$$Q_p = \omega_0 C R_p = \frac{R_p}{\omega_0 L} \tag{4.16}$$

which corresponds to the losses in R_p.

Note that the approximate values of Q_L and Q_C are only valid for small values of the corresponding resistances and can cause considerable error for Q values smaller than 5.

Problem 4.1
Calculate the error of the approximate form of (4.14) for $Q_L = 50$, $Q_L = 10$, $Q_L = 5$, and $Q_L = 3$.

It is common practice to represent the total losses of a resonance circuit with a lumped parallel resistor (effective resistance) R_{eff}, as shown in Fig. 4.5.

If we want to express the losses of the inductor branch with a parallel resistor R_{Lp}, from (4.14) and (4.16) we can obtain

$$R_{Lp} = r_L + \frac{\omega_0^2 L^2}{r_L} \cong r_L\left(1 + Q_L^2\right) \cong \frac{L}{r_L C} \tag{4.17}$$

Similarly, the parallel resistor representing the losses of the capacitor branch is

$$R_{Cp} = r_C + \frac{1}{\omega_0^2 C^2 r_C} \cong r_C\left(1 + Q_C^2\right) \cong \frac{L}{r_C C} \tag{4.18}$$

Then the effective resistance can be expressed as the parallel equivalent of R_{Lp}, R_{Cp}, and R_p:

$$R_{eff} = \left(R_{Lp}//R_{Cp}//R_p\right) \tag{4.19a}$$

or in terms of parallel conductances:

$$G_{eff} = G_{Lp} + G_{Cp} + G_p \tag{4.19b}$$

Then the effective quality factor containing all of the losses becomes

$$Q_{eff} = \frac{1}{\omega_0 L G_{eff}} = \frac{\omega_0 C}{G_{eff}} \quad \text{or} \quad Q_{eff} = \frac{R_{eff}}{\omega_0 L} = \omega_0 C R_{eff} \tag{4.20}$$

This equivalence considerably decreases the complexity of the expressions and extensively used in the literature, but it also hides some properties of the resonance circuit. For example, although the natural frequency (f_0), the frequency where the impedance is real ($f_{(Re)}$) and the frequency corresponding to the maximum magnitude of the impedance ($f_{(max)}$) are not the same for the original circuit shown in Fig. 4.5a, they are all the same for Fig. 4.5b and the skew of the frequency characteristics exemplified in Example 4.1 is now hidden.

4.1.1.2 The Quality Factor from a Different Point of View

As previously mentioned, if an oscillatory system is excited, it starts to oscillate, but due to the losses the amplitude of oscillation gradually decreases. Since the rate of decrease of the amplitude and the quality factor of the system are both defined by the losses of the system, there must be an interrelation between them.

Consider a parallel resonance circuit simplified as shown in Fig. 4.5b. The impedance of this circuit in the s-domain can be arranged as

$$Z = \frac{1}{C} \frac{s}{s^2 + s\frac{1}{R_{eff}C} + \frac{1}{LC}} = \frac{1}{C} \frac{s}{(s - s_{p1})(s - s_{p2})} \tag{4.21}$$

The poles of the impedance function are

$$s_{p1,p2} = -\frac{1}{2R_{eff}C} \mp \sqrt{\left(\frac{1}{2R_{eff}C}\right)^2 - \frac{1}{LC}}$$

or

$$s_{p1,p2} = \sigma \mp j\sqrt{\omega_0^2 - \sigma^2} = \sigma \mp j\overline{\omega}_0 \tag{4.22a}$$

where

$$\sigma = -\frac{1}{2R_{eff}C}, \quad \omega_0 = \frac{1}{\sqrt{LC}} \quad \text{and} \quad \overline{\omega}_0 = \sqrt{\omega_0^2 - \sigma^2} \tag{4.22b}$$

If the circuit is excited with a current pulse, the voltage in the s-domain between the terminals of the circuit can be written as

$$V = \left(\frac{I}{s}\right) \frac{1}{C} \frac{s}{(s - s_{p1})(s - s_{p2})}$$

and the voltage in the t-domain,

$$v(t) = I \frac{1}{C} \frac{1}{(s_{p1} - s_{p2})} \left(e^{s_{p1}t} - e^{s_{p2}t}\right)$$

which can be arranged as

$$v(t) = I \frac{1}{C} \frac{1}{2j\overline{\omega}_0} e^{\sigma t} \left(e^{j\overline{\omega}_0 t} - e^{-j\overline{\omega}_0 t}\right)$$

$$v(t) = I \frac{1}{C} \frac{1}{\overline{\omega}_0} e^{\sigma t} \sin \overline{\omega}_0 t$$

and with $\sigma = -(\omega_0/2Q_{eff})$ and $\omega_0 \cong \overline{\omega}_0$,

$$v(t) = V_{(ini)} e^{-(\overline{\omega}_0/2Q_{eff})t} \sin \overline{\omega}_0 t \tag{4.23}$$

The variation of the voltage of an excited parallel resonance circuit is shown in Fig. 4.6. From (4.23), two important properties of this damped oscillation can be extracted:

- At the end of $n = Q_{eff}$ swings, the amplitude decreases to $V_{ini}e^{-\pi}$.
- The number of swings[5] corresponding to $\hat{v} = (1/e)V_{ini}$ is equal to (Q_{eff}/π).

---

[5] In the 1960s, the company Rohde & Schwarz introduced a Q-meter (QDM) based on these interesting relations.

Fig. 4.6 The damped oscillation of an excited parallel resonance circuit

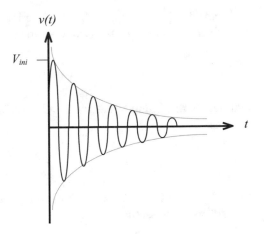

4.1.1.3 The "Q Enhancement"

There is a possibility for increasing the low-quality factor of an on-chip resonance circuit: connecting a "negative conductance" in parallel to the resonance circuit. We know that the Q factor of the resonance circuit at its resonance frequency (ω_0) can be expressed as $Q_{eff} = 1/L\,\omega_0 G_{eff}$. If we connect a negative conductance ($-G_n$) parallel to G_{eff}, the enhanced effective parallel conductance becomes $G'_{eff} = (G_{eff} - G_n)$. This decrease of the parallel conductance consequently increases the Q factor to $Q'_{eff} = 1/L\,\omega_0(G_{eff} - G_n)$.[6]

Although this negative conductance that is introduced by an appropriate electronic circuit adds an additional noise and nonlinearity, it can be effectively used to increase the quality factor of a resonance circuit [1–3]. One of the simplest negative resistance circuits is a capacitive loaded source follower. In Sect. 3.2, we have seen that the input resistance of a source follower can be negative under appropriate conditions, and can be used to enhance the quality factor of a resonance circuit.

Example 4.2
The problem that we are going to solve is of increasing the $Q = 10$ of an on-chip parallel resonance circuit to 20. The resonance frequency and the inductance (and consequently the capacitance) are given as $f_0 = 2$ GHz, $L = 10$ nH, and $C = 0.633$ pF. The effective parallel conductance of the resonance circuit can be calculated as

$$G_{eff} = \frac{1}{L\omega_0 Q} = \frac{1}{\left(10 \times 10^{-9}\right) \times \left(2\pi \times 2 \times 10^9\right) \times 10} = 7.96 \times 10^{-4}\ \text{S}$$

which corresponds to $R_{eff} = 1256.6$ ohm. It is obvious that to increase the Q factor to 20, the effective conductance must be decreased to $G'_{eff} = 3.98 \times 10^{-4}$ S and that the necessary negative conductance is $G_n = -3.98 \times 10^{-4}$ S.

---

[6] Note that for $G_n = G_{eff}$ the quality factor becomes infinite. According to (4.23), once the circuit is excited, the magnitude of the oscillation remains constant at its initial value. It means that the circuit operates as a sinusoidal oscillator. It can be shown that in all types of the L-C oscillators there exists a negative conductance parallel to the L-C circuit introduced by a positive-feedback circuit or by a device, like a tunnel diode, that inherently exhibits a negative conductance (See Chap. 6).

From (3.17), we see that the negative input conductance of a source follower is maximum for $g_m C \gg G C_{gs}$ and for $\omega \gg \omega_p$, and becomes equal to

$$g_i(\infty) = \frac{C C_{gs}}{\left(C + C_{gs}\right)^2} g_m$$

and has its maximum value for $C = C_{gs}$. Under this condition, the pole frequency given in (3.15) becomes

$$\omega_p = \frac{g_m + G}{2C_{gs}} \simeq \frac{g_m}{2C_{gs}} \simeq \frac{\mu(V_{GS} - V_T)}{L^2}$$

for $g_m \gg G$, which can be satisfied by using a current source instead of R, as shown in Fig 2.15. From the last two expressions, we obtain

$$g_i(\infty) \cong -\frac{g_m}{4}$$

which offers the first design hint. In addition, this equation can be used to calculate the channel length, L.

Inspecting Fig. 3.12 we see that, to obtain a high negative conductance with a small sensitivity, ω_0 must be chosen where the slope of the curve is small, or in other words, on (or close to) the asymptote. It can be seen that for $\omega > 3\omega_p$ the slope of the curve (or the sensitivity of the negative conductance against the frequency) is sufficiently small. This means that for $\omega_0 > 3\omega_p$, the value of the negative conductance becomes approximately equal to $g_i(\infty)$.

The aspect ratio of the transistor can be calculated in terms of G_n and the gate bias voltage. From (3.17) and (1.33)

$$g_m = 4|G_n| \cong \mu_n C_{ox}(W/L)(V_{GS} - V_T)$$

$$(W/L) = \frac{4|G_n|}{\mu_n C_{ox}(V_{GS} - V_T)}$$

which indicates a trade-off between gate overdrive and the aspect ratio. $(V_{GS} - V_T) = 1$ V is a reasonable value for AMS 035 micron technology, for which the maximum supply voltage is given as 3.3 V, and the aspect ratio can be calculated as

$$(W/L) = \frac{4|G_n|}{\mu_n C_{ox}(V_{GS} - V_T)} = \frac{4\left(3.98 \times 10^{-4}\right)}{374\left(4.54 \times 10^{-7}\right)1} = 9.37$$

For this aspect ratio and $(V_{GS} - V_{TH}) = 1$ V, the drain current is

$$I_D \cong \frac{1}{2}\mu_n C_{ox}\frac{W}{L}(V_{GS} - V_{TH})^2 = \frac{1}{2}374 \times 9.3\left(4.54 \times 10^{-7}\right) = 0.79 \text{ mA}$$

Now L can be calculated as follows, with $\omega_0 = 4\omega_p$ as a value that satisfies $\omega_0 > 3\omega_p$

$$L = \sqrt{\frac{4\mu_n(V_{GS} - V_{TH})}{\omega_0}} = \sqrt{\frac{4 \times 374 \times 1}{2\pi \times \left(2 \times 10^9\right)}} = 3.45 \times 10^{-4} \text{ cm} \cong 3.5 \ \mu m$$

and

$$W = 9.37 \times 3.45 \cong 32 \ \mu m$$

The diagram of the circuit together with the current source is given in Fig. 4.7. The V_{DD} supply is 3.3 V. The gate of M1 is directly biased from V_{DD}. To satisfy $(V_{GS} - V_T) = 1$ V (or V_{GS}) = 1.5 V, the

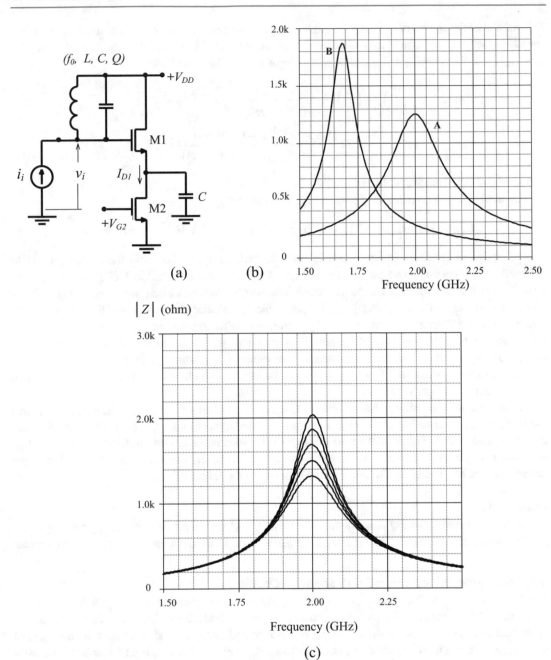

Fig. 4.7 (a) The complete circuit diagram of the Q enhancement circuit. (b) The frequency characteristic of the original resonance circuit (A) and the Q-enhanced circuit (B). (c) Q adjustment of the resonance circuit. (Curves correspond to $V_{G2} = 0.6$ V to 1.4 V with 0.2 V intervals)

DC voltage of the source of M1 must be 1.8 V, which is the drain-source voltage of the current source transistor, M2. M2 must be biased such that its current is also 0.79 mA and is in the saturation region. For $V_{GS2} = 1$ V, since $I_{D1} = I_{D2}$, the aspect ratio of M2 can be calculated as

$$\left(\frac{W}{L}\right)_2 = \left(\frac{W}{L}\right)_1 \frac{(V_{GS1} - V_{TH})^2(1 + \lambda_n.V_{DS1})}{(V_{GS2} - V_{TH})^2(1 + \lambda_n.V_{DS2})}$$

$$= 9.37 \frac{(1.5 - 0.5)^2(1 + 0.073 \times 1.5)}{(1 - 0.5)^2(1 + 0.073 \times 1.8)} \cong 37$$

and for $L_2 = 0.35$ μm, $W_2 = 13$ μm.

The final value to be calculated is C, which is equal to C_{gs1}:

$$C_{gs1} \cong W_1 L_1 C_{ox} = \left(32 \times 10^{-4}\right)\left(3.5 \times 10^{-4}\right)\left(4.54 \times 10^{-7}\right) = 508.48 \text{ fF}$$

The designed Q enhancement circuit is simulated with PSpice. To bring the current of M1 to 0.79 mA, it is necessary to adjust the gate voltage of the current source M2 to 1.22 V. The frequency characteristics of the original resonance circuit and the Q-enhanced circuit are given in Fig. 4.7b as curve A and curve B, respectively. The quality factor calculated from curve B is 17.8, which corresponds to a 78% increase of the quality factor, not 100% as targeted. One of the reasons is the losses associated with the body resistances of the devices that were not taken into account in the analytical expressions. The second reason is the value of g_m, which is usually smaller than the calculated value as discussed in Chap. 1. To compensate it, the channel width can be increased in order to reach the target value.

Another valuable possibility for this simple Q-enhancement circuit shown in Fig. 4.7a is to control the transconductance of M1 (and hence the negative conductance) with the gate bias voltage of the current source, M2. In Fig. 4.7c, the frequency characteristics of the circuit for different values of V_{G2} are shown. Note that the value of C has been decreased to compensate the effect of the input capacitance of M1 and bring the resonance frequency to the target value.

Problem 4.2
Make transient simulations of the original circuit and the Q-enhanced circuit in Design Example 4.1 for sinusoidal input currents with an amplitude of 1 mA and 5 mA. Compare and interpret the results.

4.1.1.4 Bandwidth of a Parallel Resonance Circuit

We know that a parallel resonance circuit has frequency-selective behavior. Its impedance is maximum at its resonance frequency and decreases below and above this frequency. We already know that due to the series resistances of the inductance and/or capacitance branches there may exist a skew, as seen in Fig. 4.3a. We also know that this asymmetry is associated with a shift of the zero-crossing frequency of the phase, from the top of the magnitude curve. Keeping in mind these imperfections, we will prefer the conventional approach for obtaining universal expressions for the bandwidth, assuming that the quality factor of the circuit is sufficiently high, or the series resistances of the inductance and capacitance branches sufficiently balance each other, as seen in Sect. 4.1.1.1. Since the equivalent circuit seen in Fig. 4.5b can represent both of these cases, we will base our derivations on this circuit.

The impedance of the circuit was obtained as (4.21), which can be written in the frequency domain as

$$Z = R_{eff} \frac{1}{1 + j\frac{R_{eff}}{L\omega}\left(\omega^2 LC - 1\right)}$$

With $LC = 1/\omega_0^2$ and $Q = R_{eff}/L\omega_0$, this expression can be arranged as

$$Z = R_{eff}\, \frac{1}{1 + jQ\frac{\omega_0}{\omega}\left(\frac{\omega^2}{\omega_0^2} - 1\right)} \cong R_{eff}\, \frac{1}{1 + j\beta Q} \tag{4.24a}$$

where

$$\beta = 2\Delta\omega/\omega_0 \tag{4.24b}$$

and $\Delta\omega$ is the difference from the resonance frequency. From (4.24), the magnitude and phase of the impedance can be written as

$$|Z| = Z(\omega_0)\, \frac{1}{\sqrt{1 + (\beta Q)^2}} \qquad \Phi = -\arctan{(\beta Q)} \tag{4.24c}$$

It can be seen from these expressions that:

- At resonance, the magnitude of the impedance is maximum and equal to R_{eff}, and from (4.20)

$$Z(\omega_0) = R_{eff} = Q_{eff}L\omega_0$$

- If the effective quality factor is known, the higher the inductance value is, the higher the resonance impedance. This is an important fact that is useful to obtain higher voltage gain from a tuned amplifier, as will be shown later on.
- For $\beta Q = \pm 1$, the magnitude of the impedance decreases to $Z(\omega_0)/\sqrt{2}$, which corresponds to the –3 dB frequencies of the magnitude curve:

$$f_{(-3dB)} = f_0 \pm \frac{f_0}{2Q} \quad \text{and the bandwidth} \quad B = \frac{f_0}{Q} \tag{4.25}$$

- The phase angle of the impedance is zero at the resonance frequency (the impedance is resistive), as expected.
- Below the resonance frequency, the phase angle is positive (impedance is inductive) and above f_0 the impedance is capacitive. The phase angles corresponding to the band ends (-3 dB frequencies) are $\mp\pi/4$. The normalized magnitude and phase characteristics of Z are given in Fig. 4.8.

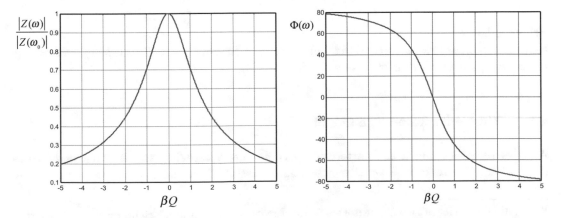

Fig. 4.8 The normalized magnitude and phase characteristics of a parallel resonance circuit

Fig. 4.9 The branch currents of a parallel resonance circuit under resonance

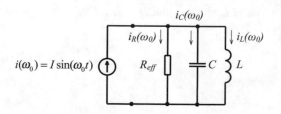

4.1.1.5 Currents of L and C Branches of a Parallel Resonance Circuit

Assume that the input current of the parallel resonance circuit shown in Fig. 4.9 under resonance is $i(\omega_0)$. The voltage between the terminals of the circuit is $v(\omega_0) = i(\omega_0)R_{eff}$. The current of the capacitor branch is

$$i_C(\omega_0) = v(\omega_0)(jC\omega_0) = i(\omega_0)R_{eff}(jC\omega_0)$$
$$i_C(\omega_0) = j\left[i(\omega_0) \times Q_{eff}\right] \tag{4.26}$$

Note that:

- The current flowing through the capacitance branch leads the input current by $90°$.
- The magnitude of the current flowing through the capacitance branch is Q_{eff} times greater than the input current.

Similarly, the current of the inductance branch can be written as

$$i_L(\omega_0) = \frac{v(\omega_0)}{jL\omega_0} = \frac{i(\omega_0)R_{eff}}{jL\omega_0}$$
$$i_L(\omega_0) = -j\left[i(\omega_0) \times Q_{eff}\right] \tag{4.27}$$

This means that:

- The current flowing through the inductance branch lags behind the input current by $90°$.
- The magnitude of the current flowing through the inductance branch is Q_{eff} times larger than the input current. This phenomenon can lead to a long-term effect called "electromigration", which means the weakening of a conductor due to the momentum transfer between the electrons and the metal atoms where the current density exceeds a certain value (approximately 2 mA/μm^2 for aluminum[7]). Note that a similar problem also exists for the interconnection of the capacitor of the resonance circuit.

Problem 4.3

Calculate the branch currents of the circuit shown in Fig. 4.5a, assuming realistic values for the on-chip passive components and their resistive parasitics.

4.1.2   The Series Resonance Circuit

A series resonance circuit can be formed by connecting a capacitor and an inductor in series (Fig. 4.10a). The series resistances representing the losses of the inductor and the capacitor are

---

[7] This value is given for DC. It is known that the electromigration is less effective at high frequencies. But the current crowding due to the skin effect and the proximity effects must be considered.

Fig. 4.10 (a) The series
resonance circuit, (b) its
simplified form

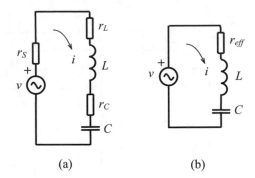

(a)   (b)

shown as r_L and r_C, and r_S is the internal resistance of the voltage source. The circuit can be simplified as shown in Fig. 4.10b, where r_{eff} is the sum of r_L, r_C, and r_S. The impedance of the circuit is

$$Z = \left(r_{eff} + sL + \frac{1}{sC}\right) \tag{4.28}$$

and the admittance

$$Y = \frac{1}{r_{eff} + sL + \frac{1}{sC}} = \frac{1}{L}\frac{s}{s^2 + s\frac{r_{eff}}{L} + \frac{1}{LC}}$$

$$Y = \frac{1}{L}\frac{s}{(s - s_{p1})(s - s_{p2})} \tag{4.29}$$

The poles of this admittance function are the same as the poles of the impedance of a parallel resonance circuit:

$$s_{p1,p2} = \sigma \mp j\sqrt{\omega_0^2 - \sigma^2} \tag{4.30}$$

where

$$\omega_0^2 = \frac{1}{LC}, \quad \sigma = -\frac{r_{eff}}{2L} = -\frac{\omega_0}{2Q_{eff}}, \quad Q_{eff} = \frac{\omega_0}{2\sigma} = \frac{L\omega_0}{r_{eff}} \tag{4.31}$$

Now the impedance of the circuit can be written as

$$Z(\omega) = Z(\omega_0)\left(1 + j\beta Q_{eff}\right) \tag{4.32}$$

where $\beta = 2\Delta\omega/\omega_0$ and $\Delta\omega$ is the difference from the resonance frequency, similar to the parallel resonance circuit. The magnitude and phase of the impedance are

$$|Z(\omega)| = Z(\omega_0)\sqrt{1 + \left(\beta Q_{eff}\right)^2} \qquad \Phi(\omega) = \arctan\left(\beta Q_{eff}\right) \tag{4.33}$$

The normalized variations of the magnitude and the phase of the impedance are given in Fig. 4.11. From expressions (4.33) and Fig. 4.11 it can be seen that:

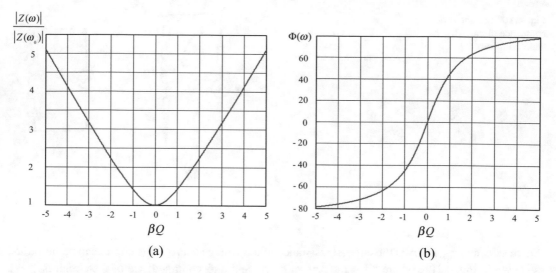

(a)   (b)

Fig. 4.11 The normalized magnitude and phase characteristics of a series resonance circuit

Fig. 4.12 The voltages in
a series resonance circuit at
resonance

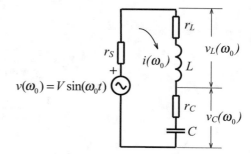

- The magnitude of the impedance is minimum and equal to r_{eff} at ω_0.
- For $\beta Q = \pm 1$, the magnitude of the impedance increases to $\sqrt{2}.Z(\omega_0)$. This can be expressed as the fact that the +3 dB frequencies of the magnitude are

$$f_{(+3dB)} = f_0 \pm \frac{f_0}{2Q} \quad \text{and the bandwidth} \quad B = \frac{f_0}{Q} \tag{4.34}$$

- The phase angle of the impedance is zero at the resonance frequency (the impedance is resistive), as expected.
- Below the resonance frequency the phase angle is negative (impedance is capacitive), and above f_0 the impedance is inductive. The phase angles corresponding to the band ends (+3 dB frequencies) are $\mp \pi/4$.

4.1.2.1 Component Voltages in a Series Resonance Circuit

Consider a series resonance circuit as shown in Fig. 4.12. The current at resonance is $i(\omega_0) = v/r_{eff}$, where $r_{eff} = r_S + r_L + r_C$. Then the voltage between the terminals of the inductor can be written as

$$v_L(\omega_0) = i(\omega_0)(r_L + jL\omega_0) = v(\omega_0)\frac{r_L + jL\omega_0}{r_{eff}}$$

$$= j.v(\omega_0)\left(\frac{r_L}{r_{eff}} + Q_{eff}\right) \cong j.v(\omega_0) \times Q_{eff}$$

(4.35)

and similarly,

$$v_C(\omega_0) \cong -j.v(\omega_0) \times Q_{eff}$$

(4.36)

This means that:

- The voltage between the terminals of the inductor is approximately Q_{eff} times larger than the input voltage and leads the input voltage by 90°.
- The voltage on the capacitor is approximately Q_{eff} times larger than the input voltage and lags behind the input voltage by 90°. This is an important property that has to be taken into account to prevent the breakdown of the capacitor dielectric, especially if the capacitor is a MOS capacitor or MOS varactor.

4.2   Tuned Amplifiers

In all wireless applications, we need "tuned" or "narrow-band" amplifiers that provide gain at a certain frequency and a narrow band around this frequency. The tuning frequency and the bandwidth depend on the area of application. Since the early days of the radio resonance circuits have been the main components of tuned amplifiers, due to their frequency-selective nature.

We have seen that the voltage gain of any amplifier is proportional to the total load impedance, which is the parallel equivalent of the external load and the output impedance of the amplifier. We also know that the impedance of a parallel resonance circuit is maximum at a certain frequency that is approximately equal to the natural frequency (or the resonance frequency) of the circuit. Then the easiest way to form a tuned amplifier is to use a parallel resonance circuit as the load of the amplifier such that the resonance frequency of the load together with the output impedance of the amplifier is equal to the desired tuning frequency. In some applications, the tuning frequency must be fixed as in the intermediate frequency (IF) amplifier of a receiver. But in some other applications, such as the input amplifier of a radio, the tuning frequency must be adjustable in a certain frequency band. To tune the frequency of an amplifier, the most commonly used way is to incorporate a suitable varactor into the resonance circuit.[8]

The bandwidth of the amplifier is certainly determined by the frequency characteristic of the overall (effective) load. For some applications, the bandwidth and the shape of the frequency characteristic in this band do not fulfill the needs of the application. There are several techniques to improve the shape of the frequency characteristic and to increase the relative bandwidth: using coupled resonance circuits instead of the simple parallel resonance circuits as the load, and the "staggered tuning" of the stages of an amplifier are well-known and extensively used solutions.

In principle, any high output impedance amplifier configuration can be used as the gain block of a tuned amplifier. The simplest and one of the most frequently used configuration, especially at the lower end of the RF spectrum, is the common-source tuned amplifier.

---

[8] Formerly, mechanically controlled "variable capacitors" were being used in non-integrated tuned amplifiers.

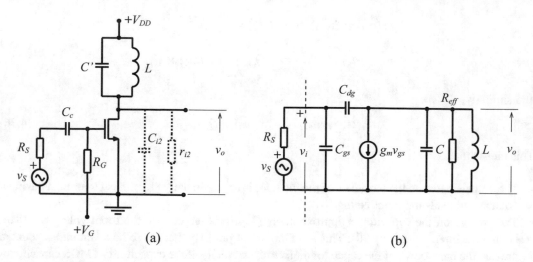

Fig. 4.13 (a) Schematic diagram of a tuned common-source amplifier. (b) The small-signal equivalent circuit. C represents the total capacitance (the sum of the external capacitor C', the output capacitance of the transistor, the input capacitance of the following stage and the parasitics). R_{eff} is the parallel equivalent of the output resistance of the transistor, the input resistance of the following stage and the parallel resistance corresponding to the losses of the resonance circuit

4.2.1 The Common-Source Tuned Amplifier

The circuit diagram of a common-source tuned amplifier is given in Fig. 4.13a. The parallel L-C circuit is connected between the drain of the transistor and V_{DD}. The DC load of the transistor is the DC resistance of the inductor, and the AC load is the impedance of this load at the frequency of operation of the circuit. The input signal source is represented with a non-ideal voltage source having an internal resistance R_S. It is connected to the gate of the transistor via a coupling capacitor C_c, which exhibits a low reactance at the frequency of operation, and can be considered short-circuited for the signal. The gate DC bias voltage of the transistor is applied via a high-value resistance R_G that can be considered open-circuited for the signal. The input impedance of the following stage at the frequency of operation is represented with a parallel r_{i2},C_{i2} combination.

The small-signal equivalent circuit of the amplifier is given in Fig. 4.13b, where the signal voltage at the input of the transistor is shown with v_i. This voltage certainly depends on the input resistance (in general, impedance) of the signal source and the input impedance of the amplifier, for the frequency of operation.

Therefore, to characterize the amplifier, the voltage gain alone is not sufficient; the input impedance (or admittance) also must be investigated. The voltage gain (v_o /v_i) of a common-source amplifier for any type of load was found to be (see Eq. (3.2)):

$$A_v = -\frac{g_m - sC_{dg}}{Y_o}$$

where Y_o represents the total output load admittance, which is the parallel equivalent of C, L, and R_{eff} for our circuit[9]:

---

[9] It is obvious that for $\omega_0 C_{dg} \ll g_m$, the voltage gain can be written as $A_v = -g_m/Y_0 = -g_m Z_0$. Consequently, the frequency characteristic and the −3 dB frequencies of the amplifier are the same as those of the total load impedance, Z_0. Here a less straightforward approach will be used to enable to discuss the effects of the "zero" of the gain function and to prepare the reader for the concepts that will be used to investigate staggered tuning in Sect. 4.3.

$$A_v = -\frac{g_m - C_{dg}}{sC + \frac{1}{sL} + G_{eff}} \qquad (4.37)$$

To derive the variations of the magnitude and the phase of the gain with frequency, it is possible to write (4.37) in the frequency domain, or to use the pole-zero diagram of the gain function. In the following, we will prefer the second method:

(4.37) can be arranged in terms of its poles and zero as

$$A_v = \frac{C_{dg}}{C} \frac{s(s - s_0)}{(s - s_{p1})(s - s_{p2})} \qquad (4.38)$$

where

$$s_0 = +\frac{g_m}{C_{dg}} \quad , \quad s_{p1,p2} = -\frac{G_{eff}}{2C} \mp j\sqrt{\frac{1}{LC} - \left(\frac{G_{eff}}{2C}\right)^2} \qquad (4.39a)$$

and with

$$\sigma = -\frac{G_{eff}}{2C} = -\frac{\omega_0}{2Q_{eff}} \quad \text{and} \quad \omega_0^2 = \frac{1}{LC}$$

$$s_0 = +\frac{g_m}{C_{dg}} \quad , \quad s_{p1,p2} = \sigma \mp j\sqrt{\omega_0^2 - \sigma^2} \qquad (4.39b)$$

The pole-zero diagram of the gain function is given in Fig. 4.14a. For any ω value the magnitude and phase of the gain can be obtained as

$$|A| = \frac{C_{dg}}{C} \frac{|s||s - s_0|}{|s - s_{p1}||s - s_{p2}|} \qquad (4.40)$$

$$\Phi = \Phi_s + \Phi_{(s-s_0)} - \Phi_{(s-s_{p1})} - \Phi_{(s-s_{p2})} \qquad (4.41)$$

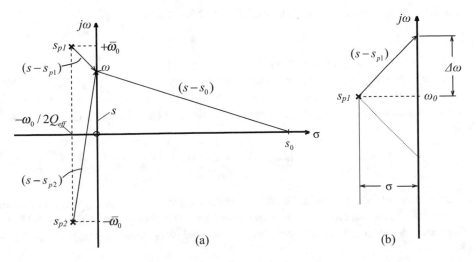

Fig. 4.14 (a) The pole-zero diagram of the gain function of a tuned amplifier, (b) its simplified form for the vicinity of the resonance frequency (note that $(s-s_{p1})$ is drawn for the upper 3 dB frequency of the gain)

Provided that $|s_0| \gg \omega_0$ and $\omega_0 \gg \sigma$, which are valid for most practical cases, for the vicinity of ω_0 the magnitude and phase can be approximated as

$$|A| \cong \frac{C_{dg}}{C} \frac{|\omega||s_0|}{|s - s_{p1}|2\omega_0} = \frac{g_m}{2C|s - s_{p1}|} \tag{4.40a}$$

$$\Phi \cong \frac{\pi}{2} + \pi - \Phi_{(s-s_{p1})} - \frac{\pi}{2} = \pi - \Phi_{(s-s_{p1})} \tag{4.41a}$$

The simplified form of the pole-zero diagram of the amplifier valid for and around the resonance frequency is shown in Fig. 4.14b. From this figure, the 3 dB frequencies and the bandwidth of the amplifier can be found as

$$f_{(-3dB)} = f_0 \pm \Delta f = f_0 \pm \frac{f_0}{2Q_{eff}}, \quad B = 2\Delta f = \frac{f_0}{Q_{eff}} \tag{4.42}$$

identical to those of a parallel resonance circuit.

The magnitude of the gain corresponding to the resonance frequency can be calculated from (4.40a) with $|s - s_{p1}| = \sigma = \omega_0/2Q_{eff}$

$$|A(\omega_0)| = g_m \frac{Q_{eff}}{\omega_0 C} = g_m R_{eff} \tag{4.43}$$

and similarly the phase angle for ω_0

$$\Phi(\omega_0) = \pi \tag{4.44}$$

Throughout the calculations above we have assumed that the zero of the voltage gain related to the drain-gate capacitance is far away on the right half-plane and is therefore negligible. This assumption corresponds to $\omega_0 C_{dg} \ll g_m$, which is usually valid. Under this assumption, the voltage gain (2.28) can be written as

$$A(\omega) = -\frac{g_m}{Y_0} = -g_m Z_0 \tag{4.45}$$

where Z_0 is the effective impedance of the parallel resonance circuit.

Example 4.3

Check the validity of the assumption of $\omega_0 C_{dg} \ll g_m$ for a typical 0.13-micron NMOS transistor operating in the velocity saturation regime. The operating frequency of the amplifier is 3 GHz.

Under the velocity saturation, (which is the case for a 0.13 micron transistor, as shown in Chap. 1) the transconductance is $g_{m(v-sat)} = kWC_{ox}v_{sat}$ and the drain-gate capacitance is $C_{dg} = W \times$ CDGW. Therefore

$$\frac{\omega_0 C_{dg}}{g_m} = \frac{\omega_0 \times \text{CDGW}}{C_{ox}v_{sat}}$$

which is independent of the gate width. The related parameter values for this 0.13 micron technology are $T_{ox} = 2.3 \times 10^{-9}$ [m] (which corresponds to $C_{ox} = 15 \times 10^{-3}$ F/m^2), and CDGW $= 5.18 \times 10^{-10}$ [F/m]. The saturation velocity of electrons in the channel was given in Chap. 1 as 6.5×10^4 [m/s]. Therefore

$$\frac{\omega_0 C_{dg}}{g_m} = \frac{2\pi \times \left(3 \times 10^9\right) \times \left(5.18 \times 10^{-10}\right)}{\left(15 \times 10^{-3}\right) \times \left(6.5 \times 10^4\right)} \cong 10^{-2}$$

which corresponds to an error of 1% on the magnitude of the gain and an excess phase shift of only 0.57°.

Problem 4.4
An amplifier tuned to 1GHz is designed using an AMS 035 NMOS transistor with $W = 100$ μm and $L = 0.35$ μm. The drain current is 5 mA. Calculate the gain and phase errors of this amplifier if the effects of C_{dg} are neglected.

The numerical results of this example show that the effects of the drain-gate capacitance are negligibly small. The question can then arise as to why this capacitance is notorious for its adverse effects on tuned amplifiers. The answer to this question is related to the effects of C_{gs} on the input admittance of the amplifier and the risk of oscillation under certain conditions.

The input admittance of a common source amplifier was found as

$$y_i = s\left(C_{gs} + C_{dg}\right) + y_{mi} = s\left(C_{gs} + C_{dg}\right) + sC_{dg}\frac{g_m - sC_{dg}}{Y_o} \tag{3.3a}$$

The first term is apparently capacitive, but the second term (the Miller component) needs to be investigated. Let us write the Miller admittance in the frequency domain, and then calculate the real and imaginary parts:

$$y_{mi}(\omega) = j\omega C_{dg}\frac{g_m - j\omega C_{dg}}{G + j\omega C + \frac{1}{j\omega L}} = \omega^2 LC_{dg}\frac{-g_m + j\omega C_{dg}}{(1 - \omega^2 LC) + j\omega LG}$$

$$\mathrm{Re}\left\{y_{mi}\right\} = \frac{(\omega/\omega_0)^2\frac{C_{dg}}{C}}{\left(1 - (\omega/\omega_0)^2\right)^2 + \omega^2 L^2 G^2}\left(-g_m\left(1 - (\omega/\omega_0)^2\right) + (\omega/\omega_0)^2\frac{C_{dg}}{C}G\right) \tag{4.46}$$

$$\mathrm{Im}\left\{y_{mi}\right\} = \frac{(\omega/\omega_0)^2\frac{C_{dg}}{C}}{\left[1 - (\omega/\omega_0)^2\right]^2 + \omega^2 L^2 G^2}\omega\left\{C_{dg}\left[1 - (\omega/\omega_0)^2\right] + LGg_m\right\} \tag{4.47}$$

From (4.47) it is possible to see that:

- At the resonance frequency (for $\omega = \omega_0$) the input conductance is

$$\mathrm{Re}\left\{y_{mi}(\omega_0)\right\} = \frac{\omega_0^2 C_{dg}^2}{G}$$

which strongly depends on C_{dg}.

- Above the resonance frequency (where the load impedance is capacitive), the input conductance is positive and is a function of frequency.
- Below the resonance frequency (where the load impedance is inductive), the input conductance has a dominant negative component and varies with frequency.

This frequency-dependent input conductance and its being negative below the ω_0 resonance frequency of the output load is important for several different reasons:

- In the case of a non-ideal input signal source (which is the realistic case), the signal voltage on the gate of the transistor changes with frequency. Therefore the overall frequency characteristic is determined not only by the output load, but also by the internal impedance of the signal source.
- If a tuned circuit exists in parallel to the input, due to the positive parallel conductance above ω_0 and the negative parallel conductance below ω_0, the quality factor of this circuit decreases above ω_0 and increases below ω_0. What results is the skew of the frequency characteristic of the input resonance circuit, which affects the overall frequency characteristic of the circuit.
- The negative conductance component of the input admittance can over-compensate the losses of the input resonance circuit and can cause the circuit to oscillate.

According to (4.47) the imaginary part of the input admittance of a tuned amplifier also depends on the frequency. However, this is not as severe as the varying, even negative, input conductance. It only acts on the tuning of the input resonance circuit, if there is any.

To illustrate these observations the PSpice simulation results of a simple tuned amplifier are given in Fig. 4.15, using an AMS 035 NMOS transistor with $L = 0.35$ μm and $W = 200$ μm. The DC supplies are $V_{DD} = 3$ V and $V_G = 0.8$ V. $L = 10$ nH and C is trimmed to 2.34 pF to tune the circuit to $f_0 = 1$ GHz. The parallel resistance representing the total losses of the resonance circuit is 1 kohm, which corresponds to $Q = 15.9$.

Curves A and B show the variations of the input capacitance and the input conductance, respectively. The negative input conductance below f_0 and positive input conductance above f_0 are clearly seen from curve B. The maximum values of the input conductance are 1.2 mS and correspond approximately to the 3 dB frequencies of the gain that is plotted as curve C.

These dramatic variations of the input admittance of a tuned MOS amplifier are obviously due to the drain-gate capacitance of the device, which is unavoidable and whose adverse effects increase with frequency. Consequently, a circuit as shown in Fig. 4.13 can be used only at the lower end of the RF spectrum. For high frequency RF amplifiers, the extensively used solution is the "cascode" circuit that was investigated in general in Sect. 3.4.

4.2.2 The Tuned Cascode Amplifier

A cascode circuit loaded with a parallel resonance circuit is shown in Fig. 4.16. M1 and M2 are biased in the saturation region with V_{G1} and V_{G2}. The load of M1 is the input impedance of M2, operating as a common gate circuit.

We know that the input impedance of a common gate circuit is approximately equal to the parallel equivalent of $1/g_m$ and C_{sg}. Consequently, the voltage gain of M1 is low and equal to $(-g_{m1}/g_{m2})$ up to the frequencies close to g_{m2}/C_{sg2}. Therefore the Miller component of the input admittance of M1 is considerably smaller compared to that of a high-gain common source amplifier. In addition, since the output resistance of a cascode circuit is higher than the output resistance of a common source circuit, the effective Q of the load becomes higher.

To visualize the benefits of the cascode configuration, the simulation results of a cascode amplifier are given in Fig. 4.17. The parameters of the circuit are the same as the parameters of the common source amplifier, whose simulation results were given in Fig. 4.15:

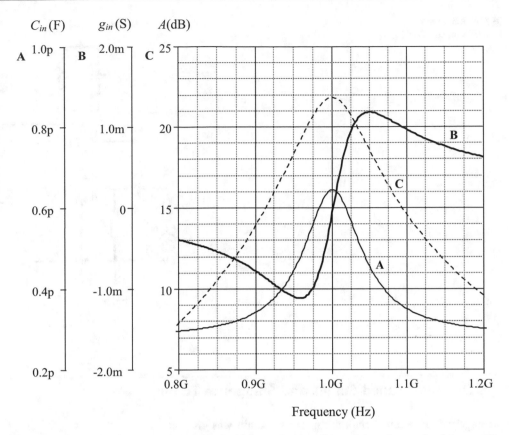

Fig. 4.15 Variations of the input capacitance (**a**), the input conductance (**b**) and the voltage gain of the amplifier (**c**). Note the fluctuations of the input capacitance and the input conductance, and especially the negativity of the input conductance below the resonance frequency

M1 and M2:   AMS 035 NMOS transistor. $L = 0.35$ μm, $W = 200$ μm.
DC supplies: $V_{DD} = 3$ V and $V_{G1} = 0.8$ V, $V_{G2} = 1.5$ V.
$L = 10$ nH, $C' = 2.34$ pF ($f_0 = 1$ GHz).
Parallel resistance representing the total losses of L and C is 1 k ohm.

To facilitate the comparison, the vertical axes in Fig. 4.17 are intentionally chosen to be the same as those of Fig. 4.15. The obvious advantages of the cascode circuit can be summarized as follows:

- The input capacitance is almost constant in the entire frequency band and equal to the input capacitance of M1.
- The input conductance is positive and almost constant in the entire frequency band. This means that the input of the circuit is a well-defined load for the driving signal source (or the previous stage) and has no adverse effect if there is another tuned circuit parallel to the input.
- The bandwidth of the gain is smaller (the effective Q is higher) compared to that of the reference common source circuit. This is the result of the high output resistance of the common gate output transistor M2, as expected.

Fig. 4.16 Schematic
diagram of a tuned cascode
amplifier

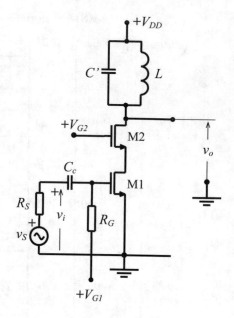

4.3   Cascaded Tuned Stages and Staggered Tuning

The voltage gain of a tuned amplifier in the s domain was given as

$$A_v = \frac{C_{dg}}{C} \frac{s(s - s_0)}{(s - s_{p1})(s - s_{p2})} \tag{4.38}$$

From Fig. 4.14a it can be seen that in the vicinity of the resonance frequency and for $|s_0| \gg |s_{p1}|$, $(s - s_{p2}) \cong 2s$ and $(s - s_0) \cong -s_0$. Therefore, (4.38) can be simplified as

$$A_v \cong -\frac{C_{dg}}{2C} \frac{s_0}{(s - s_{p1})} \tag{4.48}$$

and with $A_v(\omega_0) = -g_m R_{eff}$ and $Q_{eff} = \omega_0 C R_{eff}$,

$$A_v \cong A_v(\omega_0) \frac{\omega_0}{2Q_{eff}} \frac{1}{(s - s_{p1})} = A_v(\omega_{p1}) \frac{\omega_{p1}}{2Q_{eff}} \frac{1}{(s - s_{p1})} \tag{4.49}$$

If n identical stages are connected in cascade, the total voltage gain becomes

$$A_{vT} \cong [A_v(\omega_{p1})]^n \left(\frac{\omega_0}{2Q_{eff}}\right)^n \frac{1}{(s - s_{p1})^n}$$

and the bandwidth of the amplifier shrinks to

$$B_T = B\sqrt{2^{1/n} - 1} \tag{4.50}$$

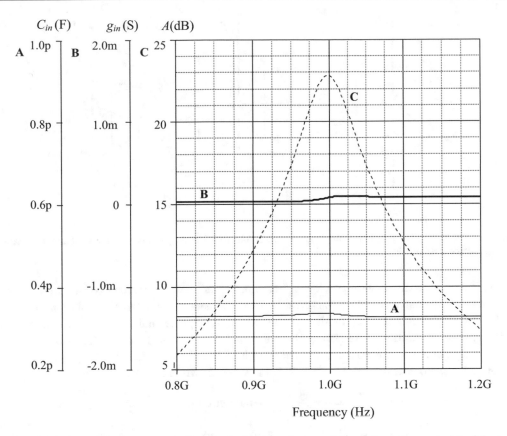

Fig. 4.17 Variations of the input capacitance (**a**), the input conductance (**b**) and the voltage gain (**c**) of the tuned cascode amplifier. Note the almost constant input capacitance and the input conductance

This is the appropriate solution if a high-gain and narrow-band amplifier is needed. But in some cases, a relatively broad bandwidth and a flat frequency characteristic in this band are needed. It can be understood intuitively that tuning the stages of this multistage amplifier to slightly different frequencies around the center frequency of the band can lead to the solution. In this case the gain of this multistage amplifier can be written as

$$A_{vT} \cong A_{v1}(\omega_{p1})\cdots A_{vn}(\omega_{pn}) \frac{\omega_{p1}}{2Q_{eff1}} \cdots \frac{\omega_{pn}}{2Q_{effn}} \frac{1}{(s-s_{p1})\cdots(s-s_{pn})}$$

which has n poles.

The appropriate positions of the poles of the transfer function (the voltage gain in our case) to obtain a desired frequency characteristic are investigated in depth in classical filter theory [4]. It is known that the number of poles and their relative positions determine the bandwidth and the shape of the frequency characteristics. Among several possibilities for the distribution of the poles in the s-domain, the Butterworth distribution and the Chebyshev distribution have prime importance and are extensively used in practice.

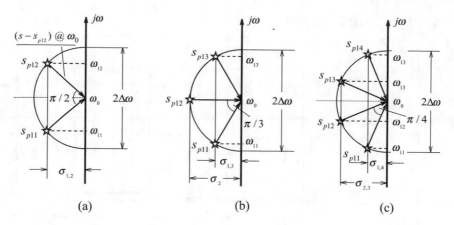

Fig. 4.18 The appropriate positions of the poles of a (**a**) 2-pole, (**b**) 3-pole, and (**c**) 4-pole circuit that has a maximally flat frequency characteristic

The Butterworth distribution provides a "maximally flat" frequency characteristic in the band. It has been shown that to obtain a Butterworth type frequency characteristic, poles must be on a semicircle whose center is at ω_0 on the vertical axis of the s-plane, and they must be symmetrically positioned equidistantly with respect to the horizontal diameter of the circle. The diameter of the circle on the $j\omega$ axis corresponds to the bandwidth of the circuit in angular frequency, $2\Delta\omega$. The appropriate positions of the poles for $n = 2$, 3 and 4 are shown in Fig. 4.18,[10] where

$$\sigma_{1,2} = -\frac{\omega_{11}}{2Q_1} = -\frac{\omega_{12}}{2Q_2} \text{ for a 2-pole circuit,}$$

$$\sigma_2 = -\frac{\omega_0}{2Q_2} \text{ and } \sigma_{1,3} = -\frac{\omega_{11}}{2Q_1} = -\frac{\omega_{13}}{2Q_3} \text{ for a 3-pole circuit,}$$

$$\sigma_{1,4} = -\frac{\omega_{11}}{2Q_1} = -\frac{\omega_{14}}{2Q_4} \text{ and } \sigma_{2,3} = -\frac{\omega_{12}}{2Q_2} = -\frac{\omega_{13}}{2Q_3} \text{ for a 4-pole circuit,}$$

In the case of small relative bandwidths, where $2\Delta\omega \ll \omega_0$, the sigmas can be written as

$$\sigma_{1,2} \cong -\frac{\omega_0}{2Q_1} = -\frac{\omega_0}{2Q_2} \Rightarrow Q_1 \cong Q_2 \quad \text{for a 2-pole circuit,}$$

$$\sigma_2 = -\frac{\omega_0}{2Q_2}, \; \sigma_{1,3} \cong -\frac{\omega_0}{2Q_1} = -\frac{\omega_0}{2Q_3} \Rightarrow Q_1 \cong Q_3 \quad \text{for a 3-pole circuit,}$$

$$\sigma_{1,4} \cong -\frac{\omega_0}{2Q_1} = -\frac{\omega_0}{2Q_4}, \; \sigma_{2,3} \cong -\frac{\omega_0}{2Q_2} = -\frac{\omega_0}{2Q_3} \Rightarrow Q_1 \cong Q_4, \; Q_2 \cong Q_3 \quad \text{for a 4-pole circuit.}$$

---

[10] It must not be overlooked that these diagrams are the simplified versions of the pole-zero diagrams, for the vicinity of the center frequency as shown in Fig. 4.14. The full pole-zero diagrams contain the complex conjugates of the poles shown in Fig. 4.18.

Example 4.4

A three-stage staggered tuned amplifier having a Butterworth type frequency characteristic will be designed. The center frequency of the frequency characteristic is 1GHz. The maximum possible effective Q value for the resonance circuits – without any Q enhancement feature – is given as 20.

(a) What is the realizable bandwidth?
(b) Calculate the tuning frequencies of the stages.
(c) Calculate the appropriate effective Q values of the resonance circuits.

Solution:

The pole-zero diagram of the voltage gain function of the amplifier is shown in Fig. 4.18b. From the geometry of the figure it can be easily seen that the magnitude of the real part of the center pole, s_{p12}, must be equal to half of the bandwidth:

$$|\sigma_2| = \left|\frac{\omega_0}{2Q_2}\right| = \Delta\omega \quad \Rightarrow \quad Q_2 = \frac{\omega_0}{2\Delta\omega} = \frac{f_0}{2\Delta f}$$

Similarly, the real parts of s_{p11} and s_{p13} must be equal in magnitude to $\Delta\omega/2$:

$$|\sigma_1| = \left|\frac{\omega_0}{2Q_1}\right| = \frac{\Delta\omega}{2} \quad \Rightarrow \quad Q_1 = \frac{\omega_{01}}{\Delta\omega} = \frac{f_{01}}{\Delta f}$$

$$|\sigma_3| = \left|\frac{\omega_0}{2Q_3}\right| = \frac{\Delta\omega}{2} \quad \Rightarrow \quad Q_3 = \frac{\omega_{03}}{\Delta\omega} = \frac{f_{03}}{\Delta f}$$

ω_{01} and ω_{03} can be calculated from the geometry:

$$\omega_{01} = \omega_0 - \Delta\omega \cos(\pi/6)$$

$$\omega_{03} = \omega_0 + \Delta\omega \cos(\pi/6)$$

Since ω_{03} is the highest among the three tuning frequencies, the quality factor corresponding to this resonance circuit is the highest and must be equal to the possible maximum Q value, which is 20:

$$Q_3 = \frac{\omega_{03}}{\Delta\omega} = \frac{\omega_0 + \Delta\omega \cos(\pi/6)}{\Delta\omega} = \frac{\omega_0}{\Delta\omega} + \cos(\pi/6)$$
$$= \frac{f_0}{\Delta f} + 0.866 = 20$$

which yields $\Delta f = 52.26$ MHz ($\Delta\omega = 328.36$ rad/s). Now the tuning frequencies and the quality factors can be calculated as

$$f_{01} = f_0 - \Delta f \cos(\pi/6) = 1000 - 52.26 \times 0.866 = 954.74 \ [\text{MHz}]$$

$$Q_1 = \frac{f_{01}}{\Delta f} = \frac{954.74}{52.26} = 18.27$$

$$f_{02} = f_0 = 1000 \ [\text{MHz}], \quad Q_2 = \frac{f_0}{\Delta f} = \frac{1000}{2 \times 52.26} = 9.57$$

$$f_{03} = f_0 + \Delta f \cos(\pi/6) = 1000 + 52.26 \times 0.866 = 1045.26 \ [\text{MHz}]$$

$$Q_3 = \frac{f_{03}}{\Delta f} = \frac{1045.26}{52.26} = 20$$

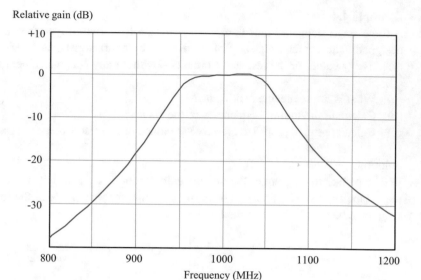

Fig. 4.19 The calculated frequency characteristic of the circuit, calculated with MATLAB

Note that since the quality factors are not high and the relative bandwidth ($2\Delta f/f_0$) is not small, Q_1 and Q_3 are not equal.

The normalized frequency characteristic of the circuit calculated and plotted with MATLAB is given in Fig. 4.19.

Problem 4.5

The center frequency and the bandwidth of a four-stage, staggered tuned amplifier are 2GHz and 80 MHz, respectively. Calculate the tuning frequencies and the quality factors to obtain a Butterworth type frequency characteristic.

The second important type of pole distribution provides a Chebyshev type frequency characteristic. The side-walls of a Chebyshev type (or equiripple) characteristic is steeper than those of a same order Butterworth type characteristic, but exhibit a typical ripple on the top of the curve, as shown in Fig. 4.20a. The number of ripples depends on the order of the circuit.

The poles of a Chebyshev type circuit are positioned on an ellipse, whose vertical axis is on the $j\omega$ axis and the length of the longer axis corresponds to the bandwidth of the circuit. It has been shown that the appropriate positions of the poles for a certain amount of ripple can be obtained from the positions of a Butterworth type circuit that has the same bandwidth. As shown from Fig. 4.20b, the tuning frequencies of the resonance circuits are same, but the real parts of the poles of the Chebyshev type circuit are smaller. For an nth order Chebyshev type circuit with r (dB) ripple, the magnitude of σ of the Chebyshev pole can be calculated in terms of the corresponding Butterworth pole from

$$(\sigma_i)_C = \tanh \alpha \times (\sigma_i)_B \tag{4.51a}$$

where

$$\alpha = \frac{1}{n} \sinh^{-1} \frac{1}{\sqrt{\varepsilon}}, \qquad \varepsilon = \log^{-1} \frac{r(\text{dB})}{10} - 1 \tag{4.51b}$$

For convenience, the values of $(\tanh \alpha)$ for $n = 2$, 3, and 4 and for several ripple values are given in Table 4.1.

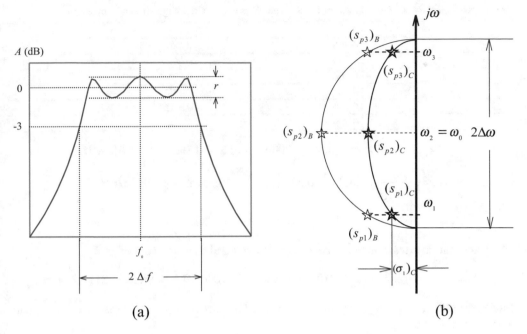

Fig. 4.20 (a) The frequency characteristic of a third-order C-type circuit. (b) The positioning of the poles of a Butterworth type circuit and a Chebyshev type circuit that have the same bandwidth

Example 4.5

A three-stage staggered tuned amplifier for 2 GHz will be designed, with a voltage gain of 40 dB and a Chebyshev type frequency characteristic with 0.5 dB ripple and 380 MHz bandwidth. The design will be made for a technology similar to the 0.35 micron AMS technology, but with an additional thick metal layer, allowing 10 nH inductors with an efficient quality factor of 10 at 2 GHz are available. The Q values can be further increased with a Q-enhancement circuit similar to the circuit given in Example 4.1

From Fig. 4.20b, we see that the bandwidth of the circuit is $B = 2\Delta\omega = 2|(\sigma_2)_B|$, where $(\sigma_2)_B$ is the negative real center pole of a Butterworth type circuit having the same center frequency and bandwidth. The center (real) Chebyshev pole can be calculated from Fig. 4.18b and Table 4.1:

$$\left|(\sigma_2)_C\right| = 0.524\left|(\sigma_2)_B\right| = 0.524 \times \Delta\omega$$

On the other hand,

$$\left|(\sigma_2)_C\right| = \frac{\omega_0}{2Q_2}$$

Now Q_2 can be calculated as

$$Q_2 = \frac{\omega_0}{2\Delta\omega}\frac{1}{0.524} = \frac{f_0}{2\Delta f}\frac{1}{0.524} = 10.04 \cong 10$$

This means that the center pole can be realized without any Q-enhancement.

From Fig. 4.18b and Fig. 4.20b, the tuning frequency and the quality factor corresponding to $(s_{p1})_C$ can be calculated as

Table 4.1 $\text{Re}[(s_{pi})_C]/\text{Re}[(S_{pi})_B]$ values for different numbers of cascaded stages (n) and ripples (r)

r (dB)	$n = 2$	$n = 3$	$n = 4$
0.05	0.898	0.750	0.623
0.1	0.859	0.696	0.567
0.2	0.806	0.631	0.505
0.3	0.767	0.588	0.467
0.4	0.736	0.556	0.439
0.5	0.709	0.524	0.416

$$f_1 = f_0 - \Delta f \times \cos(\pi/6) = 2000 - 190.8 \times 0.866 = 1834.76\,[\text{MHz}]$$

$$Q_1 = \frac{\omega_1}{|2(\sigma_1)_C|}, \quad (\sigma_1)_C = 0.524\,(\sigma_1)_B, \quad (\sigma_1)_B = \Delta\omega \times \cos(\pi/6)$$

$$Q_1 = 18.3$$

which correspond to a resonance impedance (effective parallel resistance) of

$$R_{1(eff)} = L_1\omega_1 Q_1 = \left(10 \times 10^{-9}\right) \times \left(2\pi \times 1834.76 \times 10^6\right) \times 18.35 = 2114.3\ \text{ohm}$$

Similarly, the tuning frequency, the quality factor, and the effective parallel resistance corresponding to $(s_{p3})_C$ can be found as

$$f_3 = 2165.2\ \text{MHz}, \quad Q_3 = 21.65, \quad R_{3(eff)} = 2945\ \text{ohm}$$

Since the resonance circuit tuned to 2000 MHz was intended to be used without any Q enhancement, its quality factor and the effective parallel resistance are $Q = 10$ and $R_{2(eff)} = 1256$ ohm, respectively.

The basic circuit diagram is given in Fig. 4.21a. The individual stages of this amplifier use the cascode configuration, which is the natural choice as explained in Sect. 4.2.2. The tuning frequencies of these stages will be $f_1, f_2 = f_0$ and f_3 in turn. To obtain the targeted voltage gain of 100 at the center frequency ($f_0 = f_2$), as well as frequencies corresponding to $(s_{p1})_C$ and $(s_{p3})_C$, the gains of the individual stages must be properly determined.

The total voltage gain of the amplifier at $f_0 = 2000$ MHz is equal to the multiplication of the voltage gains of the individual stages at this frequency. The DC currents and consequently the transconductances of the stages are chosen identical for convenience. The voltage gains of the individual stages at f_0 are

$$|A_{v1}(f_0)| = g_m|Z_1(f_0)|, \quad |A_{v2}(f_0)| = g_m|Z_2(f_0)| \quad \text{and} \quad |A_{v3}(f_0)| = g_m|Z_3(f_0)|$$

respectively. The load impedance of the second stage, which is tuned to f_0, is equal to the parallel equivalent resistance of the resonance circuit:

$$|Z_2(f_0)| = R_{2(eff)} = L\omega_0 Q_2 = \left(10 \times 10^{-9}\right) \times \left(2\pi \times 2 \times 10^9\right) \times 10 = 1257\ \text{ohm}$$

and the voltage gain

$$|A_{v2}(f_0)| = g_m \times 1257$$

To find the gain of the first stage at f_0, first $|Z_1(f_0)|$ must be calculated. From (4.24a)

$$|Z_1(f_0)| = |Z_1(f_1)| \frac{1}{\sqrt{1 + (\beta_1 Q_1)^2}}$$

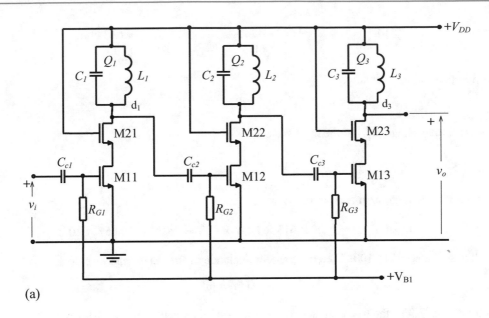

(a)

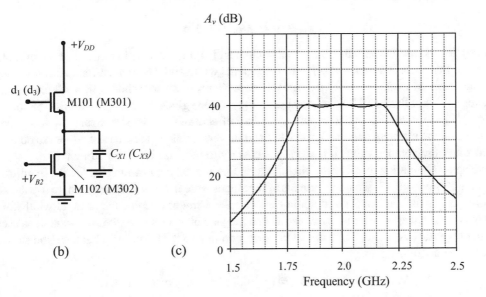

(b)   (c)

Fig. 4.21 (a) Schematic diagram of the amplifier. (b) The Q-enhancement circuits. (c) The frequency characteristic of the amplifier after a fine-tuning with PSpice $V_{DD} = 3.3$ V, $V_{B1} = V_{B2} = 1.15$ V, $I_{tot} = 9.23$ mA

where

$$\beta_1 = 2\Delta f_1/f_1 \quad \text{and} \quad \Delta f_1 = f_0 - f_1 = 2000 - 1834.76 = 165.24 \text{ MHz}$$

and

$$|Z_1(f_1)| = R_{1(eff)} = 2114.3 \text{ ohm}.$$

Hence

$$|Z_1(f_0)| = 2108.5 \times \frac{1}{\sqrt{1 + \left(\frac{2 \times 165.24}{1834.76} \times 18.3\right)^2}} = 613 \text{ ohm}$$

And the gain of the first stage

$$|A_{v1}(f_0)| = g_m \times 613$$

Similarly, the voltage gain of the third stage at f_0

$$|A_{v3}(f_0)| = g_m \times 853$$

Now the total voltage gain can be written as

$$|A_{vT}(f_0)| = (g_m)^3 \times 613 \times 1257 \times 853 = (g_m)^3 \times (0.657 \times 10^9)$$

which must be equal to 100. Then the transconductance of the stages can be calculated as

$$g_m = 5.34 \text{ mS}$$

According to (1.33), the transconductance of a non-velocity saturated transistor is

$$g_m = \sqrt{2\mu_n C_{ox}(W/L)I_D}$$

which implies the use of the minimum possible channel length, 0.35 μm in our case. Equation (1.33) also implies that there is a trade-off between the channel width and the drain current. Keeping in mind that a high channel width increases the parasitic capacitances, we will choose $W = 40$ μm. With the parameter values of the AMS 0.35 micron technology, (1.33) gives the drain current as 0.58 mA. But it must be kept in mind that due to certain secondary effects such as the series parasitic resistance of the source and the mobility degradation due to the transversal field in the channel region, to obtain the targeted transconductance value, it is usually necessary to increase the drain current.

The PSpice simulation result of the circuit after a fine-tuning is shown in Fig. 4.21c. To adjust the total voltage gain to 40 dB, the drain currents of the stages were increased. In addition, to use the same bias voltages for the gain stages and both of the Q-enhancement circuits, the channel widths of the gain stages were changed to appropriate values and the gate bias resistor of the second stage is used to adjust the Q of the first stage. The parameters corresponding to the frequency characteristic shown in Fig. 4.21c are as follows:

$$M11, M21, M12, M22, M13, M23 : W = 44\mu\text{m}, L = 0.35\mu\text{m}$$
$$M101, M301 : \qquad\qquad\qquad W = 46\mu\text{m}, L = 2.4\mu\text{m}$$
$$M102, M302 : \qquad\qquad\qquad W = 20\mu\text{m}, L = 0.35\mu\text{m}$$

$$L_1 = L_2 = L_3 = 10 \text{ nH}, Q = 10$$

$$C_1 = 445 \text{ fF}, C_2 = 520 \text{ fF}, C_3 = 235 \text{ fF}$$

$$R_{G1} = R_{G3} = 40 \text{ k}, R_{G2} = 14 \text{ k}$$

$$C_{C1} = C_{C2} = C_{C3} = 1 \text{ pF}$$

$$C_{x1} = 250 \text{ fF}, C_{x3} = 410 \text{ fF}$$

4.4   Amplifiers Loaded with Coupled Resonance Circuits

To obtain a reasonably flat frequency characteristic in a band, amplifiers loaded with a pair of "coupled" resonance circuits, individually tuned to the center frequency of the band, have been extensively used since the early days of radio. Although the resonance circuits are tuned to the same frequency, the transfer function of the circuit exhibits two pairs of conjugate poles having different imaginary parts, similar to Fig. 4.18a.

In earlier realizations, the main application area of these "double tuned amplifiers" was intermediate frequency (IF) amplifiers of all types of super-heterodyne receivers. The standard IF frequency was around 450 kHz with a bandwidth of 9 kHz for AM receivers and 10.7 MHz with a bandwidth of 150 kHz for FM receivers. The advantages of this approach were a reasonably flat response within this relatively narrow band with only one amplifying stage, and the ease of the tuning procedure. Magnetic coupling is usually preferred to couple the resonance circuits, but it is possible to show that any approach that provides interaction among the resonance circuits gives similar results and it is possible to use the same basic equations, after an appropriate parameter conversion.

The basic form of a magnetically coupled circuit is shown in Fig. 4.22a. The resonance frequencies of the two resonance circuits are equal and will be shown with ω_0, but L_1 and L_2 are not necessarily equal. The coupling coefficient is shown with k and is equal to $M/\sqrt{L_1L_2}$, where L_1 and L_2 are the values of the self-inductances of the coupled inductors and M is the mutual inductance. The losses of the input side and the output side were lumped as r_1 and r_2.

In a typical double-tuned amplifier, the first resonance circuit is connected as the load of the transistor, and the signal at the output of the second resonance circuit is applied to the input of the succeeding stage. To investigate the behavior of such a circuit, it is therefore appropriate to drive it with a current source and calculate the voltage of the output port, in other words, to calculate the transfer impedance of the circuit.[11]

The transfer impedance of the circuit given in Fig. 4.22a can be written as

$$\frac{V_2}{I_1} = \frac{sM}{(1 + sr_1C_1 + s^2L_1C_1)(1 + sr_2C_2 + s^2L_2C_2) - s^4M^2C_1C_2} \tag{4.52}$$

with

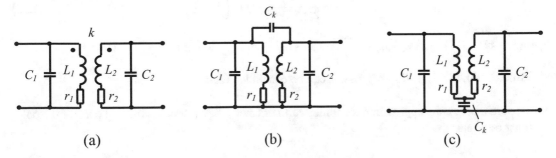

(a)   (b)   (c)

Fig. 4.22 The basic form of (a) a magnetic coupled, (b) capacitive voltage coupled, (c) capacitive current coupled double-tuned circuit

---

[11] Note that the output internal resistance of the driving transistor must be appropriately incorporated into r_1, the resistance representing the total losses of the first resonance circuit.

$$\omega_0^2 = \frac{1}{L_1 C_1} = \frac{1}{L_2 C_2}, \quad Q_1 = \frac{1}{\omega_0 C_1 r_1}, \quad Q_2 = \frac{1}{\omega_0 C_2 r_2}$$

(4.52) can be arranged as

$$\frac{V_2}{I_1} = \frac{s M \omega_0^4 Q_1 Q_2}{\left(s^2 Q_1 + s\omega_0 + \omega_0^2 Q_1\right)\left(s^2 Q_2 + s\omega_0 + \omega_0^2 Q_2\right) - s^4 k^2 Q_1 Q_2} \tag{4.53}$$

To ease the calculation of the poles, the fourth-order term of the denominator can be arranged as

$$s^4 Q_1 Q_2 \left(1 - k^2\right) = s^4 Q_1 Q_2 (1 - k)(1 + k)$$

which permits us to write the denominator as

$$\left[\left(s^2 Q_1 (1 - k) + s\omega_0 + \omega_0^2 Q_1\right)\right]\left[\left(s^2 Q_2 (1 + k) + s\omega_0 + \omega_0^2 Q_2\right)\right]$$

which yields two pairs of conjugate poles as

$$s_1, s_1' = -\frac{\omega_0}{2Q_1(1 - k)} \mp j \frac{\omega_0}{2Q_1(1 - k)} \sqrt{4Q_1^2(1 - k) - 1}$$

$$s_2, s_2' = -\frac{\omega_0}{2Q_2(1 + k)} \mp j \frac{\omega_0}{2Q_2(1 + k)} \sqrt{4Q_2^2(1 + k) - 1}$$

Since $4Q_1^2(1 - k) \gg 1$ and $4Q_2^2(1 + k) \gg 1$ for small values of k, the poles can be written as

$$s_1, s_1' \cong -\frac{\omega_0}{2Q_1(1 - k)} \mp j \frac{\omega_0}{\sqrt{1 - k}}$$

$$s_2, s_2' \cong -\frac{\omega_0}{2Q_2(1 + k)} \mp j \frac{\omega_0}{\sqrt{1 + k}}$$

Since usually $k \ll 1$, these expressions can be even further simplified as

$$s_1, s_1' \cong -\frac{\omega_0}{2Q_1(1 - k)} \mp j\omega_0 \left(1 + \frac{k}{2}\right) \tag{4.54}$$

$$s_2, s_2' \cong -\frac{\omega_0}{2Q_2(1 + k)} \mp j\omega_0 \left(1 - \frac{k}{2}\right) \tag{4.55}$$

with$(1 - k) \cong 1, (1 + k) \cong 1, 1/\sqrt{1 - k} \cong 1 + (k/2)$ and $1/\sqrt{1 + k} \cong 1 - (k/2)$.

From (4.54), (4.55), and Fig. 4.18a, it can be concluded that:

- To obtain pole pairs appropriate for Butterworth or Chebyshev responses the real parts of the poles must be equal:

$$\sigma = -\frac{\omega_0}{2Q_1(1 - k)} = -\frac{\omega_0}{2Q_2(1 + k)} \tag{4.56a}$$

and for $k \ll 1$;

$$Q_1 \cong Q_2 = Q \tag{4.56b}$$

Fig. 4.23 Positions of the poles (**a**) for a maximally flat (Butterworth type), (**b**) for an equiripple (Chebyshev type) frequency characteristic

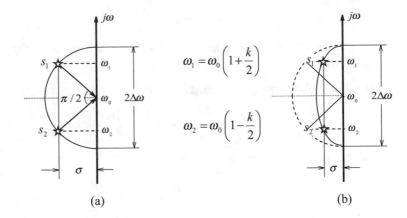

$$\omega_1 = \omega_0\left(1 + \frac{k}{2}\right)$$

$$\omega_2 = \omega_0\left(1 - \frac{k}{2}\right)$$

- To obtain a second-order Butterworth type frequency characteristic the poles must be positioned as shown in Fig. 4.23a. The corresponding k values and the bandwidth can be calculated from the geometry of the figure:

$$|\sigma| = \frac{\omega_0}{2Q} = \omega_0\frac{k}{2} \quad \Rightarrow \quad k = \frac{1}{Q} \tag{4.57}$$

$$2\Delta\omega = 2 \times \left(\sqrt{2}|\sigma|\right) = \sqrt{2}\frac{\omega_0}{Q} \quad \Rightarrow \quad B = 2\Delta f = \sqrt{2}\frac{f_0}{Q} \tag{4.58}$$

From the comparison of (4.58) and (4.25), we see that the bandwidth of a double-tuned circuit with maximally flat frequency response is $\sqrt{2}$ times larger than that of a single tuned circuit, having the same Q value.

The positioning of poles for a Chebyshev type frequency characteristic is shown in Fig. 4.23b. The shift of the poles in the horizontal direction with respect to the poles of a Butterworth type response having the same bandwidth, corresponding to a certain ripple value, was given in Sect. 4.3. For example, to obtain a Chebyshev type frequency characteristic with 0.5 dB ripple, from the geometry of the figure,

$$(\sigma)_C = \frac{\omega_0}{2Q} = 0.709 \times \omega_0\frac{k}{2} \quad \Rightarrow \quad k = \frac{1}{0.709}\frac{1}{Q} = \frac{\sqrt{2}}{Q} \tag{4.59}$$

$$2\Delta\omega = 2 \times \sqrt{2} \times \omega_0\frac{k}{2} = \sqrt{2} \times \omega_0 \times \frac{1}{0.709}\frac{1}{Q} \quad \Rightarrow \quad B = 2\Delta f = 2 \times \frac{f_0}{Q} \tag{4.60}$$

From the comparison of (4.60) and (4.25), we see that the bandwidth of a double-tuned circuit with 0.5 dB ripple Chebyshev type frequency response is twice as large as that of a single-tuned circuit, having the same Q value.

Example 4.6

A double-tuned amplifier has a load as shown in Fig. 4.24a. The center frequency of the response is 2 GHz. Effective quality factors of both sides were adjusted to 20 with an appropriate Q-enhancement circuit, which corresponds to a parallel effective resistance of $R_{p1} = R_{p2} = 2512$ ohm.

(a) Calculate the magnetic coupling coefficient for a maximally flat response and corresponding bandwidth.

The value of the coupling coefficient can be calculated from (4.57);

$$k = \frac{1}{Q} = \frac{1}{20} = 0.05$$

According to (4.58) the bandwidth is

$$B = \sqrt{2}\frac{f_0}{Q} = \sqrt{2}\frac{2 \times 10^9}{20} = 141 \text{ MHz}$$

(b) Calculate the magnetic coupling coefficient for a 0.5 dB ripple Chebyshev type response and corresponding bandwidth.

From (4.59) $k = \frac{\sqrt{2}}{Q} = \frac{\sqrt{2}}{20} = 0.0707$

From (4.60) $B = 2 \times \frac{f_0}{Q} = 2 \times \frac{2 \times 10^9}{20} = 200 \text{ MHz}$

(c) Plot the frequency responses with PSpice simulation.

Results that fit the calculated values are given in Fig. 4.24b.

Problem 4.6

A cascode amplifier stage with a double-tuned circuit load will be designed. The target values are:
- *Center frequency: $f_0 = 1$ GHz*
- *Voltage gain at 1 GHz: 20 dB*
- *Bandwidth: $2\Delta f = 120$ MHz*
- *Type of the response: 0.4 dB ripple, Chebyshev*
- *Technology: AMS035*
- *Available on-chip inductors: 5 nH ($Q = 6$ @ 1GHz), 10 nH ($Q = 5.7$ @ 1GHz),*
  15 nH ($Q = 4.7$ @ 1GHz)
- *Quality factors of the on-chip capacitors: $Q = 50$*
- *Power supply: +3.2 V*

(a) *Determine the positions of the poles of the gain function and then calculate the effective quality factors corresponding to these poles.*
(b) *Calculate the value of the negative resistances necessary to enhance the actual Q values to the calculated Q values (assume the output resistance of the cascode circuit is negligibly high).*
(c) *Calculate the magnetic coupling coefficient, k.*
(d) *Calculate the resonance capacitances, the parasitics included.*
(e) *Calculate the transconductance of the cascode stage.*
(f) *Simulate the circuit and plot the frequency characteristic (the idealized equivalent circuit of the cascode amplifier can be used).*
(g) *Check the effects of the ± 10% and ± 20% spread of the k value.*

4.4.1 Capacitive Coupling

There are other ways of establishing an interaction between two resonance circuits, as shown in Fig. 4.22b, c. The circuit in Fig. 4.22b is known as a capacitive voltage coupling. The expressions related to this case have similar forms to the expressions derived for magnetic coupling. The coupling capacitance value corresponding to a certain value of the magnetic coupling coefficient k is given in [5] as:

$$C_k = -\frac{k}{\omega_0^2 \sqrt{L_1 L_2}} \tag{4.61}$$

The negative sign indicates that the capacitive voltage coupling is equivalent to a negative magnetic coupling.

For the capacitive current coupling shown in Fig. 4.22c, C_k can be calculated as

$$C_k = -\frac{1}{k\omega_0^2 \sqrt{L_1 L_2}} \tag{4.62}$$

It must be noted that:

- The expressions given for capacitive coupling were obtained for high-Q circuits. Therefore a fine-tuning with Spice simulation is needed for low-Q circuits.
- The center frequency shifts downward for the capacitive voltage coupling and upward for the capacitive current coupling, due to the effect of the coupling capacitor.

Example 4.7

Let us calculate the coupling capacitance value for the circuit shown in Fig. 4.24a to obtain a 0.5 dB ripple Chebyshev type response. Assume that there is no magnetic coupling.

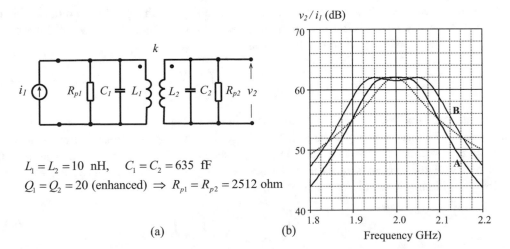

$$L_1 = L_2 = 10 \text{ nH}, \quad C_1 = C_2 = 635 \text{ fF}$$
$$Q_1 = Q_2 = 20 \text{ (enhanced)} \Rightarrow R_{p1} = R_{p2} = 2512 \text{ ohm}$$

(a) (b)

Fig. 4.24 (a) The equivalent circuit of the output side of the amplifier. (b) Frequency characteristics of the double-tuned circuit for (A) $k = 1/Q$ (maximally flat response) and (B) $k = 1.41/Q$ (0.5 dB ripple Chebyshev response). The response of a single-tuned circuit (C) is also given for comparison (It can be shown that the magnitude of the transfer impedance corresponding to $kQ = 1$ at ω_0 is half of the resonance impedance of the single-tuned circuit. To ease the comparison, the response of the single-tuned circuit is shifted by -6 dB)

From (4.61)

$$C_k = \frac{0.0707}{\left(2\pi \times 2 \times 10^9\right)^2 \sqrt{\left(10^{-8}\right)\left(10^{-8}\right)}} = 45 \text{ fF}$$

4.5   The Gyrator: A Valuable Tool to Realize High-Value On-Chip Inductances

A tuned amplifier can be interpreted as a "band-pass filter" that amplifies (passes) the frequencies in a certain band and does not amplify (stops) all other frequencies. Several types of filters, especially band-pass filters and low-pass filters, are extensively used in telecommunication systems. Filters were one of the main subjects of network theory during the 1940s and 1950s, and investigated in depth. Several methods have been developed to design a filter that fulfills the requirements related to the pass-band, stop-band, and the impedances. These filters were "passive" filters that were composed of passive components; capacitors, inductors, and resistors.

Analog and digital "active" filters were the result of developments in IC technology. For one of the types of analog active filters the methodology of the well-investigated passive filter theory can be directly applied; the inductors in these filters are not "real" inductors, but are "emulated" inductors with an electronic circuit known as a "gyrator" and a capacitor. With this approach, relatively high-value inductances (that are not possible to realize as an on-chip component) can be integrated, and filters can be realized at relatively high frequencies that could not be reached with op-amp based active filters.

As will be shown below, a gyrator can be easily realized with a number of transconductance amplifiers (OTAs). That is why the gyrator-based filters are also called "G_m-C filters," as a specific member of the broader "G_m-C filters" family.

The "gyrator" was described (or invented) by B. D. H. Tellegen with a purely theoretical approach in 1948 [6]. After a systematic classification of the linear two-ports, Tellegen noticed that there must exist a yet-unknown two-port satisfying the expression

$$\begin{pmatrix} i_1 \\ i_2 \end{pmatrix} = \begin{pmatrix} 0 & G \\ -G & 0 \end{pmatrix} \begin{pmatrix} v_1 \\ v_2 \end{pmatrix} \tag{4.63}$$

He called it a "gyrator" and offered a symbol shown in Fig. 4.25a.

It can be shown that a gyrator can be realized with two ideal voltage-controlled current sources, as shown in Fig. 4.25b, or with two ideal OTAs, as shown in Fig. 4.25c, and in addition, the G parameters of the current sources do not have to be equal [7]. Consequently, the expression describing the circuit shown in Fig. 4.25c becomes

$$\begin{pmatrix} i_1 \\ i_2 \end{pmatrix} = \begin{pmatrix} 0 & g_{m2} \\ -g_{m1} & 0 \end{pmatrix} \begin{pmatrix} v_1 \\ v_2 \end{pmatrix} \tag{4.64}$$

The input impedance of a gyrator loaded with a load admittance Y_2 can be easily calculated:

$$i_2 = -Y_2 v_2 = -g_{m1} v_1, \qquad i_1 = g_{m2} v_2 \qquad \Rightarrow \qquad Z_1 = \frac{v_1}{i_1} = \frac{Y_2}{g_{m1} g_{m2}} \tag{4.65}$$

Fig. 4.25 (a) The original
symbol of the gyrator. (b)
Realization of a gyrator
with two voltage-
controlled current sources,
and (c) with two OTAs

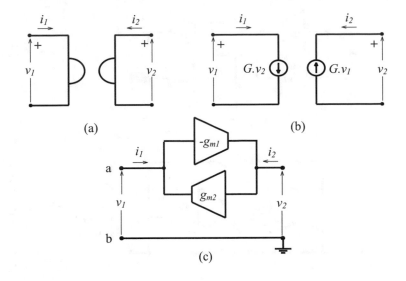

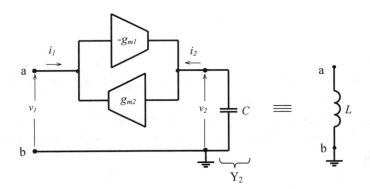

Fig. 4.26 Gyrator-based inductance realization

The practical importance of the gyrator can be best appreciated by connecting a capacitor to the
output port as shown in Fig. 4.26, and then calculating the input impedance of the circuit. In this case,
since $Y_2 = sC$, (4.65) can be arranged as

$$Z_1 = s\frac{C}{g_{m1}g_{m2}} \quad \Rightarrow \quad L = \frac{C}{g_{m1}g_{m2}} \tag{4.66}$$

Eq. (4.66) tells us that the input impedance of a capacitance-loaded ideal gyrator is equal to the
impedance of an ideal inductor. Therefore, the input port of the circuit can be used as an inductor.
This means that the high inductance values that are not possible to realize with on-chip inductors can
be realized with a gyrator and an appropriate on-chip load capacitor.

Example 4.8
A 10 μH inductance is needed. The transconductance values of the OTAs that will be used to
implement the gyrator are 0.5 mS each. Calculate the value of the capacitor to be connected to the
output port of the gyrator.

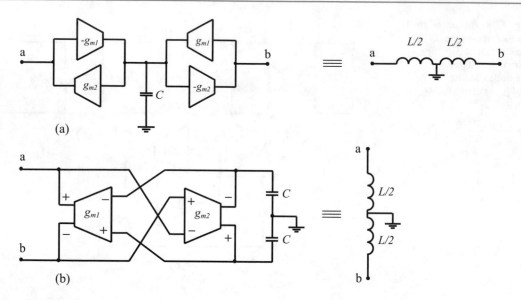

Fig. 4.27 Realization of floating inductances: (**a**) with single-ended input, single-ended output OTAs, (**b**) with symmetrical OTAs

From (4.66)

$$C = L g_{m1} g_{m2} = \left(10 \times 10^{-6}\right)\left(0.5 \times 10^{-3}\right)\left(0.5 \times 10^{-3}\right) = 2.5 \text{ pF}$$

It must be noted that one terminal of this "emulated" inductor is connected to ground. In many applications, however, floating inductors are needed. A floating inductor can be realized as shown in Fig. 4.27a, b [8].

4.5.1 Parasitics of a Non-Ideal Gyrator

Up until this point we assumed that the OTAs used to compose a gyrator were ideal. In reality, they have a finite input capacitance (C_i), a finite output capacitance (C_o), and a non-zero output conductance. In addition, the transadmittance function does not have a flat frequency characteristic but rolls off at high frequencies, which can be expressed as a one-pole gain function as given in (3.66). Consequently, all of these non-idealities must be taken into account to characterize the realized inductance.

In Fig. 4.28, a "real" gyrator is represented with an "ideal" gyrator and the input and output parasitics of the circuit. For convenience, we will assume that OTAs are in the form of the circuit given in Fig. 3.32. The output capacitance C_0 is the sum of the input capacitance of OTA-2 (the gate capacitance of the input transistor) and the sum of the drain capacitances of the transistors connected to the output node of OTA1. The output conductance g_o is the sum of the output conductances of the output transistors of OTA1. Therefore, the total load admittance of the gyrator, Y_2, is the sum of these components and the external load capacitance, C_L.

The input parasitics, C_i and g_i, are similar to their output counterparts. Since they are directly parallel to the input impedance Z_1 to be calculated, they will be excluded during the derivation to ease the calculations and will be added at the end.

Fig. 4.28 Gyrator with its input and output parasitics. Note that $g_2 = g_o$ and $C_2 = (C_o + C_L)$

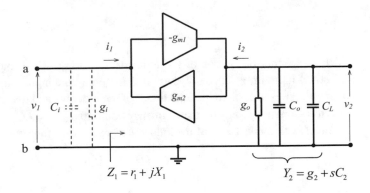

The gain of a typical OTA can be expressed with a one-pole gain function as given with (3.66):

$$y_m = g_m s_p \frac{1}{(s - s_p)}$$

Now with $Y_2 = g_2 + sC_2$ (4.65) can be written as

$$Z_1 = \frac{Y_2}{y_{m1} y_{m2}}$$

$$Z_1 = \frac{1}{g_{m1} g_{m2} s_{p1} s_{p2}} (s - s_{p1})(s - s_{p2})(g_2 + sC_2)$$

$$Z_1(\omega) = \frac{1}{g_{m1} g_{m2} \omega_{p1} \omega_{p2}} (j\omega + \omega_{p1})(j\omega + \omega_{p2})(g_2 + j\omega C_2)$$
(4.67a)

It is convenient to simplify this expression assuming that the poles of the two OTAs are the same:

$$Z_1(\omega) = \frac{1}{g_{m1} g_{m2} \omega_p^2} (j\omega + \omega_p)^2 (g_2 + j\omega C_2)$$
(4.67b)

The imaginary part of the input impedance can be calculated as

$$\text{Im}\{Z_1\} = \omega \left(\frac{C_2}{g_{m1} g_{m2}} + 2\frac{g_2}{\omega_p g_{m1} g_{m2}} \right)$$
(4.68)

and for $\omega \ll \omega_p$

$$\text{Im}\{Z_1\} \cong \omega \left(\frac{C_2}{g_{m1} g_{m2}} \right) = \omega L$$
(4.68a)

The real part of the input impedance, which corresponds to the series resistance of L, is

$$\text{Re}\{Z_1\} = \frac{g_2}{g_{m1} g_{m2}} - \frac{\omega^2}{\omega_p^2} \frac{1}{g_{m1} g_{m2}} (g_2 + 2\omega_p C_2) = r_L$$

since usually $2\omega_p C_2 \gg g_2$,

$$r_L \cong \frac{g_2}{g_{m1}g_{m2}} - \frac{\omega^2}{\omega_p} \frac{2C_2}{g_{m1}g_{m2}} \tag{4.69}$$

which represents the losses of the inductor appearing at the input port of the gyrator. The total (effective) input impedance of the gyrator, including the input parasitics, is shown in Fig. 4.29a. It is apparent that g_i adds on to the total losses of the inductor and decreases the quality factor, whereas C_i leads to resonance at an angular frequency equal to $\omega_0 = \sqrt{LC_i}$

The effective resistance of the inductor together with the effect of g_i can be calculated as

$$r_{eff} \cong r_L + \omega^2 L^2 g_i$$

$$r_{eff} = \frac{g_2}{g_{m1}g_{m2}} - \omega^2 L^2 \left(\frac{2}{\omega_p L} - g_i \right) \tag{4.70}$$

In Fig. 4.30, the imaginary part of Z_1 from (4.68a) that is proportional with L and the effective series resistance of L from (4.70) for $g_i < \frac{2}{\omega_p L}$ are plotted as a function of ω. From these plots, it can be seen that:

- The slope of $\text{Im}\{Z_1\}$ is constant and equal to L.
- For $g_i < \frac{2}{\omega_p L}$, $r_{eff}(\omega)$ decreases with frequency and crosses zero at ω_x, which can be calculated as

$$\omega_x = \sqrt{\frac{g_2}{g_{m1}g_{m2}} \frac{1}{L^2 \left(\frac{2}{\omega_p L} - g_i \right)}} \tag{4.71}$$

and then becomes negative. If L resonates with a capacity at a frequency higher than ω_x, the circuit oscillates.

Fig. 4.29 (a) The equivalent circuit of the input impedance of a gyrator, loaded with a capacitor, the input parasitics included. (b) The simplified equivalent circuit

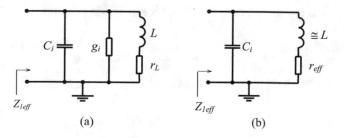

(a)   (b)

Fig. 4.30 Variations of r_{eff} (solid line) and the imaginary part of Z_1 (dashed line) with frequency for $g_i < (2/\omega_p L)$. Note that the frequency axis was chosen as linear to visualize the linear dependence of the imaginary part. The slope of the dashed line is equal to L

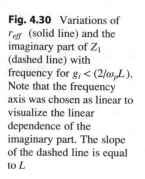

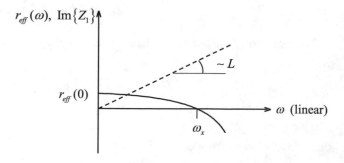

- For $g_i > \frac{2}{\omega_p L}$, $r_{eff}(\omega)$ is always positive; therefore there is no risk of instability.
- The quality factor of the inductor for any ω is

$$Q_{eff}(\omega) = \frac{L\omega}{r_{eff}(\omega)} = \frac{L\omega}{\frac{g_2}{g_{m1}g_{m2}} - \omega^2 L^2 \left(\frac{2}{\omega_p L} - g_i\right)}. \tag{4.72}$$

From (4.72) it can be concluded that:

- The effective quality factor of the emulated inductor increases with the transconductances of OTAs and decreases with the input and output conductances.
- For $g_i < \frac{2}{\omega_p L}$ the circuit has an inherent Q-enhancement feature. For frequencies approaching ω_x, the effective quality factor sharply increases and reaches infinity at $\omega = \omega_x$, where the effective resistance of the inductor becomes zero.
- Due to the increased sensitivity of Q_{eff}, operating with this excessive "Q-enhancement" (at frequencies close to ω_x) is not convenient.

Example 4.9
A gyrator as shown in Fig. 4.28 is formed with two OTAs similar to the circuit given in Example 3.3. OTAs are used in single-ended input, single-ended output form. The transconductances are $g_{m1} = g_{m2} = 1$ mS, $g_{o1} = g_{o2} = 84\,\mu$S and $f_{p1} = f_{p2} = 2.62$ GHz. The external output load capacitance is $C_L = 1$ pF.

In Fig. 4.31, the PSpice simulation results for the real and imaginary parts of the input impedance are given. The imaginary part linearly increases with frequency according to (4.68)

Fig. 4.31 The variations of the real part (solid line) and the imaginary part (dashed line) of the input impedance of the gyrator

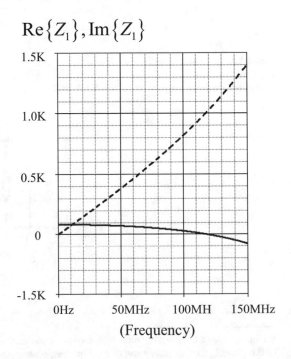

as expected.[12] The inductance value calculated from the initial slope of the imaginary part curve is 1.21 µH. According to (4.68a), if there were no parasitic capacitances (i.e. $C_2 = C_L$), the value of the inductance should be 1 µH. The difference arises from the output parasitic capacitance of the gyrator, which can be easily calculated as 0.21 pF.

The real part of the input impedance for $f = 0$ that corresponds to $r_{eff}(0)$ is 83.4 ohm, which is in agreement with the value calculated from (4.69).

Problem 4.7

(a)   *Design a 10 µH inductor with the OTAs used in Example 4.9.*
(b)   *What is the frequency limit of usability for this inductor?*
(c)   *Calculate the effective quality factor at 20 MHz and 50 MHz.*
(d)   *Compare the results with the results of pSpice simulation.*

4.5.2   Dynamic Range of a Gyrator-Based Inductor

The maximum voltage swing between the terminals of a gyrator-emulated inductor is limited by the maximum output voltage of OTA-1, which depends on (and is usually approximately equal to) the supply voltage, or the maximum input voltage of OTA-2, whichever is smaller. A similar limitation holds for the voltage of the output terminal. To ensure appropriate operation, the input port swing must not be limited due to the output port dynamic range, and vice-versa. This means that the maximum output swings of the input and output ports must be equal, i.e., $\widehat{V}_1 = \widehat{V}_2 = \widehat{V}$ (See Fig. 4.32).

From Fig. 4.32, the output port voltage can be written in terms of the input port voltage as

$$v_2 = -\frac{i_2}{Y_2} = -i_2 \frac{1}{sC_2 + g_2} = v_1 g_{m1} \frac{1}{sC_2 + g_2}$$

Fig. 4.32 Input and output port voltages of a capacitive loaded gyrator

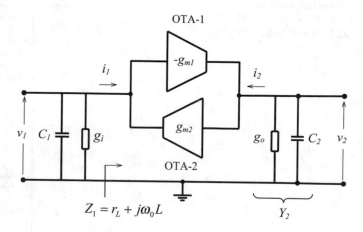

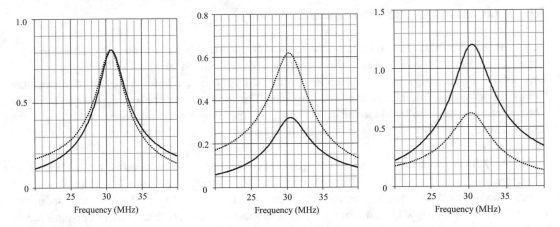

Fig. 4.33 The input (solid line) and output port (dashed line) voltages (in volts) of the gyrator for different C_2/C_1 ratios: (**a**) $C_1 = C_2 = 5$ pF, (**b**) $C_1 = 10$ pF, $C_2 = 2.5$ pF, (**c**) $C_1 = 2.5$ pF, $C_2 = 10$ pF

and the magnitude in the ω domain is:

$$V_2 = V_1 \frac{g_{m1}}{\sqrt{\omega^2 C_2^2 + g_2^2}} \cong V_1 \frac{g_{m1}}{\omega C_2} \tag{4.73}$$

Similarly, the input port voltage can be written as

$$v_1 = i_1 Z_1 = i_1 (sL + r_L) = v_2 g_{m2}(sL + r_L)$$

and the magnitude in the ω domain:

$$V_1 = V_2 g_{m2} \sqrt{\omega^2 L^2 + r_L^2} \cong V_2 g_{m2} \omega L \tag{4.74}$$

The inductance is usually used in resonance with a parallel capacitor, C_1. (4.73) and (4.74) can be written in terms of the resonance frequency, $\omega_0 = 1/\sqrt{LC_1}$.

$$V_2 \cong V_1 \frac{g_{m1}}{\omega_0 C_2} \tag{4.75}$$

and

$$V_1 \cong V_2 \omega_0 L = V_2 \frac{g_{m2}}{\omega_0 C_1} \tag{4.76}$$

From the condition of equality of the magnitudes of the input and output voltages as stated above and from (4.75) and (4.76),

$$\frac{g_{m1}}{g_{m2}} = \frac{C_2}{C_1} \tag{4.77}$$

To verify this expression, a pSpice simulation is performed. The gyrator given in Example 4.9 is driven with a 10 μA signal current source from the input port. The input and output port voltages are plotted for different C_2/C_1 ratios. It is clearly seen that the input and output port voltages at resonance are equal for $C_2/C_1 = 1$ as expected according to (4.77), since $g_{m1} = g_{m2}$ in the example.

4.6 LNAs and Input Impedance Matching

"LNA" is the acronym for a class of amplifiers, "Low Noise Amplifiers," commonly used as the input stages of wireless receivers. Since the incoming signal from an antenna is usually weak, the unavoidable noise generated in the amplifier must be as low as possible, to obtain an acceptable "signal to noise ratio" at the output of the amplifier. Therefore, one of the key design goals for the LNA is a low noise contribution to the input signal, together with a good impedance matching to the signal source, a sufficiently large output signal dynamic range and – certainly – a low power consumption.

Since LNAs are being used as the input stage of receivers, they must be tuned (or be tunable) to the carrier frequency of the transmitter that we want to receive. Therefore, they are "tuned amplifiers" by nature. The bandwidth of the amplifier must be large enough to cover the side-bands of the modulated carrier. But due to the low Q values of the on-chip inductors, the bandwidth usually becomes larger than necessary and the signals that remain outside of the modulation bandwidth must be eliminated by the succeeding stages of the amplifier.

4.6.1 Input Impedance Matching

The input signal source of the LNA is usually an antenna with the appropriate shape and dimension to provide the desired radiation pattern (omni-directional or directional with a specified beam-pattern) at the frequency of interest. The internal impedance of the antenna at this frequency (which is normally the resonance frequency of the antenna) has a dominant resistive component corresponding to the radiation losses[13] and a usually small reactive component corresponding to the connection parasitics.

If the antenna is located at a distance from the input terminal of the amplifier, a low-loss transmission line has to be used to connect the antenna to the amplifier. In this case, to maximize the signal power transfer from the antenna to the amplifier, we must satisfy:

(a) Impedance matching between the antenna and the line
(b) Impedance matching between the line and the input of the amplifier

Coaxial lines with 50-ohm characteristic impedance[14] are the most frequently used transmission lines[15] for single-ended antennas such as monopole antennas. For symmetrical antennas such as dipole antennas, folded dipole antennas, etc., symmetrical transmission lines can be used to connect to the inputs of a differential LNA. Another solution is to use a "balun"[16] to connect the output of a symmetrical antenna to the input of a coaxial line. If the antenna impedance is not resistive and not

---

[13] From antenna theory, it is known that an antenna is reciprocal in one sense; it exhibits similar behavior to a transmitting or receiving antenna. Since the concept of "radiation loss" is more understandable, this term has been used even for receiving antennas. It must be noted that the radiation loss is not constant; it is maximum when the radiation is maximum, which corresponds to the maximum of the current or voltage, in other words, to resonance. On both sides of the resonance peak, the radiation, and correspondingly, the antenna loss decreases, leading to further narrowing of the resonance curve.

[14] The characteristic impedance (Z_o) of a transmission line is the impedance for which no signal reflection occurs when the line is terminated with this impedance. The characteristic impedance of a low-loss transmission line is equal to $\sqrt{L/C}$ where L and C correspond to the series inductance and parallel capacitance of the line per unit length, and is resistive. To underline its resistive character, R_o will be used for the characteristic impedance, whenever necessary.

[15] Other standard characteristic impedance values for coaxial lines are 60 ohm and 75 ohm.

[16] "balun" is an abbreviation for "balanced-to-unbalanced" converting circuit.

Fig. 4.34 Impedance matching of a single-ended antenna when the real part of the antenna impedance is equal to the characteristic impedance of the coaxial line (**a**) For a capacitive antenna, (**b**) For an inductive antenna

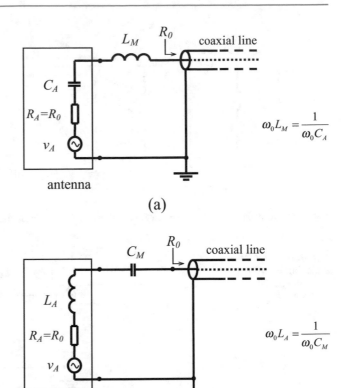

equal to the characteristic impedance of the line at the operation frequency of the amplifier, an impedance-matching circuit has to be used to provide a resistive impedance equal to Z_o (or R_o) to maximize the power transfer from the antenna to the transmission line. If the resistive component of the antenna impedance (R_A) is equal to R_o, reactive components can be easily compensated with a series matching inductor or matching capacitor, as shown in Fig. 4.34a, b.

For $R_A < R_o$ and $R_A > R_o$, there are several simple solutions. Fig. 4.35 shows the circuit diagram, the corresponding phasor diagram and the design formulae for one of the solutions pertaining to $R_A < R_o$. Note that this solution is especially advantageous for antennas having an inductive component that can be considered as a part of L_M.

Problem 4.8
Derive matching circuits similar to that given in Fig. 4.35, *which are suitable for an antenna*

(a)   *having a series capacitive component and $R_A < R_o$,*
(b)   *having a parallel capacitive component and $R_A > R_o$,*
(c)   *having a parallel inductive component and $R_A > R_o$.*

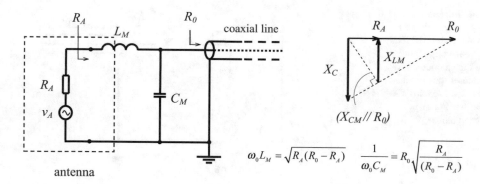

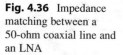

$$\omega_0 L_M = \sqrt{R_A(R_0 - R_A)} \qquad \frac{1}{\omega_0 C_M} = R_0 \sqrt{\frac{R_A}{(R_0 - R_A)}}$$

Fig. 4.35 Impedance-matching circuit for $R_A < R_0$, suitable for antennas having a small series inductive component

Fig. 4.36 Impedance matching between a 50-ohm coaxial line and an LNA

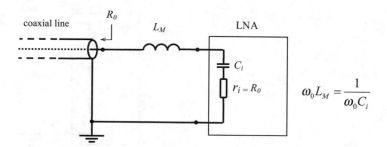

$$\omega_0 L_M = \frac{1}{\omega_0 C_i}$$

The use of impedance converting transformers is another possibility to match the real part of the antenna impedance to the line.

To provide impedance matching between the transmission line and the input of the amplifier, the amplifier can be designed such that the real part of the input impedance is equal to R_o, which is usually 50 ohms.

The reactive component of the input impedance (which is usually capacitive) can be easily compensated with a series inductance as shown in Fig. 4.36. This is the classical approach for the design of LNAs.

In mobile system applications such as mobile telephones, GPS receivers, etc., the antenna is built on the same board as the LNA chip and close enough to eliminate the need for any transmission line in between. This means that the adherence to the 50-ohm standard for the input impedance of LNAs is no longer critical (except for the ease of measurement and characterization). Therefore, LNAs must be designed to match the impedance of the antenna that is intended to be used. An even more realistic approach is the co-design (or, at least interactive design) of the antenna and the amplifier. The following discussion serves well to illustrate the importance of such considerations.

The most frequently used on-board antennas in mobile systems are folded-dipole and folded-loop antennas to drive the differential LNAs, and inverted-F antennas to drive the single-ended LNAs. There are many publications about these types of antenna, but they are mostly focused on the radiation pattern and the return-loss (or the SWR) when the antenna is driven with a 50-ohm line. The variation of the impedance with frequency is investigated only in a few publications.

One of the extensively used types of onboard single-ended antenna is the "planar inverted-F antenna." They are suitable to be realized on the multilayer board of the receiver and they can provide a fairly omni-directional radiation pattern. The resonance frequency and the impedance of inverted-F antennas strongly depend on the structure and dimensions. For one of the published inverted-F antennas [9], the real and imaginary parts of the impedance are given in Fig. 4.37a.

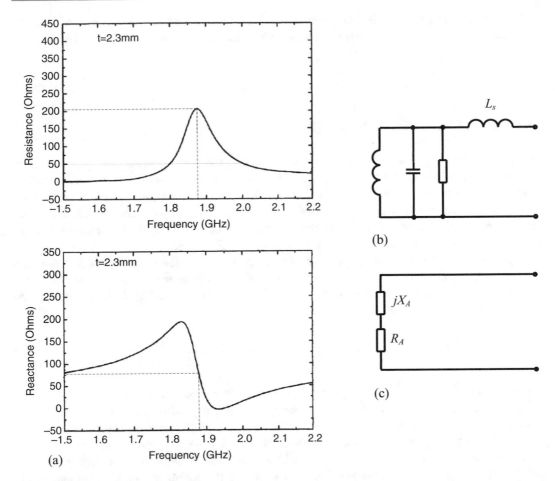

Fig. 4.37 (**a**) The variation of the real and imaginary parts of the impedance of the inverted-F antenna given in [9]. (**b**) The model suitable to represent the behavior of the antenna. L_s represents the positive shift of the reactance curve. (**c**) The model representing the resistive and reactive parts of the antenna impedance

It can be seen that there is an apparent resemblance between these curves and the curves of the real and imaginary parts of the impedance of a parallel resonance circuit. The positive shift of the reactive part indicates that the model includes a series inductance, as shown in Fig. 4.37b. It must be noted that the parallel resistance representing the radiation losses is not constant; it is minimum at the resonance frequency and increases with de-tuning on both sides of the peak. The equivalent circuit of the antenna in terms of its resistive and reactive components is given in Fig. 4.37c.

From the curves given in Fig. 4.37a, it can be seen that:

- The resonance frequency of the antenna is 1.87 GHz, where the antenna is most effective.
- The resistive and reactive components of the antenna impedance at resonance frequency are $R_A = 210$ ohm and $X_A = +75$ ohm, respectively.
- Therefore, for maximum signal transfer from the antenna to the input of the LNA, the input impedance of the LNA at 1.87 GHz must be $r_i = 210 - j75$ ohm, in other words, 210 ohm in series with a 1.147 pF capacitance.
- If f_0, the center frequency of the LNA, is not equal to 1.87 GHz, the antenna must be modified to shift the resonance frequency to f_0.
- From the resistance curve it can be seen that at approximately 2 GHz the antenna impedance is purely resistive. If the antenna is driven with a 50-ohm line, the return loss certainly becomes

minimum at these frequencies, as given in [9]. But since the antenna is not operating at the resonance frequency, the overall signal efficiency becomes considerably lower than the efficiency of the antenna operating at resonance.

4.6.2   Basic Circuits Suitable for LNAs

The real part of the impedance of an antenna at the resonance frequency is low, usually in the range of some tens of ohms to hundreds of ohms. To enable maximum signal power transfer from the antenna to the input of the amplifier, the real part of the input impedance of the amplifier must be equal to that of the antenna. This means that the LNA must be a low input impedance amplifier. The imaginary parts can be eliminated by resonance at the frequency of operation, f_o, with the addition of an inductive or capacitive reactance, into the loop as shown in Fig. 4.38a.

It must be noted that the input impedance of an amplifier is usually a parallel combination of a capacitance and a resistance, as shown in Fig. 4.38b. To investigate the matching conditions, it is convenient to use the series equivalent of the input impedance as shown in Fig. 4.38c. The conversion formulae can be easily derived as:

$$r_i = r_p \frac{1}{1 + \left(\omega \cdot C_p r_p\right)^2}, \qquad C_i = C_p\left(1 + \frac{1}{\left(\omega \cdot C_p r_p\right)^2}\right) \qquad (4.78)$$

As seen from (4.78), r_i and C_i are frequency-dependent and must be calculated for $\omega = \omega_0$.

The output of the LNA drives the output resonance circuit that is also tuned to f_o. To prevent the loading of the output resonance circuit, the output resistance of the amplifier must be as high as possible. Consequently, the appropriate configuration for an LNA is a low input impedance-high output impedance circuit. The voltage gain of the LNA is determined by the transconductance and the load impedance. Therefore, it is appropriate to characterize the amplifier with its input and output impedances and the transconductance, g_m.

Fig. 4.38 (a) The input loop of an LNA. (b) The usually valid input impedance equivalent of a LNA, (c) the series R-C equivalent of the input impedance that can be directly used in (a)

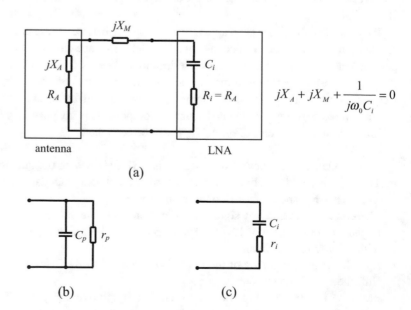

$$jX_A + jX_M + \frac{1}{j\omega_0 C_i} = 0$$

There are several circuit topologies that satisfy the low input impedance and high output impedance condition. Additional factors include high transconductance, low internal feedback, low power
consumption, and low noise.

Figure 4.39 shows the most basic circuits that satisfy the impedance condition. The basic
properties of these circuits are summarized below. Differential (symmetrical) versions of these
amplifiers are convenient for differential antennas, i.e., all versions of dipole antennas. The noise
performances of these amplifiers will be compared later on.

The input impedance of the common-gate amplifier shown in Fig. 4.39a is the parallel combination
of $1/g_m$ and the input capacitance of the transistor. According to (4.78), the real part of the input
impedance is approximately equal to $1/g_m$ for $f_0 \ll f_T$, where f_0 and f_T are the frequency of operation
of the amplifier and the high-frequency figure of merit of the transistor, respectively. The real part of
the input impedance can be made equal to the real part of the antenna impedance, with an appropriate
design of the aspect ratio and the quiescent current of the transistor. The output resistance is high, as
derived in Sect. 3.3, and its loading on the output resonance circuit is negligible. Since there are no

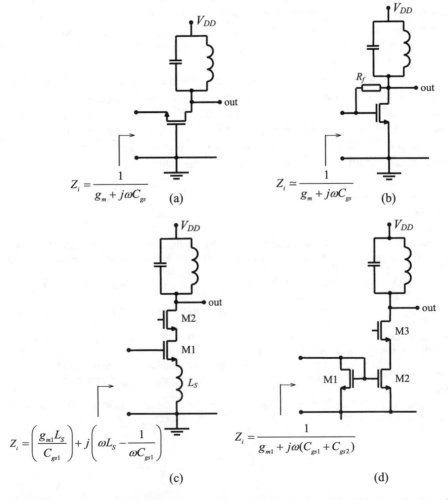

Fig. 4.39 The basic circuits suitable to use as an LNA. (**a**) The common gate amplifier, (**b**) the reduction of the input
impedance of a common source amplifier with parallel voltage feedback, (**c**) the source degenerated and cascoded LNA,
(**d**) the current mirror input cascoded circuit

direct parasitics between the output and input nodes, the internal feedback also is negligible. Consequently, the common gate amplifier must be considered as one of the usable candidates for LNAs.

It can be shown that the input impedance of a parallel voltage feedback amplifier as shown in Fig. 4.39b becomes equal to the parallel combination of $1/g_m$ and the input capacitance of the transistor, provided that $g_{ds} \ll g_m$ and $(1/R_F) \ll g_m$. This means that it is a low input impedance circuit. But it must be kept in mind that the parallel voltage feedback also decreases the output impedance of the amplifier. Therefore, due to the excessive loading of the output resonance circuit, this is not a good candidate for a tuned LNA, but can be used as a wide-band LNA with an appropriate resistive load.

The circuit shown in Fig. 4.39c and its differential versions are the circuits most widely used as LNAs. The use of the cascode configuration to reduce the output to input feedback is useful, even necessary, as explained in Sect. 3.4. Since the cascode circuit exhibits a high output resistance, the loading of the output resonance circuit is also negligible. As already derived in Sect. 3.6, the input impedance of the circuit has a real part equal to $g_{m1}L_S/C_{gs1}$ that can be designed to obtain a resistance value equal to the real part of the antenna impedance with appropriate g_m and L_S values.

The reactive components of the input impedance can be eliminated by resonance together with the reactive part of the antenna impedance and an additional series reactance, as shown in Fig. 4.38a. An additional advantage of the circuit is the voltage increase at the G-S port of the transistor due to the series resonance, as explained in Sect. 4.1.2.

The input impedance of the circuit proposed in Fig. 4.39d is determined by M1, of which the input impedance is the parallel combination of $1/g_m$ and the input capacitances of M1 and M2. M2 is the second transistor of the current mirror that is cascoded with M3 to reduce the internal feedback. The gain of the circuit can be enhanced when the current transfer ratio of the current mirror is higher than unity. The obvious disadvantage of the circuit is higher power consumption due to the second DC current path and the additional noise due to the third transistor.

References

1. R.P. Feynman, *Lectures on Physics*, vol. 1 (Addison Wesley Publishing Co., 1963), pp. 23–1 to 23–9
2. W.B. Kuhn, F.W. Stephenson, A. Elshabini-Riad, A 200 MHz CMOS Q-enhanced LC bandpass filter. IEEE J. Solid State Circ. **31**(8), 1112–1122 (1996)
3. T. Soorapanth, S.S. Wong, A 0-dB IL 2140 ± 30 MHz bandpass filter utilizing Q-enhanced spiral inductors in standard CMOS. IEEE J. Solid State Circ. **37**(5), 579–586 (2002)
4. M. E. van Walkenburg, *Analog Filter Design* (The Oxford Series in Electrical & Computer Engineering), (Oxford University Press Inc., USA, 1996)
5. Reference Data for Radio Engineers, 4th edn. (International Telephone and Telegraph Corp., 1956)
6. B.D.H. Tellegen, The gyrator, a new electric network element. Philips Res. Rep. **3**(2), 81–101 (1948)
7. L.W. Huelsman, *Active Filters: Lumped, Distributed, Integrated, Digital and Parametric* (McGraw-Hill, 1970)
8. A.G.J. Holt, J. Taylor, Method of replacing ungrounded inductances by grounded gyrators. Electron. Lett. **1**, 105 (1965)
9. D. Z, K. Gong, J.S. Fu, B. Gao, Z. Feng, A compact planar inverted-F antenna with a PBG-type ground plane for Mobile communications. IEEE Trans. Veh. Technol. **52**(3), 483–489 (May 2003)

Noise and Linearity in Amplifiers

At this point, it is worthwhile to examine the notion of noise and the modeling of noise in fundamental circuits. Amplifiers are used to enhance the signal generated by a weak signal source; for example, the signal generated by a microphone (which is proportional to the sound pressure), connected to the input of an audio amplifier, or the signal at the output of an antenna applied to the input of a receiver, which is proportional to the electromagnetic field strength at the tuning frequency of the receiver. In addition to the original signal to be amplified, there is always a physical phenomenon that disturbs the amplified signal in a certain way: for example, the high-frequency "hiss" that we hear from the loudspeaker connected to the output of the audio amplifier when the input signal level is low, or the random speckles we see on the screen of a TV receiver when we receive from a distant transmitter. This phenomenon is called "noise", and is generated in all kinds of resistors and devices like diodes and transistors, due to the random motions of electrons. In the scope of this book, we will concentrate on the noise of LNAs. But it must be kept in mind that the physical mechanisms and the basic definitions are common for all kinds of amplifiers.

The noise generated in an amplifier mainly originates from the random movements of charge carriers in resistors and devices, due to their thermal energy. Apart from this "thermal noise," there are several other types of noise: the partition noise, the multiplication noise (being the most important one), the flicker (or $1/f$) noise, etc.

Thermal noise was first observed and measured by J. B. Johnson in 1928, and interpreted by H. Nyquist who derived the following expression giving the value of the noise power in a conductor due to the random movements of electrons:

$$P_n = 4k\text{TB} \tag{5.1}$$

where k is the Boltzman constant (1.36×10^{-23} joules/K), T is the temperature of the conductor in K, and B is the bandwidth of interest. B can be placed anywhere on the frequency axis, indicating the "white" (frequency-independent) character of the noise. Expression (5.1) shows that the noise power is the same for all conductors for a certain temperature and for a bandwidth of interest, *regardless of the material and shape of the conductor.*

It must not be overlooked that the noise voltages and currents of the resistors and transistors in a circuit are processed (amplified, fed-back, mixed, etc.) just like other signals in the circuit, until they reach the output of the circuit. Hence, their contributions to the total noise power at the output must be calculated in terms of the mean square values and combined additively to find the total noise power.

The original version of this chapter was revised. The correction to this chapter is available at https://doi.org/10.1007/978-3-030-63658-6_8

D. Leblebici, Y. Leblebici, *Fundamentals of High Frequency CMOS Analog Integrated Circuits*, https://doi.org/10.1007/978-3-030-63658-6_5, corrected publication 2021

5.1   Thermal Noise of a Resistor

If the electrical resistance of a conductor is R, the noise power can be expressed in terms of the mean square noise voltage between the terminals of the resistor, or the mean square noise current flowing through the resistor as

$$P_n = \overline{i_n^2} \times R \ \text{ or } \ P_n = \overline{v_n^2} \times \frac{1}{R}$$

Consequently, the root mean square (effective) values of the noise current and the noise voltage can be written as

$$\bar{i}_n = \sqrt{4k\text{TB}\frac{1}{R}} \ \text{ and } \ \bar{v}_n = \sqrt{4k\text{TB} \times R} \tag{5.2}$$

and the noise equivalent circuit of a resistor can be drawn as seen in Fig. 5.1.

   To ease the comparison of noise behavior of different devices (or circuits) and to conform with the existing noise measurement systems that usually measure the noise in a narrow band, it is common practice to express the noise for a bandwidth of 1 Hz. Hence, the mean square noise voltage and the mean square noise current for a 1 Hz bandwidth can be written as

$$\overline{v_n^2}\Big|_{B=1Hz} = 4kT \times R = S_v \ \text{ and } \ \overline{i_n^2}\Big|_{B=1Hz} = 4kT/R = S_i \tag{5.3}$$

which is called the "spectral density" of the mean square noise voltage and noise current, respectively.

Problem 5.1

(a) *Calculate the thermal noise voltage and thermal noise current of a 50 ohm resistor (i) for 30 °C, (ii) for 100 °C (the bandwidth of interest is 10 MHz).*

(b) *Calculate the thermal noise voltage and thermal noise current of a 1000 ohm resistor for the same temperature and the same bandwidth.*

(c) *Compare and discuss the results.*

5.2   Thermal Noise of a MOS Transistor

The thermal noise of a MOS transistor was first investigated and modeled by A. van der Ziel in 1986 [1]. In the physical structure of a MOS transistor, there are several "resistances" as shown in Fig. 5.2, all of which generate noise according to (5.2).

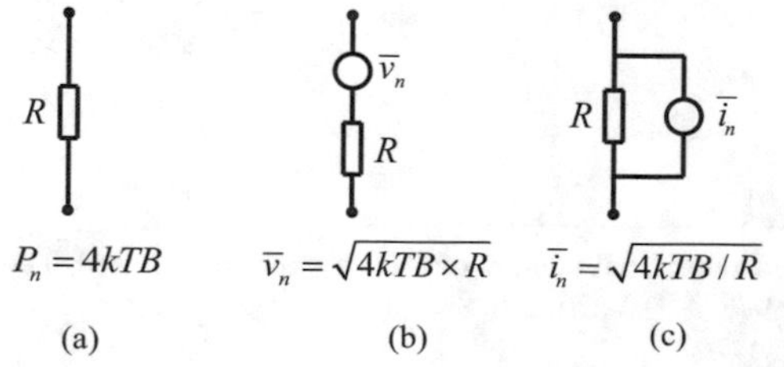

Fig. 5.1 (a) A resistor and the noise equivalent circuit with the noise voltage source (b), and with the noise current source (c)

$$P_n = 4kTB \qquad \bar{v}_n = \sqrt{4kTB \times R} \qquad \bar{i}_n = \sqrt{4kTB/R}$$

(a)   (b)   (c)

Fig. 5.2 The MOS
transistor with its noise-
generating resistances

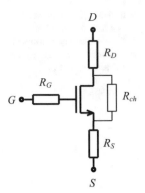

The total resistance between the external source node (S in Fig. 5.2) and the external drain node (D) of the transistor is the sum of the source series resistance (R_S), the channel resistance (R_{ch}), and the drain series resistance (R_D). The source series resistance (and similarly the drain resistance) is the sum of the intrinsic and extrinsic components and the equivalent contact resistance, as explained in Chap. 1. The source and drain series resistances are obviously technology- and geometry-dependent and can be more than one hundred ohms for small transistors and several ohms for large transistors. The gate series resistance, another noise source, is also technology- and geometry-dependent and can be minimized with appropriate finger structures [2].

It is known that the strongly dominant part of the MOS transistor noise is the thermal noise associated with the channel resistance [3]. In several publications, this noise is investigated as the sum of the noises of the pre-pinch-off region of the channel and that of the pinched-off region [4–6]. On the other hand, it has been shown that the effect of V_{DS}, and consequently the contribution of the pinched-off region on the drain current noise, is negligible [7, 8].

In this section, the noise associated with the channel resistance of a MOS transistor operating in the saturation region will be derived with a similar but more straightforward approach based on the expression derived in Chap. 1 for the inversion charge.

The cross-section of a non-velocity saturated NMOS transistor in the saturation region is shown in Fig. 5.3. According to (1.7), the effective gate voltage (and the inversion charge that is proportional to the effective gate voltage) varies with the square root of y, up to the pinch-off point L', which corresponds to a channel voltage of ($V_{GS} - V_{TH}$). Along the saturation region (from L' to L) electrons travel with the saturation velocity v_{sat}, and the inversion charge density is constant (see Fig. 1.6). It is obvious that these two sections of the channel resistance shown with r_1 and r_2 are different in nature and (as mentioned above), from the point of view of noise, the strongly dominant part of the channel resistance is r_1, as mentioned above. The value of r_1 corresponding to R_{ch} can be calculated based on the modified gradual channel approach developed in Chap. 1. The resistance of a channel element dy before the pinch-off point can be written as

$$dr(y) = \frac{dV_c}{I_D} \tag{5.4}$$

and the drain current, which is constant along the channel, is

$$I_D = \frac{d\overline{Q}_i(y)}{dt} \tag{5.5}$$

where $d\overline{Q}_i(y)$ is the inversion charge in the dy channel element that was calculated in (1.37) as

Fig. 5.3 The cross-section of a MOS transistor in saturation and components of the series resistances of the drain current path

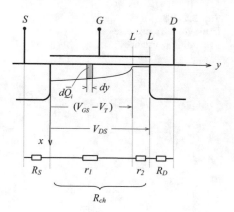

$$d\overline{Q}_i(y) = -C_{\text{ox}}W(V_{\text{GS}} - V_T)\sqrt{1 - \frac{y}{L'}}dy$$

dt in (5.5) can be written as

$$dt = \frac{dy}{v} = \frac{dy}{\mu E(y)} = \frac{dy}{-\mu(dV_c/dy)} \tag{5.6}$$

From (5.4), (5.5), and (1.37) the resistance of a channel element dy can be calculated:

$$dr(y) = \frac{dy}{\mu C_{\text{ox}}W(V_{\text{GS}} - V_T)\sqrt{1 - \frac{y}{L'}}} \tag{5.7}$$

The integral of $dr(y)$ from $y = 0$ to $y = L'$ gives the value of the first (pre-pinch-off) section of the channel resistance[1]:

$$R_{\text{ch}} = \frac{2}{\mu C_{\text{ox}}\frac{W}{L'}(V_{\text{GS}} - V_T)} \tag{5.8}$$

where L' must be considered as

$$L' = L\frac{1 + \lambda(V_{\text{GS}} - V_T)}{1 + \lambda V_{\text{DS}}} \tag{5.9}$$

according to (1.14a). It must be also noted that μ in (5.8) is a function of V_{GS} as mentioned in 1.1.2.1. and Appendix-A.

Equation (5.8) can be arranged as

$$R_{\text{ch}} = \frac{(V_{\text{GS}} - V_T)}{I_D} \tag{5.10}$$

that is – not surprisingly – the DC resistance of the inversion region. If we insert $(V_{\text{GS}} - V_{\text{TH}})$ in terms of I_D, Eq. (5.10) can be rewritten as

---

[1] Note that this is equal to $2/g_{do}$, where g_{do} is the output conductance corresponding to $V_{DS} \rightarrow 0$, in the original noise expressions of van der Ziel [1].

$$R_{\text{ch}} = \sqrt{\frac{2}{I_D \mu\, C_{\text{ox}} \frac{W}{L}}}$$ (5.11)

or in terms of g_m,

$$R_{\text{ch}} = \frac{2}{g_m}$$ (5.12)

These expressions can be interpreted as follows:

- R_{ch} decreases with the square root of the drain current, I_D.
- R_{ch} decreases with the square root of the aspect ratio.
- R_{ch} decreases with the square root of the mobility. The bias dependence of the mobility must not be ignored for small-geometry devices.
- Since the mobility of holes is considerably smaller than the mobility of electrons, the channel inversion resistance of a PMOS transistor is higher than that of an NMOS transistor having the same geometry and the same drain current.
- R_{ch} decreases with the square root of C_{ox}. This means that the channel inversion resistance is inferior for a smaller-geometry transistor with the same aspect ratio.

Now the noise current corresponding to the channel resistance of a MOS transistor can be written as

$$\overline{i^2}_{n-\text{ch}} = \frac{4kTB}{R_{\text{ch}}} = 4kTB \times \frac{1}{2} g_m \ \left[\text{A}^2\right]$$ (5.13)

It has been experimentally found and widely accepted that it is more realistic to modify this expression as:

$$\overline{i^2}_{n-\text{ch}} = 4kTB \times \gamma \times g_m$$ (5.13a)

where γ ranges from 0.5 to 1.5, depending on the depth of saturation and the channel geometry, being 0.5 for large (W/L) ratios and higher for smaller geometries.

The interpretation of (5.13) for a non-velocity saturated transistor with

$$g_m = \sqrt{2\mu C_{\text{ox}} \frac{W}{L} I_D}$$ (1.33)

shows us that the channel current noise:

- Increases with the drain DC current
- Increases with the (W/L) ratio
- Increases for smaller geometry devices due to higher C_{ox} values
- Is related to the carrier mobility, which is lower for a PMOS transistor than for an NMOS transistor having the same geometry and the same drain current.

Similarly, for a velocity saturated transistor with[2]

$$g_m = kW C_{\text{ox}} v_{\text{sat}}$$ (1.34)

the channel current noise:

---

[2] It is assumed that the expression derived for the non-velocity saturated regime is valid for the velocity-saturated regime.

- Does not depend on the drain DC current
- Increases with W
- Increases for smaller geometry devices due to higher C_{ox} values
- Is almost equal for NMOS and PMOS transistors, since the velocity saturation values of holes and electrons are approximately the same.

As an example, the variation of R_{ch} and S_{ich} for an AMS 035 ($W = 35$ μm, $L = 0.35$ μm) transistor as a function of the drain current are plotted in Fig. 5.4. Figure 5.4a shows that the inversion channel resistance acquires considerably low values for moderate to high drain currents. Therefore, if the series source and drain resistances are not sufficiently small compared to R_{ch}, they must not be ignored.

Since the noise contributions of the parasitic internal resistances (R_S, R_D, and R_G) depend not only on the transistor but also on the circuit, they must be separately investigated for different circuit configurations. R_D can be considered as part of the effective load resistance. To gain further insight on the contribution mechanisms of R_S and R_G, we will calculate their effects for the most basic common source amplifier.

The source series resistance has two effects on the noise behavior of the transistor:

(a)   The noise voltage, $\bar{v}_{nRS1}$, generated by the source series resistance of the transistor (the thermal noise of R_S). This voltage, shown in Fig. 5.5a adds a component to the drain noise current equal

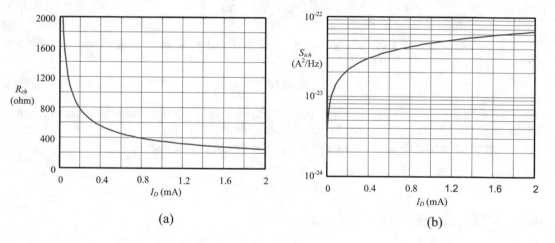

Fig. 5.4   (a) The inversion channel resistance and (b) the drain noise current spectral density of a 35 μm/0.35 μm AMS transistor. (calculated with $\gamma = 0.5$, and the V_{GS} dependence of μ is taken into account)

Fig. 5.5   (a) The noise voltage of the source series resistance. (b) The noise voltage due to the total drain noise current flowing through R_S

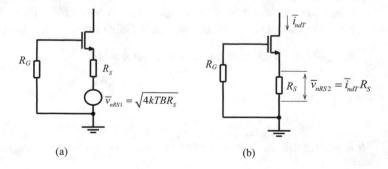

Fig. 5.6 The small-signal equivalent circuit used to calculate y_{mSD}

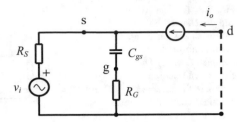

to $\bar{i}_{ndS1} = y_{mSD} \cdot \bar{v}_{nRS1}$, where y_{mSD} is the transadmittance from $\bar{v}_{nRS1}$ to the drain current (See Fig. 5.6):

$$y_{mSD} = -\frac{g_m}{R_S(g_m + j\omega C_{gs}) + (1 + j\omega C_{gs}R_G)}$$

$$= -\frac{g_m}{g_m R_S\left(1 + j\dfrac{\omega}{\omega_T}\right) + (1 + j\omega C_{gs}R_G)}$$

which is real and equal to

$$g_{mSD} \cong -\frac{g_m}{g_m R_S + 1} = -g_{m(eff)} \tag{5.14}$$

provided that $\omega \ll \omega_T$ and $\omega \ll 1/C_{gs}R_G$. The drain noise current component originating from the noise of R_S becomes

$$\bar{i}_{ndS1} = -\bar{v}_{nRS1}g_{m(eff)}$$

This noise component is totally uncorrelated to the noise originating from the channel resistance and therefore must be added to the mean square channel noise as

$$\bar{i}_{ndS1}^2 = \bar{v}_{nRS1}^2 \cdot g_{m(eff)}^2 = 4kTBR_S \cdot g_{m(eff)}^2 \tag{5.15}$$

and the total mean square noise current (with the thermal noise of R_S included) becomes

$$\bar{i}_{nd}^2 = \bar{i}_{n-ch}^2 + \bar{i}_{ndS1}^2 = 4kTB\left[\gamma g_m + R_S g_{m(eff)}^2\right] \tag{5.16}$$

(b) The additional effect of the noise voltage drop $\bar{v}_{nRS2} = \bar{i}_{nd}R_S$ on R_S due to the noise current flowing through R_S. This noise voltage provokes a drain noise current equal to

$$\bar{i}_{ndS2} = y'_{mSD} \cdot \bar{i}_{nd}R_S \tag{5.17}$$

where y'_{mSD} is the transadmittance from $\bar{v}_{nRS2}$ to the drain current.

$$y'_{mSD} = -\frac{g_m}{(1 + j\omega C_{gs}R_G)} \tag{5.18}$$

which is

$$y'_{mSD} \cong -g_m \tag{5.18a}$$

provided that $\omega \ll 1/C_{gs}R_G$. From (5.17) and (5.18a), the noise current component originating from the noise voltage drop on R_S becomes

Fig. 5.7 (a) The circuit and the channel noise current of the transistor, the thermal noise of R_S included. (b) The equivalent circuit to calculate the feedback modified value of the noise current

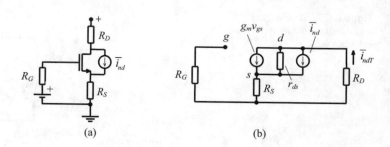

(a)   (b)

$$\bar{i}_{ndS2} = -g_m \cdot \bar{i}_{ndT} R_S \tag{5.19}$$

that is apparently related to the noise current originating from the channel resistance and is out of phase with it. Therefore, it reduces the channel noise current. This fact leads to the "noise cancelling feedback" concept that will be investigated in Sect. 5.5.3 with more detail. The channel noise current (with this feedback effect included) can be calculated from Fig. 5.7 as

$$\bar{i}_{ndT} = \bar{i}_{nd} \frac{1}{(1 + g_m R_S) + \frac{R_D + R_S}{r_{ds}}} \cong \bar{i}_{nd} \frac{1}{(1 + g_m R_S)} \tag{5.20}$$

and its mean square value is:

$$\bar{i}^2_{ndT} \cong \bar{i}^2_{nd} \frac{1}{(1 + g_m R_S)^2} \tag{5.20a}$$

It must be noted that, if there is any other noise component not correlated with the channel noise, it must be added to $\bar{i}^2_{nd}$ before the inclusion of the feedback effect.

Another important parasitic resistance of the MOS transistor is the gate resistance, whose value is strongly dependent on layout and process parameters.

The gate resistance generates a thermal noise voltage equal to $\bar{v}_{nRG1} = \sqrt{4kTBR_G}$ which is in the input loop of the transistor and provokes a drain noise current component equal to

$$\bar{i}_{ndG1} = \left| g_{m(\text{eff})} \right| \bar{v}_{nR_G} \tag{5.21}$$

where $g_{m(\text{eff})}$ is the effective transconductance defined with (2.13b). If there is an impedance (Z_S) connected in series to R_S, (2.13b) must be modified as

$$g_{m(\text{eff})} = 1/[(1 + g_m(R_S + Z_S))].$$

$\bar{i}_{ndG1}$ is a thermal (white) noise and is not correlated to the channel noise, it must therefore be added with the mean square value to the channel noise. Together with this component, (5.16) becomes:

$$\bar{i}^2_{nd} = \bar{i}^2_{n-ch} + \bar{i}^2_{ndS1} + \bar{i}^2_{ndG1} = 4kTB \left[\gamma g_m + g^2_{m(\text{eff})} (R_S + R_G) \right] \tag{5.16a}$$

It must be noted that if there is an external resistance connected in series to the gate, it must be considered together with the inherent gate resistance of the transistor.

There is another noise component related to the gate resistance. At high frequencies, a noise current owing to the noise voltage on the inversion channel flows over the gate capacitance. This noise current $(\bar{i}_{ng})$ is not "white" since it increases with frequency, and is correlated to the inversion

Fig. 5.8 (a) The noise
voltage source due to the
gate series resistance. (b)
The input noise voltage
source due to the gate noise
current

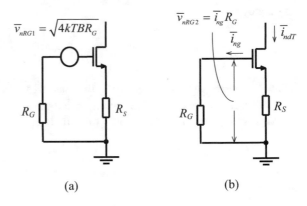

(a)   (b)

Fig. 5.9 The gate noise
current as the sum of the
incremental components

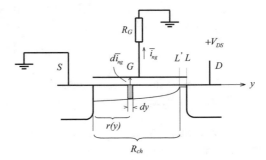

channel noise. The noise voltage drop on R_G due to this current, $\bar{v}_{nRG2} = \bar{i}_{ng}R_G$, is another noise voltage source in the input loop and adds another component to the drain noise current (see Fig. 5.8b):

$$\bar{i}_{ndG2} = g_{m(\text{eff})}\bar{v}_{nRG2} = g_{m(\text{eff})}\bar{i}_{ng}R_G \tag{5.22}$$

The approach used to calculate the gate noise current ($\bar{i}_{ng}$) is shown in Fig. 5.9. The noise voltage of a channel element dy at position y is

$$\bar{v}_n(y) = \bar{i}_{ndT}r(y)$$

where $\bar{i}_{ndT}$ is the channel noise current and $r(y)$ is the resistance of the channel segment from the source end of the channel ($y = 0$) to y, which is equal to the integral of dr given with (5.7), from zero to y:

$$r(y) = \frac{2}{\mu C_{\text{ox}}\dfrac{W}{L'}(V_{\text{GS}} - V_T)}\left(1 - \sqrt{1 - \frac{y}{L'}}\right)$$

$\bar{v}_n(y)$ induces an incremental noise current over the incremental capacitance $dC_g = C_{\text{ox}}Wdy$, which is

$$d\bar{i}_{ng} = j\omega(dC_g) \times \bar{v}_n(y) = j\omega(C_{\text{ox}}Wdy) \times r(y)\bar{i}_{ndT}$$

and hence,

$$\bar{i}_{ng} = j\omega C_{ox} W \frac{2}{\mu C_{ox} \frac{W}{L'}(V_{GS} - V_T)} \int_0^{L'} \left(1 - \sqrt{1 - \frac{y}{L'}}\right) dy \times \bar{i}_{ndT} \tag{5.23}$$

It must be noted that $\bar{i}_{ng}$ is related to the channel noise current, but there is a 90° phase difference. Therefore, it does not contribute to any noise cancelling feedback.

For $L \cong L'$, and from (1.33) and (1.42) this expression can be arranged as[3]

$$\left|\bar{i}_{ng}\right| \cong \omega \frac{C_{gs}}{g_m} \times \bar{i}_{ndT} \cong \frac{\omega}{\omega_T} \times \bar{i}_{ndT} \tag{5.24}$$

Hence $\bar{i}_{ndG2}$ becomes

$$\bar{i}_{ndG2} = g_{m(\text{eff})}\bar{v}_{nRG2} = g_{m(\text{eff})} \times \left(\omega \frac{C_{gs}}{g_m} \times \bar{i}_{ndT}\right) R_G \tag{5.25}$$

and its mean square value, which must be added to the channel noise:

$$\bar{i}^2_{ndG2} = \left(\frac{\omega C_{gs} R_G}{1 + g_m R_S}\right)^2 \times \bar{i}^2_{ndT} \tag{5.26}$$

(5.26) can be interpreted as follows.

The drain current noise component associated with the gate noise current:

- Increases with frequency,
- Increases with C_{gs},
- Is proportional to the total channel noise,
- Is a capacitive current and has a 90° phase difference with the channel noise current, and therefore does not contribute to any noise cancelling feedback.
- Must be added to the channel noise with its mean square value.

It can be seen that this noise component is usually very small and negligible except for very small oxide thicknesses and very high frequencies. Therefore, the actual output noise on the drain current (including noise feedback over R_S) can be calculated from (5.16) and (5.20a) as

$$\bar{i}^2_{ndT} \cong \bar{i}^2_{nd} \frac{1}{(1 + g_m R_S)^2}$$

and then $\bar{i}^2_{ndG2}$ can be added if it is not negligibly small.

Example 5.1

In the following, the contribution of different components of the drain noise current will be calculated for a 35 μm (7×5 μm)/0.35 μm AMS NMOS transistor at $f = 3$ GHz for 10 MHz bandwidth. The DC operating point of the transistor is $I_D = 2$ mA, $V_{DS} = 3$ V. The temperature of the transistor will be assumed as 300 K.

---

[3] It must be noted that this derivation is valid provided that $(1/\omega C_{gs}) \gg r_d$, in other words, the noise current deviated to the gate is sufficiently smaller than the drain noise current.

The parasitic source and drain series resistances of a 5 μm/0.35 μm transistor were calculated as 134 ohm each, in Example 1.3. The 35 μm width transistor is composed of seven parallel 5 μm channel regions in the form of a multi-finger transistor similar to the one shown in Fig. 1.28. Since source and drain regions are shared by the neighboring channel region and since they are connected in parallel, the source resistance of the 35 μm transistor is approximately $134/14 \approx 10$ ohm.

Since the poly gate sheet resistance for this technology is given as $R_{sh} = 7$ ohm/□, the resistance of one of the gate stripes is $7 \times (5 / 0.35) = 100$ ohm. If the gate stripes are connected in parallel as shown in Fig. 1.28, the equivalent resistance is $100/7 = 14.3$ ohm. Poly to metal contact resistance is given as 2 ohm / (0.4×0.4) micron contact. Assuming 10 contacts along the collecting stripe, the equivalent contact resistance is 0.2 ohm. Together with the resistance of the collecting stripe, the total resistance series to the gate can be taken as 15 ohm.

The gate-to-source voltage for 2 mA drain current and the mobility corresponding to this voltage can be found as 1 V and 325 cm^2/ V.s, respectively. The transconductance of the transistor for 2 mA drain current is

$$g_m = \sqrt{2 \times 325 \times (4.54 \cdot 10^{-7}) \times (2 \cdot 10^{-3})} = 7.53 \cdot 10^{-3} \simeq 7.5 \text{ mS}$$

and the effective transconductance with $R_S = 10$ ohm,

$$g_{m(\text{eff})} = \frac{7.5 \cdot 10^{-3}}{1 + 7.5 \cdot 10^{-3} \times 10} = 6.98 \cdot 10^{-3} \cong 7 \text{ mS}$$

With

$$4kTB = 4 \times (1.38 \times 10^{-23}) \times 300 \times (10 \times 10^6) = 1.656 \times 10^{-13} \ [W]$$

the inversion channel noise of the transistor from (5.13a):

$$\bar{i}^2_{n-\text{ch}} = 4kTB\gamma g_m = (1.656 \times 10^{-13}) \times 0.5 \times (7.5 \times 10^{-3}) = 6.21 \times 10^{-16} \ [A^2]$$

The drain current noise component related to the thermal noise of R_S from (5.15):

$$\bar{i}^2_{ndS1} = 4kTBR_S g^2_{m(\text{eff})} = (1.656 \times 10^{-13}) \times 10 \times (7 \times 10^{-3})^2 = 0.81 \times 10^{-16} \ [A^2]$$

The drain current noise component related to the thermal noise of R_G from (5.21):

$$\bar{i}^2_{ndG1} = g^2_{m(\text{eff})} \bar{v}^2_{nR_G} = g^2_{m(\text{eff})} \times 4kTBR_G = (7 \times 10^{-3})^2 \times (1.656 \times 10^{-13}) \times 15 = 1.217 \times 10^{-16} \ [A^2]$$

The sum of all these components according to (5.16a)

$$\bar{i}^2_{nd} = \bar{i}^2_{n-\text{ch}} + \bar{i}^2_{ndS1} + \bar{i}^2_{ndG1} = 6.21 \times 10^{-16} + 0.81 \times 10^{-16} + 1.217 \times 10^{-16} = 8.237 \times 10^{-16} \ [A]$$

The drain mean square noise current after noise feedback over R_S from (5.20a):

$$\bar{i}^2_{ndT} \cong \bar{i}^2_{nd} \frac{1}{(1 + g_m R_S)^2} = (8.237 \times 10^{-16}) \frac{1}{[1 + (7.5 \times 10^{-3}) \times 10]^2} = 7.125 \times 10^{-16} \ [A^2]$$

Now the noise component originating from the capacitive gate noise current must be calculated from (5.26). Using $A = 2.47 \times 10^{11}$ calculated for this transistor,

Fig. 5.10 The drain current noise spectral density of a AMS 35 μm/0.35 μm NMOS transistor. (**a**) of only the channel inversion resistance, (**b**) with the contributions of R_S and R_G (for $V_{DS} = 2$V, $f = 3$ GHz, and $T = 300$K)

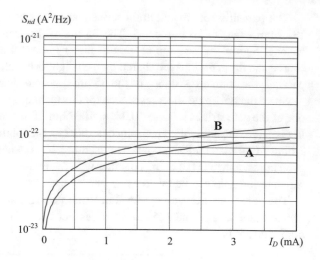

$$\overline{i^2_{ndG2}} = \left(\frac{\left(2\pi \times 3 \times 10^9\right)\left(39 \times 10^{-15}\right) \times 15}{1 + \left(7.5 \times 10^{-3}\right) \times 10} \right)^2 \times \overline{i^2_{ndT}} = 0.0001 \times \overline{i^2_{ndT}}$$

it can be seen that this is very small compared to $\overline{i^2_{ndT}}$, therefore negligible.

This example shows that the series resistances of the gate and source electrodes considerably affect the total mean square drain noise. Therefore, utmost care is necessary on the layout of the device to keep the external series resistances as low as possible.

The variation of the drain noise spectral density as a function of the drain DC current of the same transistor has been calculated, taking into account the variations of the mobility, and plotted with and without contributions of the resistive device parasitics (see Fig. 5.10).

Another aspect that must not to be overlooked is the "temperature." The "T" in the noise expressions is the temperature of the resistance, in the case of the MOS transistor the temperature of the channel region can be considerably higher than the ambient temperature and the average surface temperature of the die[4]. This fact imposes higher noise for higher power densities, and consequently higher device temperature. For example, at 400 K mean square noise currents increase approximately 33%. In Fig. 5.11, the total drain noise current spectral density of the transistor in Example 5.1 is plotted for 300K and 400K.

Problem 5.2

Derive an expression to calculate approximately the gate noise current assuming that the noise voltage along the channel is constant and equal to the noise voltage at the mid-point of the channel. Compare the result with (5.24) and discuss.

---

[4] From the publications related to the thermal simulation and mapping of ICs [9, 10], it can be seen that the temperatures of the directly heated micro-regions (hot spots) can be considerably higher than the average temperature of the die.

Fig. 5.11 The drain current noise spectral density of the transistor in Example 5.1 as a function of the drain DC current: (B) for $T = 300K$, (C) for $T = 400K$

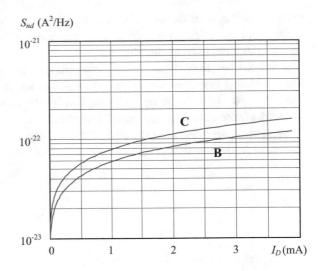

S_{nd} (A^2/Hz)

Problem 5.3

The die area of an integrated circuit is 10 mm^2 and its thickness is 0.5 mm. The die is mounted into an Amkor MLF, 44 lead miniature package, whose thermal resistance from the ambient to the bottom of the die is 24 °C/W. The power consumption of the circuit is 1W and the ambient temperature is 30 °C. Calculate the average surface temperature of the die (The specific thermal conductance of silicon is 1.5 W/cm°C).

Answer: 54.33 °C! (It is obvious that the temperature of the channel regions of the individual MOS transistors on the die are considerably higher than this value, depending on their power densities.)

5.3   The Metrics of the Noise Performance of Amplifiers

The total signal received at the input of the amplifier does not only consist of the signal sent by the transmitter[5], but also includes the unavoidable noise signal originating from the internal resistance of the signal source. To obtain a sufficiently high level of signal power with a reasonable signal-to-noise ratio (S / N) at the output of the amplifier, the noise inherently generated in the amplifier must be kept as low as possible.

The noise performance of an amplifier is usually expressed by the "noise factor," F, which is defined as

$$F = \frac{(S/N)_{input}}{(S/N)_{output}} = \frac{S_{in}/N_{in}}{S_{out}/N_{out}} \tag{5.27}$$

---

[5] In addition to the signal of interest sent by the transmitter, certain natural (atmospheric, cosmic, etc.) and man-made (originating from switching of power lines, corona discharges, etc.) noises can reach the input of the amplifier. Since these "external" noises are all sporadic and, in some cases, avoidable to some extent, they will be kept out of this discussion.

where S_{in} and S_{out} are the signal powers at the input and output of the amplifier, N_{in} is the noise power at the input originating from the internal resistance of the signal source and N_{out} is the total noise power at the output of the amplifier.

The LNA amplifies the incoming signal and the input noise generated in the signal source equally. For a noiseless amplifier, the output signal power and the output noise power are equal to the input signal and the input noise power multiplied by the power gain, respectively. Therefore, the noise factor of a noiseless amplifier is equal to unity. In a noisy amplifier, on the other hand, the output noise is the sum of the gain times the input noise and the output noise component representing the noise generated in the amplifier. From this consideration and (5.27), the noise factor can be written as

$$F = \frac{S_{\text{in}}}{S_{\text{out}}} \frac{N_{\text{out}}}{N_{\text{in}}} = \frac{S_{\text{in}}}{A_p \times S_{\text{in}}} \frac{\left(A_p \times N_{\text{in}}\right) + N_{\text{amp}}}{N_{\text{in}}} \tag{5.28}$$

where

- A_p is the power gain of the amplifier;
- N_{in} is the input noise power, originating from the resistance of the signal source;
- N_{amp} is the noise power at the output, originating solely from the amplifier itself, (i.e. excluding the noise of the resistance of the signal source).

Now (5.28) can be simplified as

$$F = 1 + \frac{N_{\text{amp}}}{A_p \times N_{\text{in}}} = 1 + \frac{N_{\text{amp}}}{N_{o(\text{in})}} \tag{5.29}$$

where $N_{o(\text{in})}$ denotes the noise power at the output related to the input noise power.

The "noise figure," NF, which is also used to express the noise performance of an amplifier, is the noise factor expressed in logarithmic scale:

$$NF(dB) = 10 \times \log F \tag{5.30}$$

The basic noise factor definition given in (5.27) is formulated in terms of signal and noise powers. But in electronic amplifiers, these quantities are not directly measurable but usually expressed in terms of voltages that can be measured. Therefore, it is useful to express the noise factor in terms of voltages instead of powers.

Figure 5.10 shows the block diagram of an amplifier whose voltage gain is A_v, input impedance is z_i, and load resistance is R_L. Note that if the amplifier has a tuned load, R_L must be replaced with the equivalent parallel resistance of the load at resonance.

The noise voltage originating from the source resistance $\bar{v}_{nA}$ is divided between R_A and z_i and then amplified. Therefore, the noise voltage and the noise power at the output originating from the source resistance are

$$\bar{v}_{noA} = \bar{v}_{nA} \frac{z_i}{R_A + z_i} \times A_v$$

$$P_{noA} = \frac{\bar{v}_{noA}^2}{R_L} = \bar{v}_{nA}^2 \left(\frac{z_i}{R_A + z_i}\right)^2 \frac{1}{R_L} A_v^2 \tag{5.31}$$

respectively. If the total noise voltage at the output of the amplifier originating only from the amplifier (the inherent thermal noise of r_L included) is $\bar{v}_{no(\text{amp})}$, the noise power originating from the amplifier at the output is

$$P_{no(\text{amp})} = \frac{\overline{v}^2_{no(\text{amp})}}{R_L} \tag{5.32}$$

Now the noise factor given in Eq. (5.29) can be written in terms of the quantities used in Fig. 5.10 as

$$F = 1 + \frac{P_{no(\text{amp})}}{P_{noA}}$$

With (5.31) and (5.32),

$$F = 1 + \frac{\overline{v}^2_{no(\text{amp})}}{\overline{v}^2_{nA}} \frac{1}{\left(\frac{z_i}{R_A + z_i}\right)^2 A_v^2} \tag{5.33}$$

In the case of impedance matching, i.e., $R_A = z_i$ and with $\overline{v}^2_{nA} = 4kTBR_A$, expression (5.33) becomes

$$F = 1 + \frac{\overline{v}^2_{no(\text{amp})}}{A_v^2 \cdot kTBR_A} \tag{5.33a}$$

and for $z_i \gg R_A$

$$F = 1 + \frac{\overline{v}^2_{no(\text{amp})}}{A_v^2 \cdot 4kTBR_A} \tag{5.33b}$$

These expressions provide some important hints for better noise performance for an amplifier:

- Higher internal signal source resistance (in case of an antenna, higher radiation resistance) reduces the noise factor.
- Voltage drive instead of input matching (whenever possible) is advantageous.
- When designing an amplifier, the inherent output noise voltage ($\overline{v}_{no(\text{amp})}$) must be kept as small as possible.
- The voltage gain must be made as high as possible.

In the case of a multistage amplifier as shown in Fig. 5.11, the total voltage gain and the total noise voltage at the output node must be calculated in terms of the individual gains of stages and the individual output noise voltages. The total voltage gain is

$$A_{vT} = A_{v1} \cdot A_{v2} \cdot A_{v3}$$

and the individual noise voltages reaching the output node are

$$\overline{v}_{no(3)}, \quad \overline{v}_{no(2)} \cdot A_{v(3)} \quad \text{and} \quad \overline{v}_{no(1)}\left(A_{v(2)} \cdot A_{v(3)}\right)$$

where $\overline{v}_{no(1)}$, $\overline{v}_{no(2)}$ and $\overline{v}_{no(3)}$ are the inherent output noise voltages of stages. Therefore, the total noise voltage at the output originating from the amplifier is

$$\overline{v}_{noT} = \sqrt{\overline{v}^2_{no(3)} + \overline{v}^2_{no(2)} \cdot A_{v(3)}^2 + \overline{v}^2_{no(1)} \cdot \left(A_{v(2)} \cdot A_{v(3)}\right)^2} \tag{5.34}$$

and the noise factor of the amplifier according to (5.33) is

$$F = 1 + \frac{\overline{v}^2_{noT}}{\overline{v}^2_{nA}} \frac{1}{\left(\frac{z_{i(1)}}{R_A + z_{i(1)}}\right)^2 A_{vT}^2} \tag{5.35}$$

Fig. 5.12 The noise
voltages at the input and
output of an amplifier

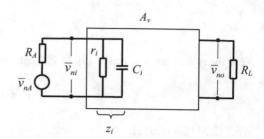

5.4   The Source Degenerated Tuned LNA and Its Noise

The circuit diagram of a typical tuned LNA is shown in Fig. 5.12. The main gain stage is a cascode circuit whose load is a parallel resonance circuit tuned to the operating frequency, ω_0. The inductor connected in series to the source (L_S) helps to obtain a low-value resistive component for the input impedance as mentioned in Sect. 3.6. The gate series inductor is used to resonate all reactive components at ω_0, and to obtain a purely resistive input impedance.

The key properties to be fulfilled by this circuit are:

- Good input matching at the tuning frequency
- Voltage gain equal to or higher than a target value
- Noise figure as low as possible at the tuning frequency
- Output swing as high as possible without the excessive nonlinear distortion that determines the usable maximum input voltage (see Sect. 5.6)
- In addition, the inductors in the circuit must be realizable in terms of their value and quality factor, for the specified fabrication technology

Another important feature of a tuned amplifier is its bandwidth. However, since the bandwidth of the tuned amplifier is imposed by the quality factor values of the on-chip inductors, setting the bandwidth as a primary design spec usually leads to unfeasible constraints. The usual approach is to accept the bandwidth of the amplifier as imposed by the quality factors of the on-chip inductors (which is usually wider than anticipated), and then rely on additional input and/or output filtering to sharpen the frequency characteristic.

Since each of these conditions mentioned above are related to the parameter values of several components in the circuit, all of these conditions influence each other. Therefore, an iterative design process becomes necessary. An analysis of the circuit leading to the calculation of the component values with some design hints is given below.

The most influential device parameter is the transadmittance g_m of the input transistor, which determines the gain of the amplifier and the noise originating from this transistor. It must be noted that the voltage gain of the circuit has two factors: the gain from the v_{gs} of the input transistor to the output, (i.e., the voltage gain of the cascode circuit, A_C), and the voltage transfer ratio from v_A to v_{gs}, which is a "series resonance gain" and equal to the value of the quality factor of the input loop (see 4.1.2.1):

$$A_R = \frac{v_{gs}}{v_A} = Q_{in} \tag{5.36}$$

In the case of impedance matching at ω_0, since the real part of z_i is equal to R_A and ($L_G + L_S$) is in resonance with the input capacitance of M1, the quality factor of the input loop is

$$Q_{\text{in}} = \frac{(L_G + L_S)\omega_0}{2R_A} \tag{5.36a}$$

which has a small value, usually slightly higher than unity.

The gain of the cascode circuit, A_C, can be written as

$$|A_C| \cong g_{m1} \times R_{\text{eff}} \tag{5.37}$$

provided that $r_{i2} \ll r_{o1}$, where R_{eff} is the impedance of the load at resonance. Since the output resistance of the cascode circuit is very high and the quality factor of the total resonance capacitance is considerably higher than that of the on-chip inductor, the impedance of the load at resonance can be written as

$$R_{\text{eff}} \cong L\omega_0 \times Q_L \tag{5.38}$$

Hence the total voltage gain becomes

$$|A_v| = A_R \cdot |A_C| \cong Q_{\text{in}}(g_{m1} \cdot R_{\text{eff}})$$
$$\cong g_{m1} \cdot Q_{\text{in}} \cdot Q_L \cdot L \cdot \omega_0 \tag{5.39}$$

Due to the high series resistance in the input loop, Q_{in} is very low and usually close to unity, as mentioned before. Therefore, the total gain can be assumed equal to A_C at the start of the design.

Since the realizable on-chip inductance values cannot exceed a certain value and the quality factor depends on the technology, the operating frequency and the value of the inductance, it is necessary to choose an appropriate L and Q_L pair to maximize A_C for a certain g_m. Using these L and Q_L values, g_m can be calculated corresponding to a target value for A_C as

$$g_{m1} = \frac{|A_C|}{L\omega_0 \times Q_L} \tag{5.40}$$

This helps to calculate W and consequently the C_{gs} parameters of the input.

For a non-velocity saturated transistor from (1.33)

$$W = \frac{g_m^2 \times L}{2\mu C_{\text{ox}} I_D} \tag{5.41}$$

and for a velocity saturated transistor from (1.34)

$$W = \frac{g_m}{k C_{\text{ox}} v_{\text{sat}}} \tag{5.41a}$$

Since g_m and the corresponding the W and C_{gs} are determined[6] (at least as a first approximation), now we can deal with the input matching:

From (3.54), the input impedance seen from the gate terminal of an inductive source degenerated transistor that is shown with z_i', can be written as

$$z_i' \cong \frac{1}{y_{\text{in}}} = \frac{1 + (g_m + sC_{gs})(sL_S + R_S)}{sC_{gs}}$$
$$= \left(\frac{g_m L_S}{C_{gs}} + R_S\right) + sL_S + \frac{1 + g_m R_S}{sC_{gs}}$$

---

[6] The calculated W values are usually very high. Therefore, a multi-finger channel structure is necessary to achieve small parasitic resistances and small source and drain parasitic capacitances.

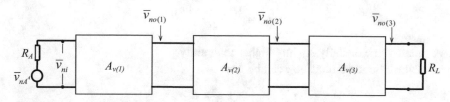

Fig. 5.13 A multistage amplifier and its noise voltages

and in the frequency domain,

$$z_i' \cong \left(\frac{g_m L_S}{C_{gs}} + R_S\right) + j\left(\omega L_S - \frac{1 + g_m R_S}{\omega C_{gs}}\right) \tag{5.42}$$

Note that $C_i' = C_{gs}/(1 + g_m R_S)$ corresponds to the input capacitance of M1, together with R_S connected in series to the source. For $g_m R_S \ll 1$, which is usually valid, C_i' is approximately equal to C_{gs}.

Now the input loop of M1 together with the signal source and the gate inductance can be drawn as shown in Fig. 5.13. In order to establish impedance matching between R_A and its load, namely z_i at the operating frequency, these two conditions must be satisfied:

$$\left(\frac{g_m L_S}{C_{gs}} + R_S + R_G\right) = R_A \tag{5.43}$$

$$\left(\omega_0(L_S + L_G) - \frac{1}{\omega_0 C_i'}\right) = 0 \tag{5.43a}$$

where $R_S = L_S \omega_0 / Q_{L_S}(\omega_0)$ and $R_G = L_G \omega_0 / Q_{L_G}(\omega_0)$.

(5.43) gives the value of L_S as

$$L_S = \frac{C_{gs}}{g_m}[R_A - (R_S + R_G)] \tag{5.44}$$

Note that $R_G = R_{LG} + R_{Gi}$, $R_S = R_{LS} + R_{Si}$, where R_{Si} and R_{Gi} are the internal parasitic source and gate series resistances and $R_{LS} = L_S \omega_0 / Q$, $R_{LG} = L_G \omega_0 / Q$.

(5.43a) indicates a series resonance between $C_i' \cong C_{gs}$ and $(L_S + L_G)$, which helps to calculate $(L_S + L_G)$ as

$$(L_G + L_S) = \frac{1}{\omega_0^2 C_i'} \cong \frac{1}{\omega_0^2 C_{gs}} \tag{5.45}$$

In many cases, the calculated $(L_S + L_G)$ is very high, in the order of tens of nH, and the L_S calculated from (5.44) is less than 1 nH. Therefore, the dominant part of the resonance inductance L_G is not realizable as an on-chip inductance. The conventional solution to reduce L_G to a realizable value is to increase C_i' with an additional capacitor C_p connected between the G and S terminals of the input transistor and return to (5.45), replace C_{gs} with $(C_{gs}+C_p)$ and continue. But related to this solution there is an interesting problem that affects the input matching.

This is related to the behavior of the C_{gs}, R_{Gi}, C_p, combination at the input of M1. Note that C_p cannot be connected directly parallel to the C_{gs} of the transistor, but before R_{Gi} as shown in Fig. 5.14a[7]. The equivalent of this combination at ω_0 is shown in Fig. 5.14b and its real and imaginary parts at ω_0 can be calculated in terms of C_{gs}, R_{Gi}, and C_p as

---

[7] Since R_{Si} is very small compared to R_{Gi} its effect is ignored.

Fig. 5.14 Most frequently
used form of the source
degenerated LNA

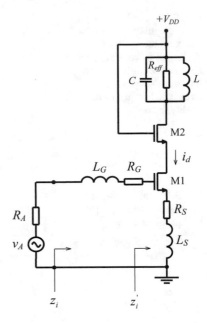

$$\text{Re}\left\{Z'_{\text{gs}}\right\} = \frac{R_{\text{Gi}}C_{\text{gs}}^2}{\left(\omega_0 C_{\text{gs}}C_p R_{\text{Gi}}\right)^2 + \left(C_{\text{gs}} + C_p\right)^2} = R_{\text{eq}} \tag{5.46}$$

$$\text{Im}\left\{Z'_{\text{gs}}\right\} = -j\omega_0 \frac{\omega_0^2 C_{\text{gs}}^2 C_p R_{\text{Gi}}^2 + \left(C_{\text{gs}} + C_p\right)}{\omega_0^2 \left[\left(\omega C_{\text{gs}}C_p R_{\text{Gi}}\right)^2 + \left(C_{\text{gs}} + C_p\right)^2\right]} \tag{5.47}$$

The corresponding equivalent capacitance is

$$C_{\text{eq}} = \frac{\left[\left(\omega_0 C_{\text{gs}}C_p R_{\text{Gi}}\right)^2 + \left(C_{\text{gs}} + C_p\right)^2\right]}{\omega_0^2 C_{\text{gs}}^2 C_p R_{\text{Gi}}^2 + \left(C_{\text{gs}} + C_p\right)} \tag{5.48}$$

that acts as the resonance capacitance of the input loop. C_p can be calculated from (5.48), which can then be inserted into (5.46) to obtain R_{eq}.

It is obvious that C_{gs} and R_{Gi} must be replaced with C_{eq} and R_{eq} in (5.45):

$$L_S = \frac{C_{\text{eq}}}{g_m}\left[R_A - \left(R_{\text{LS}} + R_{\text{Si}} + R_{\text{eq}} + R_{\text{LG}}\right)\right]$$

$$\cong \frac{C_{\text{eq}}}{g_m}\left[R_A - \left(R_{\text{Si}} + R_{\text{eq}} + R_{\text{LG}}\right)\right] \tag{5.45a}$$

provided that $R_{\text{LS}} \ll R_{\text{Si}} + R_{\text{eq}} + R_{\text{LG}}$, which is usually valid.

Fig. 5.15 The input loop
of the amplifier. R_S and R_G
represent the equivalent
series resistances of L_S and
L_G corresponding the total
losses at the operating
frequency in terms of the
quality factors

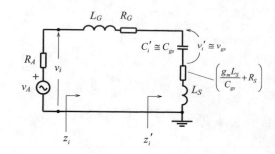

Fig. 5.16 (a) The input of
M1 together with C_p. (b)
Its equivalent at ω_0

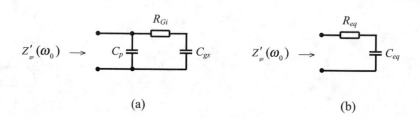

(a)   (b)

M2 is a unity gain current amplifier, insulating M1 from the load to minimize the feedback over C_{dg1}. The dimension of M2 must be determined with care, taking into consideration noise and linearity. A smaller g_{m2} (therefore a smaller gate width) reduces the contribution of M2 on the total output noise. But it must be kept in mind that the supply voltage is divided between M1 and M2, and the DC voltages of the drain nodes must be suitable for the voltage swings of these nodes without excessive nonlinear distortion. Otherwise, in addition to the self-nonlinearities of M1 and M2, the intrusion of the negative peaks of the output signal into the drain voltage of M1 will severely limit the usable output swing, as will be shown in Example 5.2.

Example 5.2

Design a tuned LNA using 0.18 micron CMOS technology. The tuning frequency is $f_0 = 2$ GHz. The internal impedance of the signal source is 50 ohm resistive and must be matched with the input impedance of the LNA at 2 GHz. The voltage gain must not be lower than 20. The DC supply voltage is 1.8 V.

The circuit diagram with all necessary components for PSpice simulation is given in Fig. 5.15. The basic parameters and restrictions for this technology are given as follows (Figs. 5.16 and 5.17):

The maximum realizable value of the on-chip inductors is 10 nH.

The quality factor of the on-chip inductors is Q = 10.

TOX = 4.2 nm, which corresponds to C_{ox}= 8.2 e−7 F/cm^2
UO = 314 cm^2/Vs, which corresponds to μ_n = 220 cm^2/Vs for $V_{GS} \approx 0.7$ V
VTHO = 0.3 V
CGSO = CGDO =2.35 e−10 F/m
The sheet resistance of gate poly = 6 ohm/□

The overall voltage gain A_v was defined as the product of the resonance gain of the input loop, A_R, and the voltage gain of the cascode stage, A_C. Since A_R is usually slightly higher than unity, $|A_v| \cong |A_C|$, is a good approximation to tackle the problem, which simplifies the calculation of g_{m1}

Fig. 5.17 The complete circuit diagram of the circuit with all parameters to be calculated, for $f_0 = 2$ GHz and voltage gain $A_v > 20$

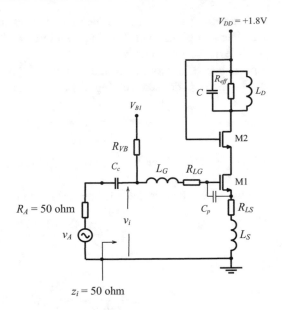

and provides a margin for the final value of the gain. Since the possible maximum value of L_D is given as 10 nH and $Q = 10$, g_{m1} of the input transistor can be calculated from (5.40) as

$$g_{m1} = \frac{20}{\left(10 \times 10^{-9}\right)\left(2\pi \times 2 \times 10^9\right) \times 10} \cong 16 \text{ mS}$$

and assuming $V_{B1} = 0.7$, the width of the transistor is calculated from (5.41)

$$W_1 = \frac{0.016 \times \left(0.18 \times 10^{-4}\right)}{220 \times \left(8.2 \times 10^{-7}\right) \times (0.7 - 0.3)} = 40 \times 10^{-4} \text{ cm} = 40\mu m$$

To ensure that the transconductance value is correct, it is useful to fine-tune the width and/or the bias voltage with PSpice simulation. In this example, we short-circuit L_S and L_G, connect the drain of M1 to ground with a high-value capacitance to prevent any feedback, apply a 2 GHz, 1 mV (peak) input signal for PSpice simulations, and adjust the signal current of M1 using W_1 and V_{B1} to 16 μA (which corresponds to $g_{m1} = 16$ mS). The appropriate values are found as $W_1 = 49$ μm and $V_{B1} = 0.8$ V.

From (1.50), the maximum channel width to prevent excessive attenuation on the gate electrode is calculated as 35 μm. This means that the overall channel width of 49 μm has to be realized using at least a 2-finger transistor with 24.5 μm finger length. But in this case, the calculated internal parasitic gate resistance is unacceptably high. To obtain a reasonable internal resistance, the number of fingers is increased to 11 with a 4.5 μm finger length that corresponds to $W_1 = 11 \times 4.5$ μm $= 49.5$ μm and provides 14 ohm internal gate series resistance and approximately 0.5 ohm internal source resistance. Another advantage of the finger structure is the reduction of the source and drain parasitic junction capacitances. The input capacitance of this 49.5 μm transistor is calculated from (1.40) as

$$C_{gs} = \frac{2}{3}C_{ox}WL + \text{CGSO} \times W$$
$$= \frac{2}{3}\left(8.2 \times 10^{-7}\right)\left(49.5 \times 10^{-4}\right)\left(0.18 \times 10^{-4}\right) + \left(2.35 \times 10^{-12}\right)\left(49.5 \times 10^{-4}\right) = 60.3 \text{ fF}$$

The sum of the source and gate inductances to resonate at 2 GHz with this capacitance can be calculated as 67 nH, which is impossible to realize on-chip. The solution is to use the maximum possible value of inductance for L_G, to estimate a small value for L_S and to connect a parallel capacitance (C_p) between G and S nodes for resonance. For $(L_G + L_S) = 11$ nH, the equivalent capacitance C_{eq} defined with Fig. 5.14 becomes

$$C_{eq} = \frac{1}{\omega_0^2(L_G + L_S)} = \frac{1}{\left(1.26 \times 10^{10}\right)^2\left(11 \times 10^{-9}\right)} = 573 \text{ fF}$$

With $C_{gs} = 60.3$ fF and $R_{Gi} = 14$ ohm, C_p from (5.48), and then R_{eq} from (5.46) can be calculated as 512 fF and 0.13 ohm, respectively.

Now L_S can be calculated from (5.45) with $R_{Si} \cong 0.5$ ohm , $R_{Gi} \cong 14$ ohm and

$$R_{LG} \simeq \frac{L_G\omega_0}{Q} \simeq \frac{\left(10 \times 10^{-9}\right)\left(1.26 \times 10^{10}\right)}{10} = 12.6 \text{ ohm}$$

as

$$L_S \cong \frac{\left(573 \times 10^{-15}\right)}{\left(16 \times 10^{-3}\right)}[50 - (0.5 + 0.13 + 12.6)] = 1.3 \text{ nH}$$

which is close enough to the previously estimated value of 1 nH.

With the resonance gain of the input loop, which is

$$A_R = \frac{v_{gs}}{v_A} = \frac{(L_G + L_S)\omega_0}{2R_A} = \frac{(10 + 1.3) \times 10^{-9} \times \left(1.26 \times 10^{10}\right)}{2 \times 50} = 1.42$$

and the gain of the cascode circuit,

$$|A_C| = \frac{v_o}{v_{gs}} \cong g_m \times R_{\text{eff}} = \left(16 \times 10^{-3}\right) \times 1260 = 20.16$$

the overall voltage gain $|A_v|$ becomes 28.6, which is higher than the target value by a considerable margin. Since the selectivity of the input resonance is very low, the bandwidth of the circuit is primarily determined by the output resonance and can be calculated from (4.25) as B = 200 MHz.

The size of M2 has limited influence on the input matching and the gain, thus we can start with $W_2 = (0.5 \times W_1)$, and fine-tune later on for noise and linearity.

The component values calculated above and used for the PSpice simulation are as follows:

$W_1 = 11 \times 4.5$ μm $= 49.5$ μm
$W_2 = 5 \times 5$ μm $= 25$ μm
$L_G = 10$ nH, $R_{LG} = 12,6$ ohm
$C_p = 512$ fF, $R_{eq} = 0.13$ ohm
$L_S = 1.3$ nH, $R_{LS} = 1.64$ ohm
$C_c = 100$ pF
$R_{VB} = 10$ k ohms
$L_D = 10$ nH, r_{eff} (@ 2 GHz)$= 1260$ ohm
$C = 633$ fF (the input capacitance of the following stage included)

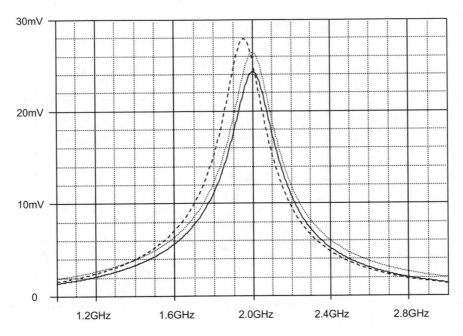

Fig. 5.18 The frequency characteristic obtained with hand-calculated values (dashed line), characteristic after fine-tuning C to 602 fF (dotted line), and the final characteristic after the improvement of the input matching and the linearity (the solid line)

The PSpice AC simulation result for the frequency characteristic of the output voltage is shown in Fig. 5.18 with a dashed line, for 1 mV input voltage amplitude and using the component values above. The resonance peak is observed at $f = 1.95$ GHz.

Usually, there is a minor discrepancy between the hand-calculated values and the simulation results, due to (a) the approximations of the expressions used for hand calculations, (b) the neglected parasitics, and (c) the differences of models and parameters used for hand calculations and in the simulator. Especially for the simulation of a tuned circuit, some "fine-tuning" is always necessary.

In our problem, it is obvious that it will be necessary to fine-tune the output resonance circuit to shift the peak frequency to 2 GHz (dotted line) by decreasing the value of C from 633 fF to 602 fF[8].

The imperfection of the input matching is not as obvious as the tuning error of the output. Nevertheless, it is necessary to check and improve that as well, if necessary.

The conventional method to check the input matching is to plot the s_{11} parameter to see if it is sufficiently low at and around the tuning frequency (see Appendix-E). But some circuit simulators (for example, PSpice) do not provide this option. On the other hand, the s_{11} plot does not give direct hints related to the adjustment of the values of the appropriate components.

As a different approach, we plot the variation of v_i with frequency (See Fig. 5.19, dashed line). If the matching were to be perfect, due to the division of v_A between R_A and the input impedance, v_i

---

[8] Note that the input capacitance value of the following stage must be subtracted from the calculated or simulated resonance capacitance value. On the realized circuit, to enable fine-tuning, an appropriate varactor must be used as part of the resonance capacitance.

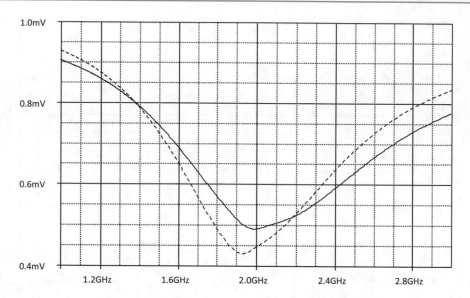

Fig. 5.19 The input voltage before the adjustments for input matching (dashed line) and after the adjustments (solid line)

should be equal to one half of the input signal voltage, i.e., to 0.5 mV in our case. The plot shows that the matching is far from perfect, not even satisfactory, such that:

(a) v_i at the resonance frequency is not 0.5 mV but 0.43 mV. This means that the input impedance is smaller than it should be. To increase its value, according to (5.43) L_S can be increased and/or C_{gs} can be decreased.

(b) On the other hand, the minimum of the series resonance curve must be shifted to 2 GHz by adjusting the values of L_S or C_{gs}. We can see that with $L_S = 1.45$ nH ($R_{LS} = 1.9$ ohm) and $C_p = 450$ fF, v_i becomes equal to 0.5 mV with good accuracy, not only for the tuning frequency but within the band of the amplifier as shown in 5.19 (solid line).

To finalize the design, it is necessary to determine the practical limit of the linearity of the amplifier, which is defined as the output signal level corresponding to the *decrease* of the gain by 1 dB (see Sect. 5.6).

To determine the −1dB level we apply a 2 GHz sinusoidal signal to the input with a low amplitude (for example 10 mV), observe the voltage waveforms at the output and on the drain of M1, measure the magnitude of the output signal and calculate the voltage gain ($v_o = 257$ mV and $A_v = 25.7$ for our circuit).

– Calculate the value of the gain corresponding to the "−1 dB point" which is $A_v /1.122$. For our circuit it is $A_{v(-1dB)} = A_v /1.122 = 23.15$.
– Increase the amplitude step by step until the voltage gain drops to 23.15. We find the input amplitude corresponding to $A_{v(-1dB)}$ as $V_A = 47.5$ mV. This is (by definition) the maximum input voltage swing for an acceptable nonlinearity.

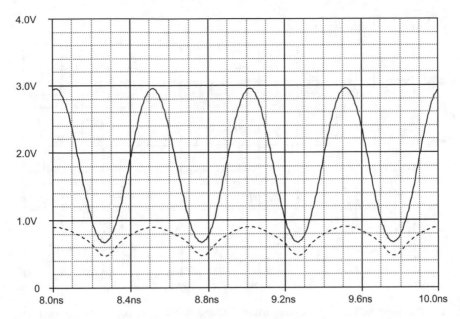

Fig. 5.20 The waveforms of the output voltage (solid line) and the drain voltage of M1 (dashed line) for 47.5 mV input voltage amplitude ($W_1 = 49.5$ μm and $W_2 = 25$ μm). Note the distortion of the signal at the drain of M1

The waveforms of the output voltage and the voltage of the drain node of M1 are shown in Fig. 5.20. The output voltage has 1.10 V amplitude with a reasonable waveform shape, but the signal at the drain of M1 (which is the input voltage of M2) is excessively distorted; this indicates severe nonlinearity. This is due to the decrease of the drain-source voltage of M2 down to the pre-saturation region around the negative peaks of the drain voltage. Since the input signal is a pure sinusoid in this simulation, the distortion terms of this signal (the harmonics of the input frequency) are filtered out by the output resonator and do not affect the output waveform. But in the case of a modulated input signal containing side-frequencies, due to the existing nonlinearity, these components may intermodulate and produce unwanted components in the band. Therefore, the maximum input amplitude must be lowered to a level that does not exhibit a considerable distortion at the (source) input of M2. For our circuit this "safe" input level is approximately 30 mV, which is well below the calculated -1dB level[9].

To increase the applicable input level up to the calculated value of 47.5 mV, we can try reducing W_2. This is due to the fact that increasing the input impedance of M2 will increase the output swing of M1, and the distortion on the output voltage of M1 appears at higher input levels. This also reduces the contribution of M2 on the total output noise. But excessive reduction of W_2 has several drawbacks:

- The increase in the input impedance of M2 increases the signal voltage on the drain of M1, which increases the unwanted feedback over C_{dg1}.
- Since $r_{i2} \ll r_{o1}$ is no longer valid, i_{d2} becomes smaller than i_{d1}, which reduces the gain.
- The decrease of the DC voltage share of M1 leads to nonlinear distortion at the output of this transistor.

---

[9] This example shows that the -1dB level definition does not reflect the nonlinearities of the previous stages, but only the output transistor.

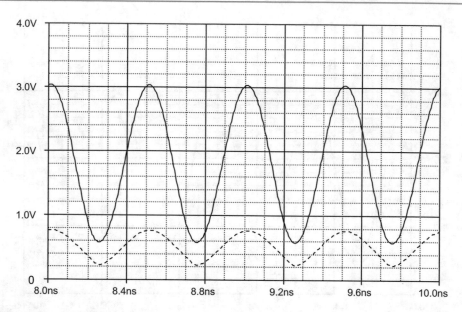

Fig. 5.21 The waveforms of the output voltage (solid line) and the drain voltage of M1 (dashed line) after final optimization, for 60 mV input signal amplitude ($W_1 = 49.5$ μm and $W_2 = 15$ μm). Note the distortion-free waveforms of both signals

Therefore, the optimization of W_2 must be done with care. To optimize W_2 in our example, after checking the performance for $W_2 = 20$ μm, 15 μm, and 10 μm, we decide to set $W_2 = 3 \times 5$ μm $= 15$ μm, which provides a voltage gain of $A_v = 22.7$ and a −1 dB gain of 20.23. The input and output signal amplitudes corresponding to a 20.23 gain are $V_{A(-1dB)} = 60$ mV and $V_{o(-1dB)} = 1.214$ V, respectively. This output signal amplitude (that corresponds to 11.7 dBm) has no remarkable distortion on the drain of M1, as seen in Fig. 5.21.

The frequency characteristic of the LNA after these adjustments is shown with a solid line in Fig. 5.21. The final values for the performance of the amplifier are as follows:

Tuning frequency: 2 GHz
Voltage gain @ 2GHz: 24.2 (27.67 dB)
The −3dB bandwidth: 200 MHz
Input impedance is matched to 50 ohm within the band
Input voltage corresponding to the −1 dB gain reduction: 60 mV (peak)
Output voltage at −1 dB point: 1.215V (peak)
Power consumption: 5.62 mW

5.4.1 Noise of a Tuned LNA

The circuit diagram of a typical source degenerated LNA and the noise sources related to its noisy components are shown in Fig. 5.22. The approach used to analyze this circuit in Sect. 5.4 helps calculate the noise contributions of the resistors placed in the input loop. The noise voltage sources situated in series in the input loop behave the same way as the signal source, v_A, i.e., the output noise components corresponding to each of these sources is equal to $A_v \times \bar{v}_n$ and, assuming that the resonance gain of the input loop is approximately 1, is equal to $g_{m1} \times \bar{v}_n$.

Fig. 5.22 Inductive source degenerated cascode LNA with its noise sources

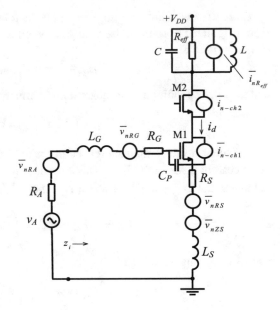

The mean square noise components of the load current for a 1 Hz bandwidth originating from each of the noisy components are as follows:

Output noise originating from M1 : $\overline{i^2_{n-\text{ch1}}} = 4kT_C \times \gamma \times g_{m1}$

Output noise originating from M2 : $\overline{i^2_{n-\text{ch2}}} = 4kT_C \times \gamma \times g_{m2}$

Output noise originating from R_A : $\overline{i}_{ndRA} = \overline{v}_{nRA} \times g_m$ $\rightarrow$ $\overline{i^2_{ndRA}} = 4kT_A R_A \times g^2_{m1}$ (5.49)

Output noise originating from R_G : $\overline{i}_{ndRG} = \overline{v}_{nRG} \times g_m$ $\rightarrow$ $\overline{i^2_{ndRG}} = 4kT_C R_G \times g^2_{m1}$

Output noise originating from R_S : $\overline{i}_{ndRS} = \overline{v}_{nRS} \times g_m$ $\rightarrow$ $\overline{i^2_{ndRS}} = 4kT_C R_S \times g^2_{m1}$

where T_A is the temperature of the antenna (the signal source) and T_C is the average temperature of the chip[10]. R_G is the sum of the R_{LG} and R_{eq} given with (5.46).

The total mean square drain noise current of M1 is

$$\overline{i^2_{ndT}} = \overline{i^2_{n-\text{ch1}}} + \overline{i^2_{ndRS}} + \overline{i^2_{ndRG}}$$

This noise is transferred to the load via M2, together with its own channel noise current, $\overline{i^2_{n-\text{ch2}}}$. Hence the total noise current flowing through R_{eff}, as well as through M1 and M2, becomes

$$\overline{i^2_{ndT}} = \overline{i^2_{n-\text{ch}}} + \overline{i^2_{ndRS}} + \overline{i^2_{ndRG}} + \overline{i^2_{n-\text{ch2}}} + \overline{i^2_{nR(\text{eff})}}$$ (5.50)

---

[10] In reality, the temperature of the channel may be considerably higher than the average temperature of the chip.

There are two additional noise components related to the total drain noise current of M1. One of them originates from the noise voltage drop on the impedance series to the source (Z_S) at ω_0, which can be written as

$$\bar{v}_{nZS} = \bar{i}_{ndT} \times Z_S(\omega_0) \cong \bar{i}_{ndT} \times (L_S\omega_0)$$

and the corresponding mean square noise component on the output is,

$$\bar{i}^2_{ndZS} = (L_S\omega_0)^2 \times g^2_{m1} \times \bar{i}^2_{ndT} \tag{5.51}$$

The other noise component related to the total drain noise current of M1 originates from the gate noise current given with (5.24). This current flows over ($R_G + R_A$) to the ground and produces a noise voltage equal to $\bar{v}_{nG} = (R_G + R_A) \times \bar{i}_{ng}$. The resulting drain noise component is

$$\bar{i}_{nG} = \left[(R_G + R_A) \times \left(\omega_0 \frac{C_{gs}}{g_m}\right)\bar{i}_{ndT}\right] \frac{g_{m1}}{(1 + g_{m1}Z_S)}$$

and its mean square value is

$$\bar{i}^2_{nG} = \left[(R_G + R_A) \frac{g_{m1}}{(1 + g_{m1}Z_S)}\right]^2 \left(\omega_0 \frac{C_{gs}}{g_{m1}}\right)^2 \times \bar{i}^2_{ndT} \tag{5.52}$$

Together with these additional components, the total mean square drain current, as well as the total output noise current flowing through the load, becomes

$$\bar{i}^2_{ndT} = \bar{i}^2_{n-ch1} + \bar{i}^2_{ndRS} + \bar{i}^2_{ndRG} + \bar{i}^2_{n-ch2} + \bar{i}^2_{nR(\text{eff})} + A \times \bar{i}^2_{ndT} + B \times \bar{i}^2_{ndT} \tag{5.53}$$

where

$$A = (L_S\omega_0 \times g_{m1})^2 \tag{5.54}$$

and

$$B = \left[(R_G + R_A)\frac{g_{m1}}{(1 + g_{m1}Z_S)}\left(\omega_0 \frac{C_{gs}}{g_{m1}}\right)\right]^2 \tag{5.54a}$$

From (5.53), $\bar{i}^2_{ndT}$ can be solved as

$$\bar{i}^2_{ndT} = \frac{\bar{i}^2_{n-ch1} + \bar{i}^2_{ndRS} + \bar{i}^2_{ndRG} + \bar{i}^2_{n-ch2} + \bar{i}^2_{nR}}{1 - (A + B)} \tag{5.55}$$

and the total output noise power dissipated on R_{eff},

$$P_{no(\text{tot})} = 4kT_CR_{\text{eff}}\left\{\frac{(\gamma \times g_{m1}) + (R_G \times g^2_{m1}) + (R_S \times g^2_{m1}) + (\gamma \times g_{m2}) + (1/R_{\text{eff}})}{1 - (A + B)}\right\} \tag{5.56}$$

Since the noise power at the output originating from R_A is

$$P_{noRA} = \bar{i}^2_{ndRA} \times R_{\text{eff}} = 4kT_AR_Ag^2_{m1}R_{\text{eff}} \tag{5.57}$$

the noise factor of the amplifier becomes

$$F = 1 + \frac{P_{no(\text{tot})}}{P_{noRA}} = 1 + \frac{T_C}{T_A} \frac{\left\{\frac{(\gamma \times g_{m1}) + (R_G \times g^2_{m1}) + (R_S \times g^2_{m1}) + (\gamma \times g_{m2}) + (1/R_{\text{eff}})}{1 - (A + B)}\right\}}{R_Ag^2_{m1}} \tag{5.58}$$

and can be rearranged as

$$F = 1 + \frac{P_{no(\text{tot})}}{P_{noRA}} = 1 + \frac{T_C}{T_A} \frac{\gamma\left(1 + \frac{g_{m2}}{g_{m1}}\right) + g_{m1}(R_G + R_S) + \left(\frac{1}{g_{m1}R_{\text{eff}}}\right)}{[1 - (A + B)]R_A g_{m1}} \tag{5.58a}$$

(5.58a) provides some hints to improve the design for better noise performance:

- It is useful to reduce the resistances connected in series to the gate and the source of M1. In order to reduce the inductor related components of these resistances, it is necessary to increase the Q values, which necessitates better inductor design. The parasitic resistances of the transistors are strongly geometry-dependent. The finger-structure and the connecting lines must be optimized for minimum resistances.
- The increase in g_{m1} improves the noise performance to some extent. This means that the width of the transistor has to be increased, which results in higher parasitic capacitances.
- g_{m2} can be carefully reduced, to prevent inconvenient operating conditions for M1 and M2, which would affect the linearity, as discussed in Example 5.2.
- The overall power consumption of the chip and the thermal resistance to the ambient temperature must be as small as possible for a small (T_C/T_A) ratio. In addition, M1 and M2 must not be close to the high power consuming parts on the chip to prevent the increase of the – already high – local temperatures of the channel regions.

Example 5.3
Calculate the noise figure of the circuit designed in Example 5.2. The parameters necessary to calculate the noise are as follows:

$R_A = 50$ ohm
$R_{(\text{eff})} = 1260$ ohm
$R_{LS} = 1.64$ ohm
$R_{LG} = 12.6$ ohm
$g_{m1} = 16$ mS
$g_{m2} = 3.4$ mS
$R_{Si} = 0.5$ ohm
$R_{eq} = 0.13$ ohm
$L_S = 1.3$ nH
$\gamma = 0.7$ (assumed)
$T_A = 300$ K
$T_C = 330$ K (assumed)

Instead of calculating the noise factor directly from (5.58a), let us calculate the individual noise powers at the output for a 1 Hz bandwidth, originating from several components from (5.55), in order to evaluate and compare the noise contribution of each component. Let us start with calculating A and B, which are common for all components, with (5.54) and (5.54a):

$$A = \left[(1.3 \times 10^{-9})(1.26 \times 10^{10})(16 \times 10^{-3})\right]^2 = 68.7 \times 10^{-3}$$

with $Z_S \cong L_S \omega_0$

$$B \cong \left[(12.73 + 50) \frac{(16 \times 10^{-3})}{[1 + (16 \times 10^{-3})(1.3 \times 10^{-9})(1.26 \times 10^{10})]} \left((1.26 \times 10^{10}) \frac{(60.3 \times 10^{-15})}{(16 \times 10^{-3})} \right) \right]^2$$

$$= 1.42 \times 10^{-3}$$

Then, $[1 - (A + B)] = 0.93$

Another factor common to all components:

$$4kT_C R_{\text{eff}} = 4 \times (1.36 \times 10^{-23}) \times 330 \times 1260 = 2.26 \times 10^{-17}$$

The noise power at the output originating from M1:

$$P_{n(\text{M1})} = \frac{4kT_C R_{\text{eff}}}{[1 - (A + B)]} \gamma g_{m1} = \frac{2.26 \times 10^{-17}}{0.93} \times 0.7 \times (16 \times 10^{-3}) = 2.72 \times 10^{-19} \text{ W}$$

The noise power at the output originating from R_G, where $R_G = R_{LG} + R_{eq}$

$$P_{n(\text{RG})} = \frac{4kT_C R_{\text{eff}}}{[1 - (A + B)]} \left(R_G \times g_{m1}^2 \right) = \frac{2.26 \times 10^{-17}}{0.93} \times 12.73 \times (16 \times 10^{-3})^2 = 0.79 \times 10^{-19} \text{ W}$$

The noise power at the output originating from R_S:

$$P_{n(\text{RS})} = \frac{4kT_C R_{\text{eff}}}{[1 - (A + B)]} \left(R_S \times g_{m1}^2 \right) = \frac{2.26 \times 10^{-17}}{0.93} \times 1.64 \times (16 \times 10^{-3})^2 = 0.1 \times 10^{-19} \text{ W}$$

The noise power at the output originating from M2:

$$P_{n(\text{M2})} = \frac{4kT_C R_{\text{eff}}}{[1 - (A + B)]} \gamma g_{m2} = \frac{2.26 \times 10^{-17}}{0.93} \times 0.7 \times (4.85 \times 10^{-3}) = 0.83 \times 10^{-19} \text{ W}$$

The noise power at the output originating from R_{eff}:

$$P_{n(\text{Reff})} = \frac{4kT_C R_{\text{eff}}}{[1 - (A + B)]} \left(\frac{1}{R_{\text{eff}}} \right) = \frac{2.26 \times 10^{-17}}{0.93} \times \frac{1}{1260} = 0.19 \times 10^{-19} \text{ W}$$

The sum of all of these components gives the total noise power at the output originating from the amplifier:

$$P_{no} = P_{n(\text{M1})} + P_{n(\text{RG})} + P_{n(\text{RS})} + P_{n(\text{M2})} + P_{n(\text{Reff})}$$

$$= (2.72 + 0.79 + 0.1 + 0.83 + 0.19) \times 10^{-19} = 4.63 \times 10^{-19} \text{ W}$$

The output noise originating from the source resistance, R_A;

$$P_{n(\text{RA})} = 4kT_A R_A \times g_m^2 \times R_{\text{eff}} = 4 \times (1.36 \times 10^{-23}) \times 300 \times 50 \times (16 \times 10^{-3})^2 \times 1260$$

$$= 2.63 \times 10^{-19} \text{ W}$$

Hence the noise factor and corresponding noise figure can be calculated as

$$N = 1 + \frac{P_{no}}{P_{n(\text{RA})}} = 1 + \frac{4.63 \times 10^{-19}}{2.63 \times 10^{-19}} = 2.76 \qquad NF = 10 \log (N) = 4.4 \text{ dB}$$

Note that if the temperature of the chip were assumed to be equal to the ambient temperature, the noise figure would be 4.15 dB.

The PSpice noise simulation gives the total output mean square noise voltage as $\overline{v}_{no}^2 = 5.81 \times 10^{-16}$ V^2/Hz, which corresponds to $P_{no} = (5.81 \times 10^{-16})/1260 = 4.6 \times 10^{-19}$ W and the mean square output noise voltage originating from R_A as $\overline{v}_{n(RA)}^2 = 3.4 \times 10^{-16}$ V^2/Hz , which corresponds to $P_{no} = (3.4 \times 10^{-16})/1260 = 2.69 \times 10^{-19}$ W. Hence the simulation results for the noise factor and the noise figure becomes $N = 2.71$ and $NF = 4.32$ dB which are in concordance with the calculated values.

5.4.2   The Differential LNA

For some applications, designing the LNA as a differential amplifier is more convenient:

(a)   If the antenna is balanced (differential), i.e., the signal voltages at both output nodes have equal magnitude and are opposite in phase with respect to the ground [11]
(b)   If the circuit following the LNA has a differential input

Another advantage of a differential LNA is that its even harmonics are small (theoretically zero), due to the symmetrical structure of the circuit.

The schematic of a typical differential LNA is given in Fig. 5.23a. The circuit is redrawn in Fig. 5.23b, which helps to consider the circuit as two single-ended LNAs and uses the expressions derived in Sect. 5.4.1.

A center-tapped inductor is composed of two identical parts and its inductance depends on the coupling between these two parts. If the coupling of the two halves of the inductor is zero or negligibly small, the total inductance is equal to $L_T = 2L$. This case corresponds to two separate, physically uncoupled and identical inductors. If these inductors are closely coupled, for example, designed as a center-tapped inductor as shown in Fig. 1.40b, the total inductance becomes $L_T = 2L$ $(1 \mp k)$, where k is the coupling coefficient[12]. If the two halves of the inductor are "wound" in the same direction the k is positive, and the value of the inductance is higher than for the case of zero coupling. This is an advantage from the point of view of the area and the effective series resistance of the inductor.

But there is another effect that must not be overlooked. The output noise voltage originating from M21, $\overline{v}_{nd1}$, which appears on the left half of the center-tapped inductor, induces a noise voltage on the right half of the inductor as an autotransformer, which is equal to $k \cdot \overline{v}_{nd1}$, and vice versa. Hence the total mean square noise voltage becomes $(1 + k)^2 (\overline{v}_{no1}^2 + \overline{v}_{no2}^2)$.

5.5   Wideband LNAs, "Noise Signal Feedback" and "Noise Cancellation"

Another approach to the LNAs, different from the conventional amplifiers tuned to a certain frequency at the output, is to design a wideband amplifier and to define the operating frequency at the input with a suitable passive band-pass filter (or directly using the tuned behavior of the antenna). With this approach, it is possible to use a wideband amplifier for any frequency, certainly in the band

---

[11] Another solution is to connect a balanced antenna to the input of a single-ended LNA via a BALUN, a simple structure that converts a balanced signal to an unbalanced signal, and vice-versa, to the expense of extra parasitics and noise.

[12] Typical values of k for center-tapped inductors are in the range of 0.6–0.8.

Fig. 5.23 (a) Schematic
of a typical differential
LNA. (b) The rearranged
schematic that helps to
clarify the expressions
derived for the single-
ended LNA

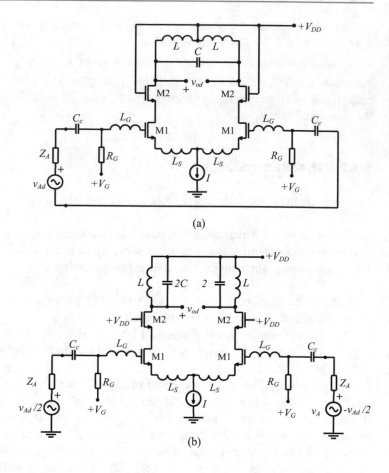

of the amplifier, only by changing the input filter or the antenna. Another advantage of this approach is that it reduces the neighboring channel interference.

The wideband amplifiers, thanks to their relatively flat phase characteristic, allow us to benefit from the negative feedback. The noise voltages and currents of the resistors and transistors in a circuit are processed (amplified, fed-back, mixed, etc.) as other signals in the circuit, until they reach the output. Therefore, the effects of the feedback on noise signals can be evaluated to improve the noise performance of amplifiers. This has been known since the early days of electronics[13], and has led to the concepts of "noise feedback" and "noise cancellation." This fact will be exemplified on several basic circuits below.

5.5.1 The Common Gate Amplifier as a Wideband LNA

In the common gate amplifier given in Fig. 5.24, there are three noisy components: the internal resistance of the input signal source R_A, the MOS transistor, and the load resistor R_L. The contributions of these noise components to the output noise are as follows:

---

[13] For example: H. J. Reich, "Theory and Applications of Electron Tubes," p. 201, McGraw-Hill, 1944.

Fig. 5.24 The principal
circuit diagram of a
common gate amplifier

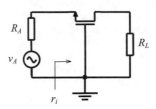

(a) Contribution of R_A: The instantaneous value of the noise voltage of the source resistance (v_{nA}) is transferred to the input of the amplifier as

$$v_{ni} = \frac{r_i}{R_A + r_i} v_{nA}$$

and then amplified by the transistor with a voltage gain of $A_v \cong g_m R_L$. Provided that the amplifier can be considered as linear for this signal level, the amplified noise voltage at the output is a replica of the noise voltage at the input, and the RMS value of this voltage is equal to

$$\bar{v}_{nAo} = A_v \times \bar{v}_{ni} = A_v \times \frac{r_i}{R_A + r_i} \bar{v}_{nA} \qquad (5.59)$$

and if the input is matched, i.e. $r_i = R_A$

$$\bar{v}_{nAo} = A_v \times \frac{1}{2} \bar{v}_{nA} = \frac{1}{2} g_m R_L \sqrt{4kTBR_A} \qquad (5.59a)$$

Since $R_A = r_i \cong 1/g_m$ for a common gate amplifier,

$$\bar{v}_{nAo} = \frac{1}{2} g_m R_L \sqrt{4kTB/g_m} \qquad (5.59b)$$

Hence the noise power dissipated on the load resistance owing to the noise of the source resistance becomes

$$P_{nAo} = \frac{\bar{v}_{nAo}^2}{R_L} = kTBg_m R_L \qquad (5.60)$$

(b) Contribution of the channel noise: The channel noise current of the transistor $\bar{i}_{n-ch} = \sqrt{4kTB\gamma g_m}$ flows through the load resistor R_L, as well as the internal resistance of the input signal source, R_A. The noise current flowing through R_A produces a voltage drop v_{ni}, which provokes a channel noise current equal to ($v_{ni} \times g_m$). This "feedback noise component" is in the opposite direction of i_{n-ch} and reduces it. In other words, there is a "noise-cancelling feedback." This qualitative explanation can be developed by including the effect of the internal resistance of the transistor, as shown in the equivalent circuit in Fig. 5.25b, from which the resulting output noise current can be calculated as

$$i_{n-ch(tot)} = i_{n-ch} \frac{r_{ds}}{2r_{ds} + R_A + R_L} = i_{n-ch} \times \delta \qquad (5.61)$$

where δ is the noise reduction parameter that approaches ½ for

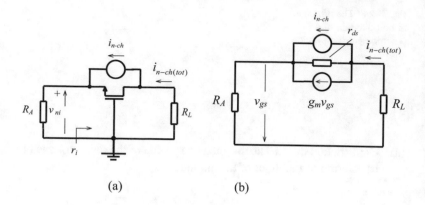

Fig. 5.25 (a). The channel noise current, (b) the equivalent circuit to calculate the internal feedback effect of this current

$$(R_A + R_L) \ll 2r_{ds}$$

The RMS value of this noise current

$$\bar{i}_{n-\text{ch(tot)}} = \delta \bar{i}_{n-\text{ch}} = \delta \sqrt{4kTB\gamma g_m} \tag{5.62}$$

The contribution of the channel noise on the output power now can be calculated as

$$P_{n-\text{cho}} = \bar{i}_{n-\text{ch(tot)}}^2 \times R_L = \delta^2 (4kTB\gamma g_m R_L) \tag{5.63}$$

(c) Contribution of the load resistance noise on the total output noise is simply

$$P_{n-\text{RL}} = 4kTB \tag{5.64}$$

Now the noise factor of a common gate amplifier can be written as

$$F = 1 + \frac{P_{n-\text{cho}} + P_{n-\text{RL}}}{P_{n-Ao}} = 1 + \frac{\delta^2 (4kTB\gamma g_m R_L) + 4kTB}{kTB g_m R_L}$$

$$F = 1 + 4\left(\delta^2 \gamma + \frac{1}{g_m R_L}\right) = 1 + 4\left(\delta^2 \gamma + \frac{1}{|A_v|}\right) \tag{5.65}$$

As an example for $R_A = 50$ ohm, $g_m = 20$ mS, $R_L = 500$ ohm, $r_{ds} = 1$ kohm and $\gamma = 0.7$ the noise factor is $F = 1.83$, which corresponds to a noise figure of $NF = 2.6$ dB.

Example 5.4

The noise reduction parameter (δ) of the common gate amplifier given in Fig. 5.26a will be determined by pSpice simulation. The dimensions of the transistor are $W = 11 \times 8$ μm and $L = 0.18$ μm. These dimensions and the DC current provide a 50 ohm input resistance and a voltage gain of 5. The channel noise current is emulated with a sinusoidal 1 μV/100 MHz current source. For $v_A = 0$, the "noise voltage" at the output node is measured as 172 μV, as demonstrated in Fig. 5.26b. To eliminate the noise feedback R_A is shunted with a high-value capacitor. In this case, the output voltage was 345 μV, showing us that the noise reduction parameter is 172/345 = 0.49, which agrees with (5.61).

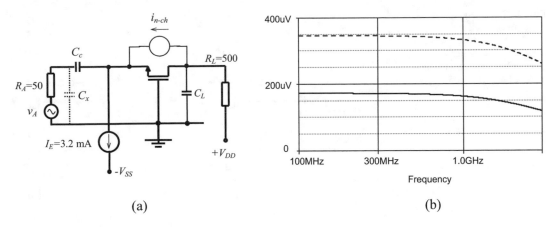

Fig. 5.26 (**a**) The circuit diagram of the simulated amplifier. (**b**) The output noise voltage of the circuit originating from the emulated channel noise current (lower trace) and the noise voltage without noise-reducing feedback (upper trace)

5.5.2   Parallel Voltage Feedback Amplifier as Wideband LNA

Figure 5.27a shows a parallel voltage feedback-applied resistance-loaded common source amplifier, which is a trans-impedance amplifier in nature. It is known that the low-frequency values of the trans-impedance, the voltage gain, and the input and output impedances of such an amplifier are as follows:[14]

$$Z_m(0) = \frac{v_o}{i_i} = -R_L \frac{g_m R_F - 1}{g_m R_L + 1} \cong -R_F \tag{5.66}$$

$$A_v(0) = \frac{v_o}{v_i} = \frac{Z_m(0)}{r_i} \cong -R_F g_m \tag{5.67}$$

$$A_{vA}(0) = \frac{v_o}{v_A} = \frac{r_i}{(r_i + R_A)} A_v(0) \tag{5.67a}$$

$$Z_i(0) = r_i = \frac{v_i}{i_i} = \frac{R_F + r_p}{1 + g_m r_p} \cong \frac{1}{g_m} \tag{5.68}$$

$$Z_o(0) = r_o = \frac{v_o}{i_o} = \frac{r_p}{1 + g_m r_p} \cong \frac{1}{g_m} \tag{5.69}$$

where r_p is the parallel equivalent of R_L and r_{ds} . These expressions can be approximated as mentioned, provided that g_m , R_F and r_{ds} are sufficiently high.

The input matching condition can be written for $r_p \cong R_L$ from (5.68) as

$$R_A = \frac{R_F + R_L}{1 + g_m R_L} \tag{5.70}$$

and with (5.70), equations (5.66) and (5.67) can be arranged as

---

[14] The effects of the reactive components and parasitics will be discussed later on.

Fig. 5.27 (a) The circuit diagram of a self-biased common source feedback amplifier. (b) Schematic to calculate the effect of the noise of R_F

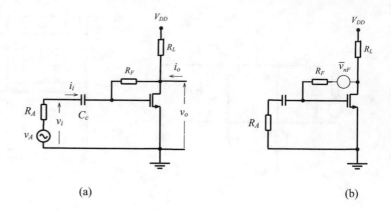

(a)   (b)

$$|Z_m| = (R_F - R_A) \tag{5.71}$$

$$|A_v| = \frac{(R_F - R_A)}{R_A}, \qquad |A_{vA}| = \frac{1}{2}\frac{(R_F - R_A)}{R_A} \tag{5.72}$$

Problem 5.4
Derive expression (5.71).

The noisy components in this circuit are the internal resistance of the input signal source R_A, the MOS transistor, the load resistor R_L, and the feedback resistor R_F.

The contributions of these noise components to the output noise are as follows:

(a) The equivalent noise voltage of the signal source internal resistance v_{nA} is transferred to the input of the amplifier as

$$v_{nAi} = v_{nA}\frac{R_A}{R_A + r_i} \tag{5.73}$$

where r_i is the input resistance of the amplifier. If the input is matched, the corresponding output noise voltage is

$$v_{nAo} = \frac{v_{nA}}{2} \times A_v$$

its RMS value is

$$\bar{v}_{nAo} = \frac{\bar{v}_{nA}}{2}|A_v| \tag{5.74}$$

and the corresponding noise power at the output is

$$P_{nAo} = \frac{\bar{v}_{nAo}^2}{R_L} = \bar{v}_{nA}^2 \frac{1}{4}\frac{|A_v|^2}{R_L}$$

Since $\bar{v}_{nA} = \sqrt{4kTBR_A}$,

$$P_{nAo} = kTB\frac{R_A}{R_L}|A_v|^2 \tag{5.75}$$

(b) The output noise power component related to the channel noise current of the transistor is

$$P_{n-\mathrm{cho}} = \bar{i}_{n-\mathrm{ch}}^2 R_L = 4kTB\gamma g_m R_L \tag{5.76}$$

(c) The noise power of the load resistor R_L itself is simply

$$P_{nLo} = 4kTB \tag{5.77}$$

(d) The contribution of R_F to the output noise is twofold: its inherent noise and the negative feedback effect over R_F.

The noise of R_F is represented with a noise voltage source, $\bar{v}_{nF}$, in Fig. 5.27b. This source provokes a noise current along R_L, R_F and R_A:

$$\bar{i}_{nF} = \frac{\bar{v}_{nF}}{\left(R'_L + R_F + R_A\right)} \tag{5.78}$$

where $R'_L = R_L // r_{ds} \cong R_L$. The contribution of this current to the output noise power can be calculated as

$$P_{nFo} = \bar{i}_{nF}^2 R_L = \left[\frac{\bar{v}_{nF}}{\left(R'_L + R_F + R_A\right)}\right]^2 R_L$$
$$\cong 4kTBR_F\frac{R_L}{\left(R_L + R_F + R_A\right)^2} \tag{5.79}$$

Hence, the total noise power dissipated on the load resistor due to the noisy components of the amplifier becomes

$$P_{no} = P_{n-\mathrm{cho}} + P_{nFo} + P_{nL}$$
$$= 4kTB\left(\gamma g_m R_L + \frac{R_L R_F}{\left(R_L + R_F + R_A\right)^2} + 1\right) \tag{5.80}$$

which corresponds to an RMS output noise voltage of

$$\bar{v}_{no} = \sqrt{P_{no}R_L} \tag{5.81}$$

The noise feedback effect of a negative feedback amplifier can be modeled using the block diagram shown in Fig. 5.28a, which is applicable to all single-loop feedback amplifiers. The input signal (a_i) is assumed zero. The noise signal at the output node of the amplifier originating from the

Fig. 5.28 (a). The block diagram of a feedback amplifier with output noise feedback. (b) Application to the parallel voltage feedback amplifier

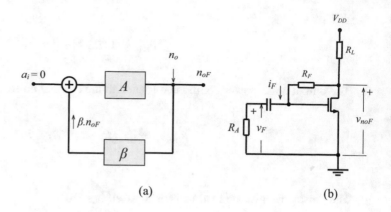

(a)

(b)

components of the amplifier is shown as n_o. The total noise output signal n_{oF} is the sum of n_o and the feedback noise signal:

$$n_{oF} = n_o + \beta A \times n_{oF}$$

which yields

$$n_{oF} = \frac{n_o}{(1 - \beta A)}$$

This expression shows that if either β or A is negative (which means that the feedback is negative) the noise signal at the output node becomes $(1 - \beta A)$ times smaller than the original noise. This means that there is a noise cancelling feedback.

For the amplifier shown in Fig. 5.28b, the voltage feedback factor and the voltage gain are

$$\beta = \frac{v_F}{v_{noF}} = \frac{R_A}{R_F + R_A} \quad , \quad A_v = -\frac{|Z_m|}{r_i} = -\frac{|Z_m|}{R_A} \tag{5.82}$$

$$v_{noF} = v_{no} \frac{1}{1 + \beta \frac{|Z_m|}{R_A}} = \frac{1}{1 + \frac{(R_F - R_A)}{(R_F + R_A)}} \tag{5.83}$$

$$\delta = \frac{v_{noF}}{v_{no}} = \frac{1}{1 + \frac{(R_F - R_A)}{(R_F + R_A)}} = \frac{(R_F + R_A)}{2R_F} \tag{5.84}$$

where δ is the output noise voltage reduction factor owing to the feedback and is always smaller than unity. The corresponding noise power reduction factor is δ^2. Hence the output noise power and the noise factor with feedback become

$$P_{noF} = \delta^2 \times P_{no} \tag{5.85}$$

$$F \cong 1 + \frac{P_{noF}}{P_{nAo}} = 1 + \frac{4\delta^2 \left(\gamma g_m R_L + \frac{R_L R_F}{(R_L + R_F + R_A)^2} + 1 \right)}{\frac{R_A}{R_L} |A_v|^2} \tag{5.86}$$

Note that the noise factor increases with g_m, strongly increases with R_L, and decreases with gain.

Now a design flow for an input-matched self-biasing parallel voltage feedback amplifier can be defined as follows:

The basic DC condition for the circuit is

$$V_{DS} = V_{GS} = V_{DD} - I_D R_L \tag{5.87}$$

The drain current for a non-velocity saturated transistor is

$$I_D \cong \frac{1}{2}\mu C_{ox}\frac{W}{L}(V_{GS} - V_T)^2 = \frac{1}{2}g_m(V_{GS} - V_T) \tag{5.88}$$

and for a velocity saturated transistor,

$$I_D \cong kC_{ox}W.v_{sat}(V_{GS} - V_T) = g_m(V_{GS} - V_T) \tag{5.89}$$

Since the amplifier to be designed is a wideband circuit, a velocity-saturated (small geometry) transistor must be preferred to extend the cut-off frequency. Therefore, the design flow must be based on (5.71). These expressions lead to

$$R_F = (|A_v| + 1)R_A \tag{5.90}$$

$$g_m = \frac{1}{R_L}\left(\frac{R_F + R_L - R_A}{R_A}\right) \tag{5.91}$$

Since $V_{DS} = V_{GS}$

$$I_D = \frac{(V_{DD} - V_T)}{R_L + \frac{1}{g_m}} \tag{5.92}$$

(5.90) directly gives the value of R_F corresponding to any source resistance and gain. (5.91) and (5.92) help to calculate the transconductance and the drain DC current corresponding to any anticipated load resistor that determines the -3dB frequency together with the load capacitance and the output node parasitics.

From (5.86), it can be seen that the noise factor increases linearly with g_m, quadratically with R_L, and decreases quadratically with A_v. On the other hand, the power consumption linearly decreases with the increase of the load resistor. The bandwidth depends on the load resistance and the transistor geometry, which affects the corner frequencies of the output as well as the input node. There is therefore a trade-off among noise, power consumption, and bandwidth.

Example 5.5

An input-matched wideband LNA as shown in Fig. 5.27a will be designed. The technology is 0.18 micron UMC CMOS technology for which $L = 0.18$ μm, $V_{TH} = 0.3$V, $C_{ox} = 8.2 \times 10^{-7}$ F/cm^2, and $V_{DD} = 2$V. The internal resistance of the input signal source is $R_A = 50$ ohm and the load capacitance is 100 fF. The required voltage gain from the source is 10 dB ($|A_v| = 3.2$), which corresponds to a voltage gain from the input node of 16 dB ($|A_v| = 6.4$). The -3dB frequency and the noise figure at 1 GHz must be higher than 3 GHz and 3 dB, respectively. It will be assumed that the transistor is operating in the velocity-saturated mode, which must be checked at the end of the design procedure. The r_{ds} of the transistor will be assumed as significantly higher than the drain load resistance.

(5.72) imposes the value of the feedback resistors

$$R_F = (|A_v| + 1)R_A = 370 \text{ ohm}$$

The g_m and I_D values calculated from (5.91) and (5.92), and *NF* from (5.86) corresponding to different R_L values are given in the table below. The width of the transistor corresponding to each (g_m, I_D) pair must be separately calculated. The -3 dB frequencies obtained from the SPICE simulations are also given. From this table, it can be seen that $R_L = 200$ ohm is the appropriate choice. The noise figure can be decreased to smaller values at the expense of a higher DC power consumption and higher transistor widths, which impairs the bandwidth.

R_L (ohm)	g_m (mS)	I_D (mA)	*NF* (dB)	*BW* (GHz) (simulation)
100	84	15	1.55	3.25
150	62.7	10.2	2.33	3.75
200	52	7.75	2.8	3.83
250	45.6	6.25	3.2	3.85
300	41.3	5.28	4.5	3.70

In Fig. 5.29, the SPICE simulation results corresponding to $R_L = 200$ is shown. The width of the transistor is optimized to obtain a good input match, i.e., $r_i = R_A = 50$ ohm, which is obtained for $W = 300$ μm. From these results, it can be seen that:

- The voltage gain is 3.17, very close to the target value
- The bandwidth is 3.83 GHz
- The noise reduction factor (δ) and the noise factor calculated from (5.84) and (5.86) with these circuit parameters are found as $\delta = 0.567$ and $F \cong 1.91$, which corresponds to $NF = 2.8$ dB
- The input matching is maintained perfectly up to 1 GHz and reasonably up to 3 GHz. The slight decrease of v_{in} at higher frequencies indicates a phase shift due to the input capacitance, which impairs the noise feedback and results in a frequency-dependent increase of δ and the consequent increase of the noise figure.

To prove the calculations of the noise reduction effect of the feedback, a simulation was performed emulating the total noise current originating from the amplifier with a sinusoidal 1 μA current source in parallel to the load and the total output noise voltage simulated for $R_A = 0$ (which corresponds to no feedback) and $R_A = 50$ ohm (which corresponds to the case of feedback). The corresponding noise

Fig. 5.29 Simulation results for $R_F = 370$ ohm and $R_L = 200$ ohm

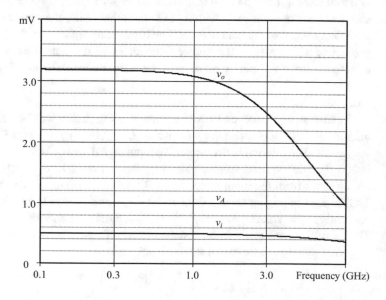

Fig. 5.30 Simulation results showing the output noise voltage (**a**) feedback effect eliminated, (**b**) with feedback for a 1 μA output noise current. The dotted curve shows the feedback voltage on R_A

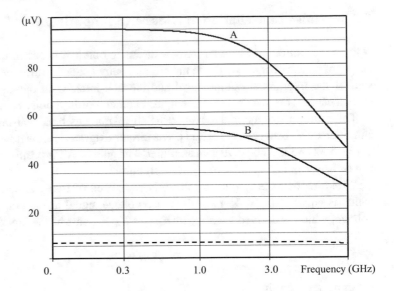

Fig. 5.31 Symmetrical version of the parallel voltage feedback amplifier

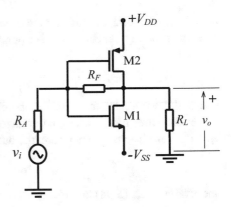

voltages at the output shown in Fig. 5.30 indicate a 0.57 noise voltage reduction thanks to the feedback, which is in very good accordance with the calculated value.

There is an obvious drawback to the self-biased amplifier shown in Fig. 5.27: the position of the DC operating point. Since $V_{DS} = V_{GS}$, the negative swing range of the output voltage is smaller than the positive swing range. This asymmetrical and limited output dynamic range impairs the linearity.

The symmetrical version of this circuit shown in Fig. 5.31 is an effective solution to increase the symmetrical output dynamic range. This circuit is none other than a CMOS inverter. The rail-to-rail symmetrical output swing capability of the circuit guarantees a good linearity around the operating point, situated in the middle of the output voltage swing range. The small-signal equivalent of this circuit is exactly the same as that of the circuit given in Fig. 2.11, where g_m and r_{ds} are the sum of the transconductances and internal resistances of M1 and M2, and all expressions derived for the original circuit are valid for the symmetrical version (See Sect.2.2).

5.5.3   Noise Cancellation in Wideband Amplifiers

Noise cancellation – in broader terms – means to obtain a negative replica of the noise signal at the output of a system and to add it to the original output signal. Since the total signal at the output is the sum of the intentional output signal and the non-intentional output noise, the generated negative replica cancels out the noise at the output. In other words, a noise cancelling feedback is established.

This concept was applied to reduce the noise of a wideband self-biased common-source feedback amplifier by F. Bruccoleri et.al. [11]. The circuit diagram is shown with the cancellation mechanism in Fig. 5.32. A3 is a negative gain amplifier. The gain of A2 is positive and small in magnitude, which can be lower than, equal to, or higher than unity.

We know that the v_o output voltage is composed of two components: the normal output signal which is equal to $(v_i \times A_{v1})$ and the noise voltage signal at the output, v_{no}. A replica of this noise voltage, which is derived from v_{no} by R_F and R_A, appears at the input node:

$$v_{noi} = v_{no} \frac{R_A}{R_A + R_F} \tag{5.93}$$

It can be seen that if

$$v_{noi} \times A_{v3} = -v_{no} \times A_{v2}$$

then the noise voltages reaching the summing node become equal in magnitude but opposite in sign, and consequently cancel each other out. The gain of A2 must fulfill

$$A_{v2} = |A_{v3}| \frac{R_A}{R_A + R_F} \tag{5.94}$$

and be "fine-tunable" for a perfect cancellation.

On the other hand, the signal voltages reaching the summing point, namely $v_i(|A_{v1}| \times A_{v2})$ and $v_i \times |A_{v3}|$ are in-phase. Therefore, the output voltage at the noise-compensated output becomes

$$v_{oNC} = v_i(|A_{v1}| \times A_{v2} + |A_{v3}|) \tag{5.95}$$

and from (5.72), (5.94) and (5.95)

$$v_{oNC} = v_i \left(A_{v2} \frac{2R_F}{R_A} \right) \tag{5.96}$$

which is almost twice as high as the voltage at the output of A1.

Fig. 5.32 Principal circuit diagram of a wideband LNA with noise cancellation feature

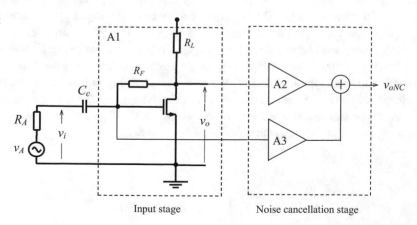

Fig. 5.33 The circuit diagram of the wideband LNA with noise cancellation feature

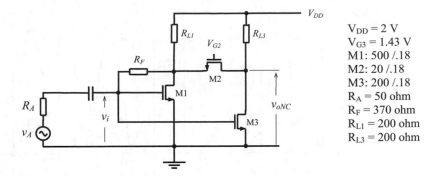

$V_{DD} = 2$ V
$V_{G3} = 1.43$ V
M1: 500 /.18
M2: 20 /.18
M3: 200 /.18
$R_A = 50$ ohm
$R_F = 370$ ohm
$R_{L1} = 200$ ohm
$R_{L3} = 200$ ohm

It must be noted that there are two realities affecting the noise cancellation advantage of this approach:

- For a perfect cancellation of the output noise of A1, the delay (or phase) characteristics of A2 and A3 must be identical along the bandwidth of the amplifier
- The noise from A2 and A3 cannot be cancelled and appear at the output of the summing circuit

The noise at the output originating from A2 and A3 depends on the structure of these amplifiers and will be demonstrated in the following example.

Example 5.6

Figure 5.33 gives the circuit diagram of a noise-cancelled wideband LNA designed with 0.18 micron CMOS technology. M1 is the feedback amplifier investigated in Example 5.5 with a low input impedance for matching. M3 is a common source amplifier providing the necessary negative voltage gain from the input node to the summing node. M2 is a grounded gate amplifier providing a positive, low voltage gain. R_{L3} is shared by M2 and M3 and serves to sum the output voltages of M2 and M3. Since the voltage gain of M2 is low (around unity) it is a low transconductance, low current, and relatively high input resistance circuit. The gate bias voltage of M2 helps to fine-tune the gain of this stage to obtain a perfect cancellation and minimum noise factor. The simulation results for the voltage gain of the amplifier from v_i is 10.66 (20.5 dB), the bandwidth is 2.25 GHz, and the power consumption is 25.2 mW.

Figure 5.34 shows the results of a SPICE simulation that observes noise cancellation. For this purpose, the total noise current generated in A1 is emulated with a 1 µA, 100 MHz sinusoidal current, parallel to R_{L1}. (a) is the noise voltage at the output of M1 while (c) is the noise voltage at the output of the circuit that indicates a perfect noise cancellation.

But it can be seen that the cancellation process is very sensitive to the bias voltage (and gain) of M2 and permits a spread of the bias voltage in the order of only a few tens of millivolts.

The simulation results given in Fig. 5.35 show the variations of the noise voltage at the output of M1, at the input of M1, and at the output of the amplifier, corresponding to a 1 µA sinusoidal noise current parallel to R_{L1}. As seen from curve (c), the noise at the output is fully cancelled out across the frequency band of the amplifier. This indicates that the signal delays of M2 and M3 are matched. The slight increase of noise at the high end of the band indicates increased discrepancy of the delays with frequency.

As previously mentioned, the cancellation process cancels out the noise component at the output port originating from M1. The noise from M2 and M3 appears at the output and dominates the noise figure of the amplifier.

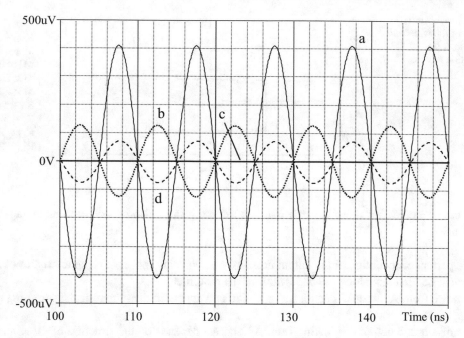

Fig. 5.34 (**a**) The emulated noise voltage at the output of M1, corresponding to a 1 μA noise current on R_{L1}. (**b**) The noise voltage at the output of the amplifier for $V_{GS2} = 1.335$ V (the phase relation with (a) indicates an unsatisfactory gain of the A1-A2 chain.) (**c**) The noise voltage at the output of the amplifier for $V_{GS2} = 1.435$ V that corresponds to a perfect noise cancellation. (**d**) Corresponds to $V_{GS2} = 1.535$ V. Note the phase reversal due to the over-increased gain of M2

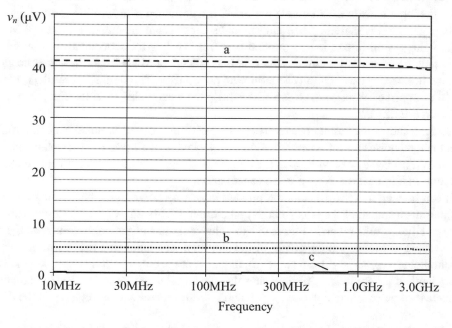

Fig. 5.35 Variation of the emulated noise originating from the first stage: (**a**) at the output of M1, (**b**) at the input of M1 fed-back over R_F, (**c**) the cancelled noise at the output of the amplifier.

Now let us calculate the noise originating from M2 and M3 for this example and find the NF of the amplifier. The transconductances of M2 and M3 are 3.37 mS and 31.47 mS, respectively, and they share the same load resistor, R_{L3}. The total noise power at the output originating from these components is

$$P_{no2,3} = 4kT\gamma(g_{m2} + g_{m3})R_L + 4kT$$

The noise voltage at the output owing to the signal source:

$$v_{nAo} = \frac{1}{2}v_{nA}(|A_{v1}||A_{v2} + |A_{v3}|)$$

$$= \frac{1}{2}v_{nA}|A_{v3}|\frac{R_F}{R_A + R_F}$$

The noise power at the output due to the signal source:

$$P_{nAo} = \frac{v_{nAo}^2}{R_L} = kTR_A|A_{v3}|^2\left(\frac{2R_F}{R_A + R_F}\right)^2\frac{1}{R_{L3}} = 4kT|A_{v3}|^2\left(\frac{R_F}{R_A + R_F}\right)^2\frac{R_A}{R_{L3}}$$

Hence the noise factor becomes

$$F = 1 + \frac{4kT(\gamma(g_{m2} + g_{m3})R_{L3} + 1)}{4kT|A_{v3}|^2\left(\frac{R_F}{R_A+R_F}\right)^2\frac{R_A}{R_{L3}}} = 1.76$$

which corresponds to $NF = 2.46$ dB, which is only slightly better than the noise figure of the non-noise-compensated first stage.

These results show that:

- The noise of the input stage is effectively cancelled out at the output.
- The voltage gain is approximately doubled.

But it must be noted that:

- The cancellation stage (for the proposed circuit M2 and M3) adds a noise comparable with the non-compensated noise of the input transistor. With a better design of the cancellation stage it is possible to reach a smaller overall noise
- For the cancellation stage, the equality of the signal delays of the paths is very important for wideband noise cancellation
- More importantly, the cancellation process is very sensitive to the gain of M2 (or M3) and needs fine-tuning
- Another drawback is the increase in the power consumption

5.6   The Limits of Usable Input Levels for LNAs

The signal voltage level delivered to the input of an LNA from the antenna may vary in a very wide interval, from very weak signals comparable to the noise level, to high-amplitude signals resulting in severe nonlinear (harmonic and intermodulation) distortion.

For weak signal levels, the signal-to-noise ratio at the output determines the lower limit of the input signal. The acceptable minimum value of the output S/N ratio depends on the area of application and the demodulation techniques applied to the incoming carrier. As mentioned in Sect. 5.4, the output S/N ratio is determined by the noise factor (F) of the amplifier, which is a measure of the internally generated noise of the amplifier, and the input S/N ratio. On the other hand, the input S/N ratio depends on the field strength of the incoming wave, the antenna, and the coupling scheme connecting the output of the antenna to the input of the amplifier.

An antenna can be considered a "transducer," delivering a voltage to its output port, proportional to the strength of the electromagnetic wave illuminating the antenna. The proportionality factor depends on the direction of the incoming wave, the structure of the antenna, and the frequency. It must be kept in mind that an antenna is a resonating system (with very few exceptions) such that the output voltage reaches its maximum at a certain frequency, depending on the structure (dimensions, shape, etc.) of the antenna. It is obvious that to maximize the input signal voltage, and correspondingly, to improve the input S/N ratio, it is necessary to operate the antenna at its resonance frequency.

The open-ended maximum output voltage of an antenna (v_o^{antenna}) can be expressed in terms of the magnitude of the electric field component of the incoming wave (E) and the maximum value of the proportionality factor (α^{antenna}), which corresponds to the resonance frequency and zero angle between the direction of the wave and the axis of the main lobe of the radiation pattern of the antenna:

$$v_o^{\text{antenna}} = \alpha^{\text{antenna}} \times E \tag{5.97}$$

Since the current of a MOS amplifier is controlled by the input signal voltage (and not the signal power), v_o^{antenna} must be transferred to the input of the amplifier with maximum possible efficiency to increase the input S/N ratio, and correspondingly, the output S/N ratio.

The signal voltage reaching the input port of the LNA depends on the interface between the output of the antenna and the input of the amplifier. If we denote the efficiency of this interface with β^{coupling}, the input signal of the amplifier corresponding to a certain value of E becomes

$$v_{\text{in}} = \alpha^{\text{antenna}} \times \beta^{\text{coupling}} \times E \tag{5.98}$$

For example, if the antenna is very close to the input of the amplifier as mentioned in Sect. 4.6.1, and if the input impedance of the amplifier is matched to the antenna at the resonance frequency, the input voltage of the amplifier is equal to one half of v_o^{antenna}, in other words $\beta^{\text{coupling}} = 0.5$. On the other hand, if the input impedance is very high compared to the internal impedance of the antenna, the input voltage of the amplifier is equal to v_o^{antenna} and $\beta^{\text{coupling}} = 1$.

If a transmission line has to be used between the output of the antenna and the input of the amplifier, the conventional approach is to match the input impedance of the amplifier to the characteristic impedance of the line and to use the antenna at a point where its impedance is real and equal to the line impedance. The β^{coupling} in this case is considerably smaller and obviously impairs the S/N ratio.

For high signal levels there is another problem – not only for LNAs but also for all kinds of amplifiers: the nonlinear distortion of the output signal, which produces harmonics of the sinusoidal components of the input signal and their intermodulation products. The amount of nonlinearity depends on the supply voltage, the load, the position of the operating point on the output characteristic curves, and obviously, on the input signal level. This basic behavior of amplifiers will be examined below on the most frequently used common source amplifier.

The nonlinearity is primarily related to the relation between the drain current and the gate-source voltage. The output voltage is equal to the voltage drop on the load. In the case of a wideband amplifier, the load is resistive up to the vicinity of the -3dB frequency. Therefore, the nonlinearity of

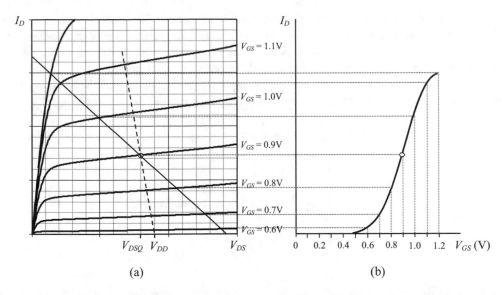

Fig. 5.36 (a) The DC load line (dashed line) and the dynamic load line (solid line) on the output characteristic of a transistor. (b) The dynamic transfer characteristic corresponding to the AC load curve shows the feedback voltage on R_A

the drain current directly reflects on the output voltage. But in the case of a tuned amplifier, the tuned load filters out the harmonics and the intermodulation products falling outside of its passband.

The drain current depends not only on the gate-source voltage but also on the drain-source voltage to some extent, and consequently, on the load. In Fig. 5.36, the "static" (DC) and the "dynamic" load lines of a common source amplifier loaded with a parallel resonance circuit are shown. The static and the dynamic load lines are determined by the DC resistance of the inductance (r_L) and the resonance impedance of the load (R_{eff}), respectively. The input-voltage-to-output-current characteristic of this circuit, called the "dynamic transfer characteristic," can be derived from Fig. 5.36a and is shown in Fig. 5.36b. The nonlinearity of the transfer curve is obvious; in order to obtain a large dynamic range for the drain current with an acceptable distortion, the operating point (Q) must be in the middle of this curve.

In the case of a resistance-loaded (wideband) amplifier, the static and dynamic characteristics are the same up to the vicinity of the upper cut-off frequency.

The nonlinearity of the drain current can be represented with a Taylor series referring to the operating point:

$$I_D = I_{DQ} + \frac{dI_D}{dV_{GS}}\bigg|_Q \Delta V_{GS} + \frac{1}{2!}\frac{d^2I_D}{dV_{GS}^2}\bigg|_Q (\Delta V_{GS})^2 + \frac{1}{3!}\frac{d^3I_D}{dV_{GS}^3}\bigg|_Q (\Delta V_{GS})^3 + \dots.$$

This expression can be converted into

$$i_d = g_1 v_{gs} + g_2 v_{gs}^2 + g_3 v_{gs}^3 + g_4 v_{gs}^4 + \dots \tag{5.99}$$

where v_{gs} and i_d are the small-signal components of the gate-source voltage and the drain current, respectively. The output signal voltage is $v_o = i_d R_{eff}$ for a tuned amplifier with a bandwidth determined by the effective quality factor of the resonance circuit.

For a wideband amplifier, the output voltage is $v_o = i_d R_D$ up to a 3 dB frequency of the circuit. The g_i coefficients of the series given in (5.99) depend on the shape of the dynamic transfer characteristic and each of them can be positive or negative. The first coefficient (g_1) in (5.99) is nothing but the g_m of the transistor for this operating point, and others are the higher order derivatives of g_m [12, 13]. Since the magnitude of the coefficients usually decreases with increasing order, the fourth and higher order terms will be neglected to keep the expressions manageable.

To evaluate the harmonic distortion, a sinusoidal input signal voltage ($v_{gs} = v_i = V_i \cos \omega t$) must be applied to the input of the amplifier. The corresponding drain current can be solved as:

$$
\begin{aligned}
i_d = & \left(\frac{1}{2} g_2 V_i^2 + \frac{3}{8} g_4 V_i^4 + \ldots \right) \\
& + \left(g_1 V_i + \frac{3}{4} g_3 V_i^3 + \ldots \right) \cos \omega t \\
& + \left(\frac{1}{2} g_2 V_i^2 + \frac{1}{2} g_4 V_i^4 + \ldots \right) \cos 2\omega t \\
& + \left(\frac{1}{4} g_3 V_i^3 + \ldots \right) \cos 3\omega t \\
& + \ldots
\end{aligned}
\tag{5.100}
$$

To evaluate the intermodulation distortion, we shall investigate the effects of two sinusoidal signals[15] with different frequencies arriving at the input of the amplifier:

$$
v_{gs} = V_1 \cos \omega_1 t + V_2 \cos \omega_2 t.
$$

For this case, the drain current becomes:

$$
\begin{aligned}
i_d = & \frac{1}{2} g_2 \left(V_1^2 + V_2^2 \right) \\
& + \left[g_1 V_1 + \frac{3}{2} g_3 \left(V_1 V_2^2 + \frac{1}{2} V_1^3 \right) \right] \cos \omega_1 t + \left[g_1 V_2 + \frac{3}{2} g_3 \left(V_1^2 V_1 + \frac{1}{2} V_2^3 \right) \right] \cos \omega_2 t \\
& + \frac{1}{2} g_2 V_1^2 \cos 2\omega_1 t \frac{1}{2} g_2 V_2^2 \cos 2\omega_2 t \\
& + \frac{1}{4} g_3 V_1^3 \cos 3\omega_1 t \frac{1}{2} g_3 V_2^3 \cos 3\omega_2 t \\
& + \frac{1}{2} g_2 V_1 V_2 \cos (\omega_1 + \omega_2) t + \frac{1}{2} g_2 V_1 V_2 \cos (\omega_1 - \omega_2) t \\
& + \frac{3}{4} g_3 V_1^2 V_2 \cos (2\omega_1 + \omega_2) t + \frac{3}{4} g_3 V_1 V_2^2 \cos (2\omega_2 + \omega_1) t \\
& + \frac{3}{4} g_3 V_1^2 V_2 \cos (2\omega_1 - \omega_2) t + \frac{3}{4} g_3 V_1 V_2^2 \cos (2\omega_2 - \omega_1) t
\end{aligned}
\tag{5.101}
$$

It is obvious that for $V_1 = 0$ or $V_2 = 0$, (5.101) reduces to (5.100).

From (5.100) it can be seen that:

- The drain current has acquired a DC term (the rectification term) due to the even harmonic coefficients. This term indicates a shift in the operating point.

---

[15] The assumption of two different signals represents the simplest case; if there are more than two signals the intermodulation occurs among all of these components.

Fig. 5.37 The linear term (**a**) and the cubic term (**b**) of the fundamental component of the drain current as a function of the input signal voltage. The solid line corresponds to the actual variation of the fundamental term of the drain current

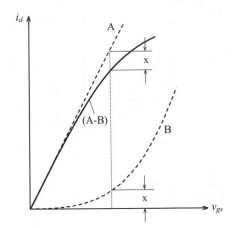

- The coefficient of the fundamental frequency has two components: a term that increases linearly with the amplitude of the signal (Fig. 5.37, line A) and a term originating from the third-order nonlinearity that increases with the third power of the amplitude of the signal (Fig. 5.37, curve B)
- For an S-shaped transfer curve, it can be seen that the sign of the component originating from the third-order nonlinearity is negative. Therefore, the actual variation of the fundamental term is equal to $(A-B)$ on (Fig. 5.37)
- This "saturation" of the drain current amplitude indicates that in the case of amplifying a modulated signal, the share of the term associated with the cubic term must be well below that of the linear term in order to maintain the relative amplitudes of the carrier and side-frequencies due to the modulation. The usually accepted criterion is to not exceed the input level corresponding to a -1dB drop at the output power, which corresponds to a factor of 1.259 reduction of the output power, or factor of 1.122 reduction of the output signal current or voltage. The input voltage corresponding to this point can be calculated from

$$\frac{g_1 V_i}{g_1 V_i - \frac{3}{4}|g_3|V_i^3} = 1.122 \quad \rightarrow \quad V_i|_{-1\text{dB}} = 0.38\sqrt{\frac{g_1}{|g_3|}} \qquad (5.102)$$

- The coefficient of the second harmonic is composed only of the even order parameters, and the coefficient of the third harmonic is composed only of the odd order parameters. These terms are not important for tuned amplifiers, since these harmonics are normally filtered out by the output tuned circuit. But for wideband amplifiers they may be in the passband of the amplifier.
- As already mentioned, symmetrical circuits eliminate the even harmonics. Another advantage of the elimination of the even harmonics is the elimination of the DC shift of the operating point due to the rectification term.

For (5.101), the above interpretations related to the rectification term as well as the second and the third harmonics are equally valid. The effects related to the intermodulation products can be interpreted as follows:

- The first order intermodulation products $(\omega_1 + \omega_2)$ and $(\omega_1 - \omega_2)$ are far from ω_1 and ω_2 and will be filtered out by the output resonance circuit for tuned amplifiers

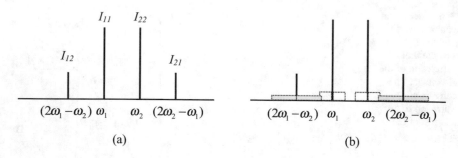

Fig. 5.38 (a) The close intermodulation products of two sinusoidal signals having equal amplitude and different frequencies. (b) Illustration of the interfering effects of one of these intermodulation products on an LNA tuned to (or to the vicinity of) this frequency

- The second-order intermodulation products ($\omega_1 + 2\omega_2$) and ($2\omega_1 + \omega_2$) are also likely to be filtered out
- The other set of the second-order intermodulation products, ($2\omega_1 - \omega_2$) and ($2\omega_2 - \omega_1$), (we shall name them the "close intermodulation products") are critical. If the difference between ω_1 and ω_2 is $\Delta\omega$, for example, $\omega_2 = \omega_1 + \Delta\omega$, these second-order intermodulation products become

$$(2\omega_1 - \omega_2) = \omega_1 - \Delta\omega$$
$$(2\omega_2 - \omega_1) = \omega_2 + \Delta\omega$$

In Fig. 5.38a, the relative positions of the fundamental components and these intermodulation products are shown for $V_1 = V_2 = V_i$. In reality ω_1 and ω_2 are modulated and have side-(modulation) bands having a certain width. Note that the side bands of $2\omega_1$ and $2\omega_2$ are twice as large as those of ω_1 and ω_2. In Fig. 5.38b, these four frequencies are shown with their modulation side-bands. It can be easily seen that the interfering effects of the intermodulation products are more severe compared to the interfering effects of the two neighboring channels, ω_1 and ω_2.

It must be noted that in the case of a wideband LNA, all these components existing in the drain current produce corresponding output voltage components up to the 3 dB frequency of the amplifier. In this case, an input filter is necessary to exclude all signals other than the carrier (and its side frequencies) intended to be received.

From (5.101), it can be seen that the magnitudes of the fundamentals and these second-order intermodulation products are affected by g_3 and rapidly increase with the amplitudes of the ω_1 and ω_2 components, $V_1 = V_2 = V_i$, as described in the following.

The magnitudes of the ω_1 and ω_2 components of the drain current:

$$I_{11} = I_{22} = g_1 V_i + \frac{9}{4} g_3 V_i^3 = g_1 V_i - \frac{9}{4} |g_3| V_i^3 \tag{5.103}$$

The magnitudes of ($2\omega_1 - \omega_2$) and ($2\omega_2 - \omega_1$) components:

$$I_{12} = I_{21} = \frac{3}{4} |g_3| V_i^3 \tag{5.104}$$

In the case of a tuned amplifier, if all these components are close to each other and are in the passband of the amplifier, the corresponding components of the output voltage become

$$V_{11} = V_{22} = R_{\text{eff}} g_1 V_i - \frac{9}{4} R_{\text{eff}} |g_3| V_i^3 \tag{5.105}$$

and

$$V_{12} = V_{21} = \frac{3}{4} R_{\text{eff}} |g_3| V_i^3 \tag{5.106}$$

Therefore, the ratio of the magnitude of one of the fundamental components to the magnitude of one of the close intermodulation products can be written as

$$\frac{V_{11}}{V_{12}} = \frac{R_{\text{eff}} g_1 V_i - \frac{9}{4} R_{\text{eff}} |g_3| V_i^3}{\frac{3}{4} R_{\text{eff}} |g_3| V_i^3} = \frac{4}{3} \frac{g_1}{|g_3|} \frac{1}{V_i^2} - 3 \tag{5.107}$$

The numerical value of this ratio for the input level corresponding to the -1 dB point can be calculated from (5.102) and (5.107):

$$\left.\frac{V_{11}}{V_{12}}\right|_{-1\text{dB}} = \frac{4}{3} \frac{g_1}{|g_3|} \frac{1}{V_i^2} - 3 = \frac{4}{3} \frac{V_i^2}{0.145} \frac{1}{V_i^2} - 3 = 6.195 \quad \rightarrow \quad 15.84 \text{ dB} \tag{5.108}$$

This relation provides a hint for a method to find the input level corresponding to the -1 dB point using a relatively easy intermodulation measurement.

Another metric to evaluate the nonlinearity of an amplifier is the input level corresponding to the so-called "third-order intercept point," or "IIP3" in short. IIP3 is defined as the intercept point of the line corresponding to the linear term of one of the fundamental components of (5.101) and the line corresponding to the magnitude of one of the close intermodulation components shown on log-log axes or in dB scales (Fig. 5.39).

The input levels corresponding to the IP3 can be calculated from (5.99):

$$g_1 V_i = \frac{3}{4} |g_3| V_i^3 \quad \rightarrow \quad V_i|_{\text{IP3}} = 1.154 \sqrt{\frac{g_1}{|g_3|}} \tag{5.109}$$

The input voltage corresponding to the -1 dB point was calculated as:

$$V_i|_{-1\text{dB}} = 0.38 \sqrt{\frac{g_1}{|g_3|}} \tag{5.110}$$

Therefore

Fig. 5.39 Definition of IP3, the third-order intercept point

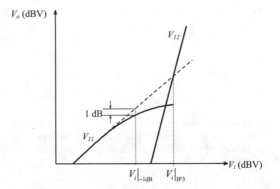

$$\frac{V_i|_{\text{IP3}}}{V_i|_{-1\text{dB}}} = 3.037 \quad \rightarrow \quad 9.65 \text{ dB} \tag{5.111}$$

This expression shows that the input level corresponding to the IP3 is well above the level corresponding to the -1 dB point, and practically not applicable due to the excessive nonlinearity[16].

References

1. A. van der Ziel, *Noise in Solid State Devices and Circuits* (Wiley, New York, 1986)
2. R.P. Jindal, Noise associated with distributed resistance of gate structures in integrated circuits. IEEE Trans. Electron Devices **31**(10), 1505–1509 (1984)
3. A.J. Scholten, L.F. Tiemeijer, R. van Langevelde, Noise modeling for RF CMOS circuit simulation (Invited paper). IEEE Trans. Electron Devices **50**(3), 618–632 (2003)
4. G. Knoblinger, P. Klein, M. Tiebout, A new model for thermal channel noise of deep submicron MOSFET'S and its application in RF-CMOS design. IEEE J. Solid State Circuits **36**(5), 831–837 (2001, May)
5. S. Asgaran, M.J. Deen, Analytical modeling of MOSFET's channel noise and noise parameters. IEEE Trans. Electron Devices **51**(12), 2109–2114 (2004, December)
6. Z.Q. Lu, Y.Z. Ye, *A Simple Model for Channel Noise of Deep Submicron MOSFETs*. IEEE Conference on Electron Devices and Solid-State Circuits, December 2005
7. C.H. Chen, M.J. Deen, Channel noise modeling of deep submicron MOSFETs. IEEE Trans. Electron Devices **49**(8), 1484–1487 (2002, August)
8. M.J. Deen, C.H. Chen, S. Asgaran, G.A. Rezvani, J. Tao, Y. Kiyota, High frequency noise of modern MOSFETs: compact modeling an measurement issues (Invited paper). IEEE Trans. Electron Devices **53**(9), 2062–2081 (2006 September)
9. V. Szekely, M. Rencz, B. Courtois, Tracing the thermal behavior of ICs. IEEE Des. Test Comput. **15**(2), 14–21 (1998)
10. K. Shadron et al., *HotSpot: Techniques for Modeling Thermal Effects at the Processor-Architecture Level*. THERMINIC Workshop, 1–4 October 2002, Madrit.
11. F. Bruccoleri, E.A.M. Klumpering, B. Nauta, Wide-band CMOS low-noise amplifier exploiting thermal noise cancelling. IEEE J. Solid State Circuits **39**, 275–282 (2004, February)
12. P.M. Jupp, D.R. Webster, Application of Derivative Superposition to low-IM3 Distortion IF Amplifiers, in *Workshop on RF Circuit Technology, A CMP Europe Conference*. (Cambridge, UK). (2002 March)
13. B. Razavi, *RF microelectronics*, 2nd edn. (Prentice Hall, Upper Saddle River, 2011)

---

[16] Although the intercept point lies outside the usable limits of amplifiers, the reason behind the widespread acceptance of the IP3 as a metric of nonlinearity is the simplicity of its determination by a relatively easy measurement procedure.

RF Oscillators

Oscillators, especially sinusoidal oscillators, are among the main components of all RF systems. They are used as pilot oscillators to generate the carrier frequencies of transmitters and as local oscillators in receivers to generate the signals that are necessary for frequency conversion purposes. The important features of sinusoidal oscillators are the frequency of oscillation and its spectral purity, the frequency stability and the phase stability. The RF sinusoidal oscillators are almost exclusively based on the resonance effect.

In certain applications, it is necessary to adjust the frequency of oscillation in a certain range, or fine-tune the frequency to a predetermined value. Since this feature is usually realized by a varactor in the resonance circuit of the oscillator, they are called voltage-controlled oscillators (VCOs). In some other applications, the frequency of oscillation must be fixed to a certain frequency with maximum possible precision and stability. For these cases, quartz crystal oscillators have to be used. If the target frequency is higher than the range supported by the crystal oscillator, a higher frequency VCO can be "locked" to the frequency of the crystal oscillator.

As already mentioned in Chap. 4, when we excite an LC circuit, it starts to swing at a frequency determined by the values of its components. The magnitude of the swing decreases in time, due to the losses of the system. If the losses of the system are compensated in some way, the magnitude of the swing remains constant; in other words, the system "oscillates." There are two well-known ways of examining the loss compensation mechanisms of the resonance circuit: the negative resistance approach and the feedback approach.

6.1 The Negative Resistance Approach to LC Oscillators

In Fig. 6.1, a series resonance circuit is shown. As it was shown in Sect. 4.1.2, the natural frequency of the circuit is

$$\omega_0 = \sqrt{\frac{1}{LC}}$$

and the effective series resistance is r_{eff}, which represents the total losses of L and C. At ω_0 the reactances of L and C cancel each other and the input impedance becomes equal to its real part, i.e.

The original version of this chapter was revised. The correction to this chapter is available at https://doi.org/10.1007/978-3-030-63658-6_8

D. Leblebici, Y. Leblebici, *Fundamentals of High Frequency CMOS Analog Integrated Circuits*,
https://doi.org/10.1007/978-3-030-63658-6_6, corrected publication 2021

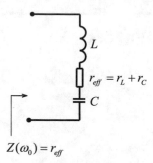

Fig. 6.1 The input impedance of a series resonance circuit at resonance

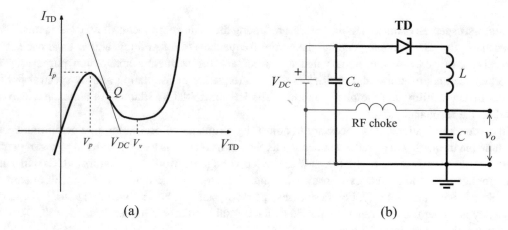

Fig. 6.2 (a) Typical current-voltage characteristic of a tunnel diode. The peak and valley voltages, V_p and V_v, depend on the semiconductor and are in the range of hundreds of millivolts. (b) Schematic of a tunnel diode oscillator. The bias circuit is drawn with fine lines. The series resistances of L and the RF choking coil determines the DC load line shown in (a)

r_{eff}. If we connect a "negative resistance" in series, equal in magnitude to r_{eff}, the total resistance of the circuit becomes zero and the circuit is now ready to oscillate under *any* excitation – for example, noise, which always exists in the system.

A "negative resistance" does not exist as a physical reality but some devices or circuits exhibit a negative resistance behavior in a limited range of their current-voltage characteristic. This means that the slope of its current-voltage characteristic is negative, in other words, if the voltage increases, the current decreases.

There are a number of two-terminal negative resistance devices: for example, the "tunnel diode" which is also known as the "Esaki diode." The tunnel diode is a semiconductor p-n junction, where both sides are doped up to the degenerate level, i.e., the Fermi levels of the p and n-type sides are shifted into the valence band and the conduction band, respectively. Due to the "quantum-mechanical tunneling" of electrons in a range of the bias voltage, the voltage-current characteristic exhibits a negative slope as shown in Fig. 6.2a.[1]

---

[1] For a clear explanation of the tunnel diode, refer to Ref. [1].

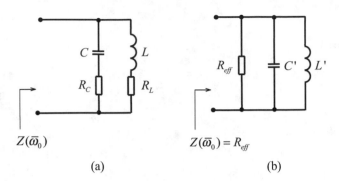

Fig. 6.3 (a) A typical on-chip parallel resonance circuit with lossy inductance and capacitance. (b)   Same circuit in terms of its parallel components at $\overline{\omega}_0$

If a tunnel diode is biased at a point in the middle of the negative resistance region, the corresponding small-signal resistance becomes negative and equal in magnitude to the inverse of the slope of the curve at this point. The value of this resistance is usually small, in the range of tens of ohms, therefore suitable to compensate for the effective series resistance of a series resonance circuit. Since the parasitics can be made very small, tunnel diodes can be used at frequencies up to several tens of GHzs.[2]

The schematic of a tunnel diode oscillator is shown in Fig. 6.2b. As already mentioned, if the negative resistance corresponding to the slope of the characteristic curve at the DC operating point Q is equal to the effective resistance of the resonance circuit, oscillation starts. But there is a risk of the oscillation ceasing owing to the variations of the parameters that may decrease the value of the negative resistance. To guarantee sustained oscillation, the magnitude of the negative resistance must be higher than the effective resistance of the resonance circuit. In this case, the amplitude of oscillation tends to increase steadily. But since the range of the negative resistance is limited, the amplitude can increase up to the points where the magnitude of the negative resistance drops to r_{eff} and stabilizes itself at the magnitude corresponding to these points.

There are a number of transistor circuits that exhibit a negative resistance port. One of them is the input port of a source follower under appropriate operating and loading conditions, which was investigated in Sect. 3.2. and used for Q enhancement in Example 4.2. Another and most frequently used one is the cross-connected differential negative resistance circuit. Since the magnitude of the negative resistance of this circuit is relatively high, it is convenient to use it to compensate for the relatively high effective parallel resistance of a parallel resonance circuit at resonance.

The general form of a parallel resonance circuit is shown in Fig. 6.3a. The frequency for which the input impedance is real is denoted with $\overline{\omega}_0$. The input impedance of the circuit can be represented in terms of its parallel components at $\overline{\omega}_0$, as shown in Fig. 6.3b, where

$$L' = L + \frac{R_L^2}{\overline{\omega}_0^2 L} \tag{6.1}$$

---

[2] Tunnel diodes are fabricated and packaged as discrete devices. But there are serious efforts to integrate the tunnel diodes into silicon ICs (see the publications of A. Seabaugh).

Fig. 6.4 The cross-connected differential negative resistance circuit

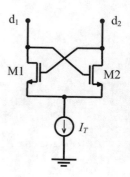

$$C' = \frac{C}{1 + \overline{\omega}_0^2 C^2 R_C^2} \tag{6.2}$$

$$\overline{\omega}_0 = \sqrt{\frac{1}{L'C'}} = \frac{1}{LC} \times \sqrt{\frac{L - CR_L^2}{L - CR_C^2}} \tag{6.3}$$

$$R_{eff} \cong \left(Q_L^2 + 1\right) \frac{R_L^2}{R_L + R_C} \tag{6.4}$$

If a negative resistance equal in magnitude to R_{eff} is connected in parallel to this circuit, it cancels out R_{eff} and the remaining parallel resonance circuit oscillates at $\overline{\omega}_0$, which is the oscillation frequency of the circuit and will be shown with ω_{osc}.

The schematic diagram of a cross-connected negative resistance circuit is given in Fig. 6.4. It can be easily seen that the circuit has two robust stable operating points; M1 is off, M2 is on and the whole I_T tail current flows through T2, and visa-versa. But there is another operating condition that is stable in theory but unstable in practice, which corresponds to $I_{D1} = I_{D2} = (I_T / 2)$. Any small (even infinitesimal) change in this condition impairs the stability of the operating points and the circuit swings to one of the robust operating points.

For this critical operating condition, the small-signal admittance seen from the d_1-d_2 output port can be calculated from the equivalent circuit shown in Fig. 6.5a as

$$y_o = -\frac{1}{2}(g_m - g_{ds}) + \frac{1}{2} C_{gs} \tag{6.5}$$

that corresponds to a negative resistance of

$$r_o = -\frac{2}{(g_m - g_{ds})} \cong -\frac{2}{g_m} \tag{6.6}$$

parallel to a capacitance equal to $C_o = C_{gs}/2$.

g_m in (6.6) depends on the technology, the dimensions of the transistors, the drain DC currents, and also whether the transistor is operating in the velocity-saturated region or not. For velocity saturation, the transconductance is considerably small and cannot be controlled with the tail current. Note that expression (1.33) must be used for non-velocity saturated operation and (1.34) for velocity saturated operation to calculate the transconductance.

This negative resistance calculated from the linearized small-signal equivalent circuit is valid only in the middle of the operating range that corresponds to $I_{D1} = I_{D2}$, and tends to infinity at both ends of

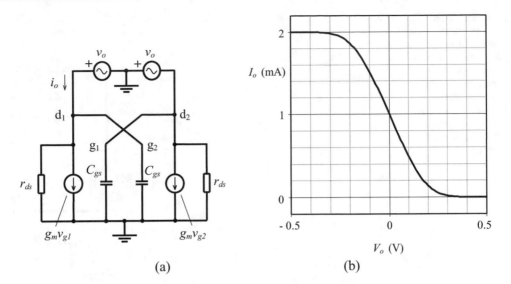

Fig. 6.5 (a) The small-signal equivalent to calculate the output admittance of a cross-connected differential negative resistance circuit. (**b**) pSpice simulation result of the output current-voltage characteristic (AMS035, 200 µm/0.35 µm, $I_T = 2$ mA)

the range. To observe this nonlinear behavior of the output resistance, a simulation result is given in Fig. 6.5b.

It is seen from this curve that the negative resistance is minimum in the mid-point that corresponds to the value calculated from the small-signal equivalent circuit, and steadily and symmetrically increases on both sides. For a parallel resonance circuit whose effective parallel resistance is higher than the negative resistance corresponding to the mid-point of the curve, the circuit oscillates. As explained for the tunnel diode oscillator, the nonlinearity of the curve limits the amplitude of the oscillation at a certain level that depends on g_m and can be controlled by I_T. The average negative resistance corresponding to a sustainable oscillation can be considerably higher than the mid-point resistance.

As an example, in Fig. 6.5b, the negative resistance at the mid-point is 200 ohms and the resistance corresponding to the on-set of limitation is about 300 ohms. Therefore, to obtain a sustained oscillation with a safety margin, the resonance impedance of the LC circuit connected in parallel to the d_1-d_2 port must not be smaller than 300 ohms.

Another aspect (and advantage) of this circuit is that due to the symmetry of the curve with respect to the mid-point, the even harmonics in the spectrum of the drain currents – and consequently on the differential output voltage – is zero,

The schematic of an oscillator that uses a cross-connected negative resistance circuit is given in Fig. 6.6, which is also known as a differential oscillator. As explained for the differential LNAs, the two halves of the inductor can be coupled or not. The total equivalent value of the inductance is equal to

$$L_T = 2L(1 + k)$$

where k is the coupling coefficient and its value is approximately 0.7 for a center-tapped inductor as shown in Fig. 1.40, and can be considered as zero for two separate inductors on the chip.

Fig. 6.6 The differential
negative resistance
oscillator. The differential
output of the oscillator is
the d1-d2 port. The load
connected to this port
certainly affects the
oscillation frequency and
the oscillation condition
and therefore must be taken
into account

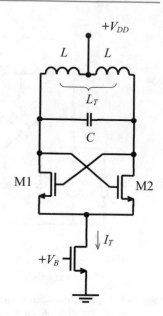

Example 6.1

A differential oscillator for $f_o = 1.6$ GHz will be designed. The inductor to be used will be selected from a library in which there are 1 nH, 2 nH, 5 nH, and center-tapped 10 and 20 nH inductors with $Q_{eff} = 7$. The transistors have characteristics similar to AMS 035 micron technology. It is assumed that a 1 V bias voltage is available on the chip. The allowed DC supply current is 2 mA.

The schematic of the circuit together with the tail current source is given in Fig. 6.6. Since the output voltage swings around V_{DD}, the supply voltage must be chosen lower than the allowed maximum value by the expected amplitude of the oscillation with a safety margin, and must be checked later on. We will use $V_{DD} = 2.5$ V that corresponds to 1 V maximum amplitude, the safety margin included.

To obtain the oscillation with a relatively high negative resistance that corresponds to a small transconductance (i.e. small supply current and small transistor width), the resonance impedance must be as high as possible. Since the resonance impedance of the resonance circuit is $R_p = L\omega_0 \times Q_{eff}$, the first choice would be to use the maximum inductance value, which is 20 nH in our case. This corresponds to a resonance capacitance equal to

$$C = \frac{1}{\omega_0^2 L_T} = \frac{1}{\left(2\pi \times 1.6 \times 10^9\right)^2 \times \left(20 \times 10^{-9}\right)} = 495 \text{ fF}$$

This capacitance is the sum of $(C_{gs}/2)$, $(C_{db}/2)$, the connected resonance capacitor (or varactor) and the input capacitance of the following circuit, which can be in the range of several hundreds of femtofarads. For our case, it is preferred to work with a higher resonance capacitance that forces us to use a 10 nH inductor that resonates at 1.6 GHz with 905 fF, which is more reasonable.

The effective parallel resistance of this resonance circuit at 1.6 GHz is

$$R_{eff} = L\omega_0 Q_{eff} = 10^{-8} \times \left(1.05 \times 10^{10}\right) \times 7 = 735 \text{ ohm}$$

Since we will work with a considerable safety margin, the output resistance of the transistor and the losses of the resonance capacitance can be neglected. The value of the negative resistance must be smaller than this theoretical value to guarantee sustained oscillation, so that the amplitude reaches the saturation end of the negative resistance curve. Let us start with $r_o = -350$ ohm, that according to (6.6), corresponds to $g_m = 5.7$ mS.

The aspect ratio to obtain this transconductance value with 1 mA drain DC current can be calculated from (1.33):

$$\frac{W}{L} = \frac{g_m^2}{2\mu C_{ox}I_D} = \frac{\left(5.7 \times 10^{-3}\right)^2}{2 \times 325 \times \left(4.54 \times 10^{-7}\right) \times 10^{-3}} = 110$$

Therefore the gate width must be 38 μm.

The gate width of the tail current source can be sized (or for better accuracy determined by simulation) to allow 2 mA under 1 V gate bias voltage, which is found as 62 μm for 0.35 μm channel length.

The PSpice transient simulation file is given below:

DIFFERENTIAL NEGATIVE RESISTANCE OSCILLATOR

```
vdd 100 0 2.5
M1 1 2 3 0 MODN w=38e-6 l=.35e-6 ad=13.3e-12 as=13.3e-12 pd=40u ps=40u nrd=.01 nrs=.01
M2 2 1 3 0 MODN w=38e-6 l=.35e-6 ad=13.3e-12 as=13.3e-12 pd=40u ps=40u nrd=.01 nrs=.01
MT 3 4 0 0 MODN w=62e-6 l=.35e-6 ad=21.7e-12 as=21.7e-12 pd=40u ps=40u nrd=.005
    nrs=.005
.LIB "ams035.lib"
VB 4 0 1
L1 1 11 5n
R1 11 100 7.5
L2 2 12 5n
R2 12 100 7.5
C1 1 2 .9P
.IC V(1)=2.6
.TRAN .03n 51N 50N .01n
.probe
.end
```

The simulation results with calculated values are shown in Fig. 6.7. It must be noted that the amplitude of the voltage swing on the drain nodes is half of the differential output voltage v_o. This means that the maximum drain voltages are well below the permitted maximum voltage, 3.5 V.

For the transient simulation of oscillators, it must be noted that:

- To start the oscillation, it is necessary to impose a suitable "initial condition" to the circuit. For example, for this circuit, the initial voltage of d_1 was given as 2.6 V.
- To be sure that the oscillation is sustained, it is necessary to observe the output signal after a reasonable time from the beginning.

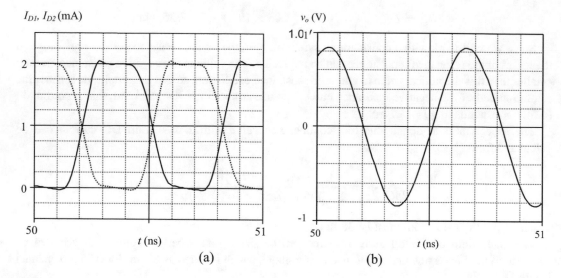

Fig. 6.7 (a) The drain current waveforms of the drain currents. (b) The differential output voltage waveform

Fig. 6.8 The block
diagram of a feedback
amplifier

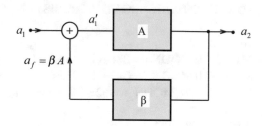

- The distortion (flattening of the peaks) on the currents of M1 and M2 indicates that the circuit is operating in the entire range of the negative resistance region. It is wise to adjust the circuit such that the peaks are slightly flattened. This means that the circuit has a safety margin against parameter changes and the distortion reflected to the output voltage waveform is not excessive.

6.2   The Feedback Approach to LC Oscillators

The classical approach to study oscillators is based on the oscillation criterion stated by H. Georg Barkhausen in 1921 for feedback amplifiers. In Fig. 6.8, the general form of a single loop feedback amplifier – which is assumed linear – is shown. Here, the gain block is usually an amplifier with a gain function A that can be a voltage gain, a current gain, a transadmittance or a trans-impedance. The block β is the feedback circuit that is usually (but not necessarily) a passive two-port. The signals a_1, a_1', a_f, and a_2 can be either voltage or current, but a_1, a_1', and a_f obviously have to have the same dimension. The basic relations among the signals are

$$A = \frac{a_2}{a_1'}, \quad \beta = \frac{a_f}{a_2}, \quad a_1' = a_1 + a_f \quad \text{and} \quad A_f = \frac{a_2}{a_1} \tag{6.7}$$

where A is the gain of the amplifier, β is the transfer function of the feedback block and A_f is the gain of the circuit as a whole, that is called the "gain with feedback" or the "gain of the feedback amplifier."

From the basic relations given in (6.7), the gain of the feedback amplifier can be easily solved as

$$A_f = \frac{A}{(1 - \beta A)} \tag{6.8}$$

where βA is called the "loop gain."

From this expression, it can be seen that:

- If the magnitude of $(1 - \beta A)$ is bigger than unity, the gain of the feedback amplifier is smaller than A, in other words, the gain is reduced by feedback. This is called "negative feedback."
- If the magnitude of $(1 - \beta A)$ is smaller than unity, the gain of the feedback amplifier is higher than A, in other words, the gain is increased by feedback. This is called "positive feedback."
- As a special case of positive feedback, if

$$(1 - \beta A) = 0 \text{ or, equally, } \beta A = 1 \tag{6.9}$$

the gain of the feedback amplifier goes to infinity.

Since the gain of the amplifier block and the transfer function of the feedback block are frequency-dependent, in other words, A and β are complex quantities in the frequency domain, (6.9) must be written as

$$\begin{aligned}
\beta A &= 1 + j.0 \quad \text{or} \\
\text{Re}\{\beta A\} &= 1 \text{ and } \text{Im}\{\beta A\} = 0 \quad \text{or} \\
|\beta A| &= 1 \quad \text{and} \quad \varphi(\beta A) = 0
\end{aligned} \tag{6.10}$$

According to Barkhausen, infinite gain at a frequency where (6.10) is satisfied means that the circuit will oscillate at that frequency. The amplitude of the oscillation will eventually be limited in a certain range by the amplifier, where the assumption of linearity is – at least approximately – valid.

As an example of this approach, we will examine the MOS transistor version of one the classical oscillator circuits: the Colpitts oscillator, which was first developed by Edwin H. Colpitts in 1919 using vacuum tubes. The simplified schematic of the circuit and its block representation are shown in Fig. 6.9a and b, respectively. The input capacitance of the transistor is included in C_1. The load of the amplifier, Y_L, is the admittance shown in Fig. 6.9b and can be calculated as

$$\begin{aligned}
Y_L &= sC_2 + \cfrac{1}{sL + r + \cfrac{1}{sC_1}} \\
&= \frac{s^3 LC_1 C_2 + s^2 C_1 C_2 r + s(C_1 + C_2)}{s^2 LC_1 + sC_1 r + 1}
\end{aligned} \tag{6.11}$$

Then the open-loop voltage gain of the transistor is

$$A_v = -g_m \frac{1}{(g_{ds} + Y_L)}$$

and in open form

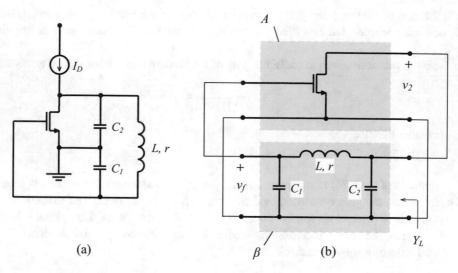

Fig. 6.9 (a) The simplified schematic of a MOS Colpitts oscillator. (b) The circuit redrawn with the amplifier block and the feedback block

$$A_v = -g_m \frac{s^2 LC_1 + sC_1 r + 1}{s^3 LC_1 C_2 + s^2(C_1 C_2 r + LC_1 g_{ds}) + s[(C_1 + C_2) + rC_1 g_{ds}] + g_{ds}} \qquad (6.12)$$

The voltage transfer ratio of the feedback block can be calculated as

$$\beta = \frac{1}{s^2 LC_1 + sC_1 r + 1} \qquad (6.13)$$

Hence the loop gain βA_v becomes

$$\beta A_v = -g_m \frac{1}{s^3 LC_1 C_2 + s^2(C_1 C_2 r + LC_1 g_{ds}) + s[(C_1 + C_2) + rC_1 g_{ds}] + g_{ds}}$$

If we convert this expression into the ω domain and apply the Barkhausen criterion, stating that the loop gain must be real and equal to unity, from $\text{Im}\{\beta A_v\} = 0$ the oscillation frequency that satisfies this condition can be calculated as

$$\omega_{osc} = \omega_0 \sqrt{1 + r \cdot g_{ds} \frac{C_2}{C_1 + C_2}} \qquad (6.14)$$

Note that the oscillation frequency is not equal to the natural frequency of the resonance circuit but slightly different, which will be discussed later on.

Similarly, the oscillation condition that gives the minimum value of the transconductance can be found from $\text{Re}\{\beta A_v\} = +1$ or $|\beta A_v| = 1$ as

$$g_m \cong \frac{1}{r_{eff}} \frac{C_1 + C_2}{C_2} \qquad (6.15)$$

It must be noted that in order to maintain the oscillations, g_m must be higher than this value to provide a safety margin. It is convenient to modify (6.15) as

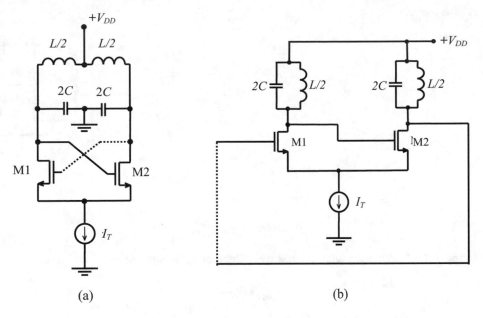

Fig. 6.10 (a) The cross-connected oscillator. (b) The same circuit drawn as a feedback oscillator

$$g_m = k_s \frac{1}{r_{eff}} \frac{C_1 + C_2}{C_2} \tag{6.15a}$$

where k_s is a safety factor, usually 1% to 10% bigger than unity (i.e. between 1.01 and 1.1).

Another interesting example is to calculate the oscillation frequency and condition of the cross-connected oscillator that was given in Fig. 6.6, with the feedback approach. The circuit is redrawn in Fig. 6.10a where the resonance circuit is divided into two identical parts as in Fig. 5.23. In Fig. 6.10b, the circuit is drawn as a feedback amplifier. The total voltage gain of the two cascaded stages is positive and equal to the multiplication of the gains of the identical stages:

$$A_{vT}(\omega_0) = \left(-g_m \cdot R'_{eff}\right)^2$$

where R'_{eff} is the effective parallel resistance of the individual loads at resonance and equal to

$$R'_{eff} = \left(\frac{L}{2}\right) \omega_0 \times Q_{eff}$$

Therefore the total gain becomes

$$A_{vT}(\omega_0) = g_m^2 \frac{1}{4} \left(L\omega_0 \times Q_{eff}\right)^2 = g_m^2 \frac{1}{4} R_{eff}^2$$

where R_{eff} is the effective parallel resistance of the circuit composed of L and C.

Since the output of the amplifier is directly connected to its input, β is equal to unity. Then the loop gain is

Fig. 6.11 The common
drain version of the
Colpitts oscillator

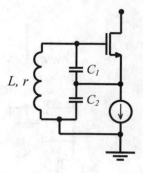

$$\beta A_{vT}(\omega_0) = g_m^2 \frac{1}{4} R_{eff}^2$$

If we apply the Barkhausen criterion and write the loop gain equal to unity, we obtain

$$g_m = \frac{2}{R_{eff}}$$

that is equivalent to the expression (6.6) obtained with the "negative resistance approach."[3]

Problem 6.1
The schematic of a common drain Colpitts oscillator is shown in Fig. 6.11.

(a) Derive expressions to calculate the oscillation frequency and the minimum value of the transconductance for sustained oscillations.
(b) Compare the circuit with the negative resistance circuit used for Q enhancement given in Fig. 4.7, and discuss.

6.3 Frequency Stability of LC Oscillators

The oscillation frequency of an LC oscillator is mainly determined by the values of the inductor and the capacitor of the resonance circuit. It is also influenced by some secondary factors, namely the losses of the resonance circuit and certain parameters of the transistor. For example, the oscillation frequency of a MOS Colpitts oscillator was given as (6.14). If we combine this expression with (6.15a), it is possible to arrange it as follows:

$$\omega_{osc} = \frac{1}{\sqrt{LC}} \sqrt{1 + k_s \frac{g_{ds}}{g_m} \frac{1}{Q^2} \frac{C_1}{C_2}} \tag{6.16}$$

where C is the series equivalent of C_1 and C_2, and Q is the quality factor of the resonance circuit.

This frequency is a function of a number of parameters and can vary depending on the tolerances or the variations of these parameters due to some effects, for example, temperature and supply voltage fluctuations. The influences of L and C tolerances are well defined and can be compensated with the

---

[3] This is an interesting example to see how different approaches converge for a certain physical reality.

trimming of frequency, for example with the aid of a varactor used as part of the resonance capacitance. The variations of other parameters are usually unpredictable and consequently more severe.

The effects of the small changes of parameters on the oscillation frequency can be expressed with an exact differential. For example, for a Colpitts oscillator, the change of the oscillation frequency due to the small variations of the parameters is

$$d\omega_{osc} = \frac{\partial \omega_{osc}}{\partial L} dL + \frac{\partial \omega_{osc}}{\partial C_1} dC_1 + \frac{\partial \omega_{osc}}{\partial C_2} dC_2 + \frac{\partial \omega_{osc}}{\partial g_m} dg_m + \frac{\partial \omega_{osc}}{\partial g_{ds}} dg_{ds} + \frac{\partial \omega_{osc}}{\partial Q} dQ$$

The change of the value of an on-chip inductance is mainly related to the dimensional changes due to temperature variations. The resistance of the inductor that affects the quality factor is also subject to change, not only due to the dimensional changes but also to the temperature coefficient of the material (aluminum or copper).

The stability of an on-chip capacitor depends on its construction. The values of the MIM and poly-1-poly2 capacitors are subject to variations due to thermal expansion. The capacitances of the varactors are sensitive to the bias voltage variations.

The small-signal output conductance and the transconductance of the transistor depend on the voltage and the current of the operating point.

As a conclusion, it can be stated that for the stability of the oscillation frequency of a Colpitts oscillator, the variation of the operating point of the transistor and the bias voltages of varactors must be well stabilized, and the variations of the temperature must be small. It can be shown that these conclusions are equally valid for other types of feedback oscillators.

(6.16) also shows that if the effective quality factor of the resonance circuit is high, the oscillation frequency approaches the natural frequency of the resonance circuits that depends only on L and C, and the overall sensitivity of the oscillation frequency decreases.[4]

If we solve C_1/C_2 from (6.15) and insert it into (6.16) then we obtain

$$\omega_{osc} = \frac{1}{\sqrt{LC}} \sqrt{1 + \frac{g_{ds}}{g_m} \frac{1}{Q^2} \left(g_m R_{eff} - k_s\right)} \tag{6.17}$$

According to this expression, if the condition $g_m R_{eff} = k_s$ is satisfied, the oscillation frequency becomes equal to the natural frequency of the resonance circuit. This corresponds to the maximum available frequency stability.

As another example, let us look at (6.3) which was derived for a negative resistance oscillator. From this expression, it can be seen that if the series resistance of the capacitance branch of the parallel resonance circuit is equal to that of the inductance branch, the oscillation frequency becomes equal to the natural frequency of the resonance circuit.[5] Since the oscillation frequency depends on only L and C in this case, the frequency stability improves,[6] despite the decrease in the effective quality factor of the resonance circuit.

From these examples, we understand that there are some measures to improve the stability of the oscillation frequency of integrated oscillators to some extent, but not sufficiently for applications that need very high frequency stability.

---

[4] This fact is applicable to the oscillator circuits constructed with discrete hi-Q inductors and high-quality capacitors, and extensively used in the past.

[5] This phenomenon was investigated in 1934 by J. Groszkowski from a different and more basic point of view [2].

[6] This behavior was underlined in Sect. 4.1.1. Since the natural frequency, the frequency where the imaginary part of the impedance is zero, and the frequency corresponding to the maximum of the magnitude of the impedance coincide, it is useful also for feedback oscillators.

The classical method to obtain oscillations with very high frequency stability is to use a quartz crystal that electrically behaves as a very high Q resonance circuit.

6.3.1   Crystal Oscillators

A thin prism cut and lapped from a native quartz crystal is shown in Fig. 6.12. The x, y, and z axes are called the electrical, mechanical, and optical axes, correspondingly. Thin metal films are vapor-deposited to make electrical connections to the faces parallel to the x-y plane. If a DC voltage is applied to the X-X' port, the crystal expands or shrinks in the y-direction, according to the polarity of the voltage. Correspondingly, if a pressure or tension is applied in the y-direction, a voltage occurs between X and X'. This effect is called the "piezo-electric effect." It can be understood that under an alternating voltage the crystal starts to vibrate. If the frequency of the signal becomes equal to the mechanical resonance frequency of the crystal[7] the amplitude of the vibration becomes maximum, in other words, the crystal is in mechanical resonance. The losses of this resonance system are the internal friction losses of the crystal, the load corresponding to the connections and the friction of the air. With appropriate precautions (for example, operating in a vacuum), the quality factor of the system can be increased to the order of hundreds of thousands.

This electro-mechanical system behaves as an electrical circuit between its X-X' connection ports, which can be represented with a circuit as shown in Fig. 6.13a. In this figure, the series branch represents the mechanical resonance. The electrical components can be derived from the value of the series resonance frequency (ω_S) and the value of the quality factor.[8] This branch becomes inductive above ω_S and resonates in parallel with C_p at ω_P, where $\omega_P > \omega_S$. The parallel resonance can be adjusted – to some extent – with an additional parallel capacitance that increases the value of C_p.

In order to give an idea about their exceptional component values (otherwise impossible to realize as electrical components), the electrical equivalents of certain crystals are given below.

f_s (MHz)	16	20	32	40
r (ohm)	75	50	30	30
C (fF)	1.5	1.4	2.5	1.5
Q ($\times 10^3$)	90	110	70	90
C_p (pF)	0.9	0.9	1.1	1.0

A quartz crystal can be used as a resonator in series resonance or parallel resonance. Since the series resonance is mechanics-based and depends only on the dimensions, it is very robust and can shift only due to temperature variations. Cutting the prism from the quartz crystal with a specific angle that minimizes the temperature coefficient, and stabilizing the temperature of the crystal, considerably helps to improve the frequency stability. Hence, the temperature coefficient can be well below 1 ppm/°C.

To make a crystal-controlled oscillator, one way is to use a suitable negative resistance circuit to compensate for the losses of the crystal. Another way is to insert the crystal into an amplifier to provide a positive feedback at the resonance frequency of the crystal.

---

[7] A solid prism has at least three resonance frequencies depending on the dimensions, the "Young modulus" and the density of the material. The resonance frequency mentioned above is the resonance frequency in the y-direction. In addition to these fundamental resonance frequencies, the crystal can resonate at their harmonics (the overtones).

[8] The inductance values are extremely high; usually in the range of 1 ...10 H, correspondingly C is very small. The resistance ranges from 100 ohm to several thousands of ohms. The parallel capacitance is in the range of picofarads.

Fig. 6.12 A quartz crystal cut, lapped, and polished to the dimension corresponding to the targeted frequency, and metal electrodes plated on the two y-z faces for electrical connections

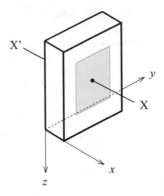

Fig. 6.13 (a) The electrical equivalent of a quartz crystal. (b) The variation of the impedance. Note that the impedance axis is not to scale; in reality, ω_S and ω_P are very close to each other

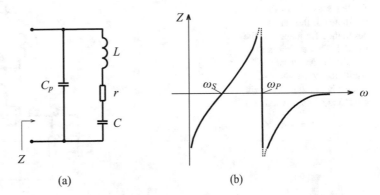

For a negative resistance crystal oscillator, since the losses of the crystal are very small compared to a conventional resonance circuit, the necessary transconductance values and the DC power consumptions are considerably small. Figure 6.14 shows an oscillator circuit based on the negative resistance circuit given in Fig. 4.7, with its source follower output buffer. This circuit can also be interpreted as a Colpitts oscillator. C_1 is totally or partially the input capacitance of M1.

Since the dimensions of crystals decrease with frequency, fabrication and precision problems increase. Usually, for frequencies higher than 10 MHz, it is useful to use the overtone resonances of crystals. The crystals originally intended to be used at a certain overtone are mounted into the case in such a manner to ease the vibration at this frequency.

The crystal oscillator circuit shown in Fig. 6.15 is a feedback oscillator. The output of M1 is connected to the input of M2 via a crystal. At the series resonance frequency of the crystal, its impedance is a resistance equal to r, which forms a positive feedback path together with R_1, to oscillate the circuit at the series resonance frequency of the crystal.

Another feature of this circuit is that it is suitable to make an overtone oscillator. If the resonance circuit is tuned to an overtone of the crystal, the gain becomes maximum at this frequency and the Barkhausen criterion can be fulfilled together with the crystal, to oscillate the circuit at this overtone frequency.

Fig. 6.14 A negative resistance crystal-controlled oscillator

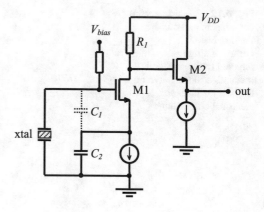

Fig. 6.15 A crystal feedback oscillator that oscillates at the series resonance frequency of the crystal. This circuit is also suitable as an overtone oscillator, when the *L-C* circuit is tuned to a certain overtone of the crystal

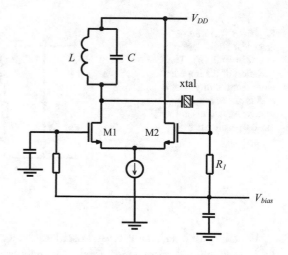

6.3.2 The Phase-Lock Technique

We have seen that crystal oscillators can serve for generating high stability oscillations with frequencies of up to about 100 MHz. For higher frequencies, it is possible to control the frequency of a voltage-controlled oscillator (VCO) with a high stability crystal oscillator, in other words, to "lock" the phase of the VCO to the phase of the crystal oscillator.

The history of the phase-lock technique goes back to the early days of electronics.[9] The concept was first developed in the analog electronics era, and the phase-locked loop (PLL) has since been one of the most important components of communication systems, with a wide range of applications such as clock and data recovery, demodulation, frequency multiplication, frequency synthesis, and synchronization. The possibilities of digital electronics further helped the development of PLL systems, after the 1970s.

---

[9] For example; Appleton, E.V., "The Automatic Synchronization of Triode Oscillators," Proc. Camb. Phil. Soc, Vol. 21, pp. 231–248, 1922–1923.

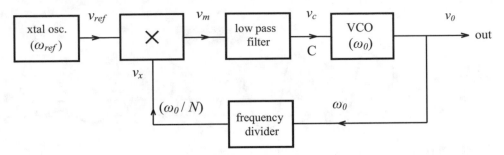

Fig. 6.16 Block diagram of a phase-locked loop. Frequencies on the figure correspond to the locked operation

Fig. 6.17 The assumed frequency-control voltage of the VCO

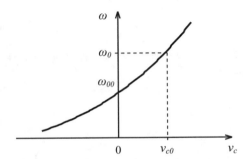

6.3.2.1 The Basic Analog PLL

The block diagram of a phase-locked loop (PLL) is shown in Fig. 6.16. The high-frequency output signal v_0, whose frequency is targeted as ω_0, will be generated by a VCO that can be any type of L-C oscillator with a varactor as its tuning element. Assume that the frequency-control voltage characteristic of the VCO is as shown in Fig. 6.17. The target frequency is ω_0 and the frequency corresponding to $v_c = 0$ is ω_{00}, which is called the "free-running frequency." A stable reference signal v_{ref} is being generated by a crystal oscillator, whose frequency must be a fraction of ω_0, such that $\omega_{ref} = \omega_0/N$, where N is an integer. The output frequency of the VCO is divided by N with a digital frequency divider and then applied to the input of an analog multiplier,[10] together with the output signal of the reference oscillator. The low-pass filter in the loop is usually a simple R-C filter.

To understand the behavior of the circuit, assume that in the beginning the loop is open at the C connection and the internal input bias voltage of the VCO is zero. For this case, if we show the output voltages of the reference oscillator, the VCO, and the frequency divider as

$$v_{ref} = V_{ref} \sin\left(\omega_{ref} t\right)$$

$$v_0 = V_0 \sin\left(\omega_{00} t + \varphi_0\right)$$

$$v_x = V_x \sin\left(\frac{\omega_{00}}{N} + \varphi_x\right)$$

the voltage at the output of the multiplier becomes

---

[10] The analog multiplier here acts as a phase detector (PD). Therefore, any other type of a phase comparator, for example, an exclusive-OR circuit, can be used instead.

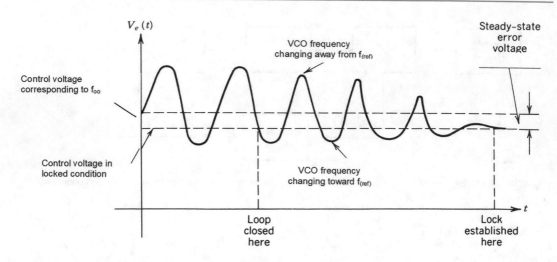

Fig. 6.18 The onset of locking upon closing the loop (After Alan B. Grebene)

$$v_c = k\left[V_{ref}\sin\left(\omega_{ref}t\right)\right] \times \left[V_x\sin\left(\frac{\omega_{00}}{N} + \varphi_x\right)\right] \tag{6.18}$$

where k is the multiplication factor of the analog multiplier. Assuming that the multiplier is linear, (6.18) can be expanded as

$$v_m = \frac{1}{2}kV_{ref}V_x\cos\left[\left(\omega_{ref} - \frac{\omega_{00}}{N}\right)t - \varphi_x\right] - \frac{1}{2}kV_{ref}V_x\cos\left[\left(\omega_{ref} + \frac{\omega_{00}}{N}\right)t + \varphi_x\right]$$

and the signal at the output of the low-pass filter becomes

$$v_c = \frac{1}{2}kV_{ref}V_x\cos\left[\left(\omega_{ref} - \frac{\omega_{00}}{N}\right)t - \varphi_x\right] \tag{6.19}$$

This means that the control voltage v_k sinusoidally varies with a frequency equal to the difference of the frequency of the reference oscillator and that of the output of the frequency divider. Now, if we close the loop at C, the circuit locks at $v_k = v_{k0}$ which corresponds to the targeted output frequency, provided that v_{k0} is in the range of the amplitude of v_c (See Fig. 6.18).[11]

Hence, the output frequency becomes equal to ω_0 and maintains this value. But it must be noted that the reference oscillator checks the frequency of the VCO not at every period but once every N period. Therefore, the control voltage at the output of the low-pass filter (that is a DC voltage when the circuit is locked) must keep its value for N periods.

6.3.2.2 The Digital PLL[12]

In this architecture, the phase detector (PD) is replaced by a phase-frequency-detector (PFD) that compares the phases of the reference signal and the divided output signal (Fig. 6.19). It produces "digital" error signals proportional to the measured phase difference. Then, the charge pump (CP) circuit sources or sinks current into a capacitive load according to these digital error signals.

---

[11] For a clear explanation with sufficient detail of the locking mechanism refer to [3].

[12] The authors would like to thank Tuğba Demirci and Gülperi Özsema for their contributions to these sections.

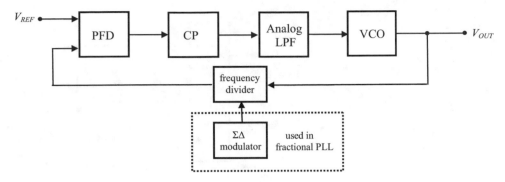

Fig. 6.19 Block diagram of a digital PLL (CP-PLL)

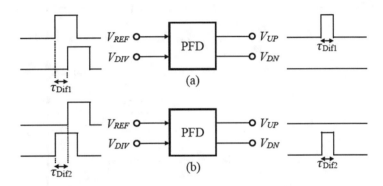

Fig. 6.20 Conceptual operation of a PFD

The output of the CP that is filtered by an analog low-pass filter (LPF) serves as the control voltage of the VCO, which serves to minimize the phase difference. The frequency divider in the feedback path divides the VCO output frequency by an integer (N). If fractional division ratios are attained, a $\Sigma\Delta$-modulator is necessary to modify the integer division factor on a regular basis. Although the internal signals are mostly analog, this architecture is called the "digital PLL" or, alternatively, the "charge pump PLL (CP-PLL)" since the error signal produced by its phase-frequency detector is a digital signal.

The critical part of a digital PLL is the PFD-CP combination. The PFD translates the sensed difference between the phases of the reference signal and the divided output signal into digital error signals at its output. The conceptual operation of a PFD block is depicted in Fig. 6.20, for two different cases. In case (a), the rising edge of the V_{REF} signal arrives earlier and the V_{UP} output goes high for a duration that is proportional to the phase difference $\varphi_{REF} - \varphi_{DIV}$, while the V_{DN} signal stays low. This operation implies that the PFD "orders" the loop to speed up the oscillator. In case (b), V_{DIV} leads the reference signal. Hence, the V_{DN} produces pulses to correct the error while the V_{UP} remains low. Thus, the PFD sends a command to slow down the oscillator.

The widely used logical realization of the described PFD functionality is illustrated in Fig. 6.21. The circuit consists of two DFFs and an AND gate. The reference signal and the divider output are connected to the flip-flop clock terminals. Assuming that the *Up* and *Down* outputs are at logic low level, a rising transition at the V_{DIV} input drives V_{DN} output to logic high. While the PFD is in this state, a rising edge of the reference signal will cause V_{UP} to rise as well. When both of the outputs are logic high, the AND gate sends a reset signal to the flip-flops. The additional delay placed in the reset

Fig. 6.21 Logical realization of a PFD with two DFFs and one AND gate

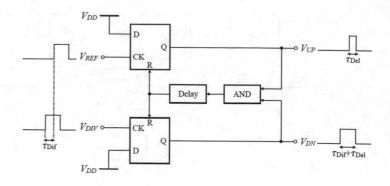

Fig. 6.22 Conventional CP model

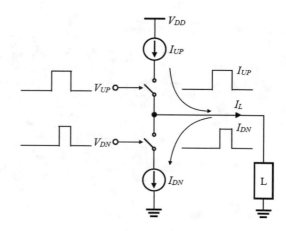

path is dedicated to eliminate the dead-zone that occurs when the PFD/CP combination cannot respond to small phase errors. The delay guarantees a minimum pulse width at the output of the PFD.

In CP-PLLs, the PFD is accompanied by a CP block that translates the *Up* and *Down* signals to charge/discharge current for the capacitive load connected to its output. The key components of a conventional CP circuit are modeled in Fig. 6.22. It is constructed from two current sources and two switches controlled by the V_{UP} and V_{DN} generated by the PFD. When the *Up* switch closes, the CP sources I_{UP} current and injects charge into the capacitive output load. On the other hand, when the *Down* switch closes, the CP sinks I_{DN} current and forms a discharge path from the output node to the ground. In both operations, the charge is proportional to the pulse duration. The voltage on the output load of the charge pump serves as the control voltage of the analog VCO.

6.3.2.3 The All-Digital PLL

The block-level schematic of the all-digital PLL is shown in Fig. 6.23. The phase-frequency detector and the charge pump are replaced by a time-to-digital converter (TDC) and the VCO is replaced by a digitally-controlled-oscillator (DCO). Also, instead of an analog LPF, a digital LPF is used for the loop filter. The TDC is usually constructed in the form of a delay line consisting of a cascaded inverter chain, where the individual inverter outputs are decoded and converted into a digital "word" representing the phase difference between the reference signal and the divided output signal. Similarly, the digitally-controlled oscillator is constructed in the form of a ring oscillator where the drive strength of each inverter cell can be incrementally increased or decreased by activating one or more parallel-connected transistors of identical size in the pull-up and pull-down path. Hence, this approach converts all internal signals in the block to digital form, with the exception of the output of

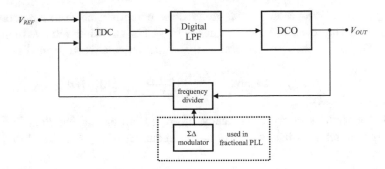

Fig. 6.23   Block diagram of an all-digital PLL

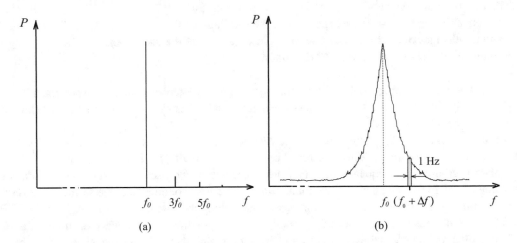

Fig. 6.24   (a) The frequency spectrum of a "noiseless" oscillator. (b) The frequency spectrum of a real oscillator, for the vicinity of f_0

the digitally controlled oscillator. The advantages of the all-digital PLLs can be summarized as follows:

–   The removal of the CP, which is a complex analog circuit, simplifies the design of all-digital PLLs.
–   Digital filters usually occupy less silicon area on the chip compared to their analog counterparts.
–   DCO offers a wider tuning range and better noise performance compared to that of an analog VCO.

6.3.3   Phase Noise in Oscillators

For an ideal sinusoidal oscillator, the frequency spectrum contains only one component at the frequency of oscillation, f_0, which is usually called the "carrier frequency." The components at the harmonics of f_0 indicate a nonlinear distortion on the waveform, which is an expected imperfection for an electronic circuit (See Fig. 6.24a). The unexpected reality is that the frequency spectrum of an oscillator contains an infinite number of side-frequencies on both sides of f_0, decreasing with the distance from f_0. The envelope of the side-frequencies exhibits a noise-like random character as shown in Fig. 6.24b. The power of a side-frequency in a 1 Hz bandwidth relative to the power of the carrier is called the "phase noise" of the oscillator for this frequency and defined as

$$L(\Delta f) = 10 \log \left(\frac{P(f_0 + \Delta f)_{B=1Hz}}{P(f_0)} \right) \quad [\text{dBc/Hz}] \tag{6.20}$$

where $L(\Delta f)$ is the phase noise for a frequency Δf apart from the carrier, $P(f_0 + \Delta f)_{B=1Hz}$ is the power density (power for 1 Hz bandwidth) of a side frequency of $(f_0 + \Delta f)$ and $P(f_0)$ is the power density of the carrier.

There are several mechanisms contributing to the phase noise, which are all interrelated in some way:

- The thermal noise on the voltages and currents in the circuit
- Random shifts of the zero-crossing points of the oscillation
- Random changes of the frequency due to the low stability of the oscillation
- Frequency modulation of f_0 with the thermal noise

Due to the noise superimposed onto the oscillation signal, the zero-crossing points on the time axis may randomly shift back and forth, which corresponds to a frequency modulation with a random and small frequency deviation [4].

Another theory is based on the random fluctuations of the limit-cycle of the oscillation, which is related to the nonlinearity of the electronic parts of the circuit [5]. For example, in a negative resistance oscillator, the frequency at which the amplitude of the oscillation stabilizes depends on the value of the transconductance. The fluctuations of the transconductance due to the noise on the drain currents of the transistor or due to the signal delay along the gate R-C line (as explained in Sect. 1.1.3.3), and the related uncertainty of g_m, result in small variations on the phase of the oscillation.

The global frequency stability is certainly another factor that acts on the phase noise of the oscillator. It is known that the ω_0, $\omega_{(Re)}$, and ω_{max} frequencies of the resonance circuit are not the same but very close to each other, as explained in Sect. 4.1.1, and that each has an influence on the oscillation frequency.

To minimize the phase noise of an oscillator, it is necessary to try to minimize the effects of all these factors.

For example, to minimize the noise reaching the resonance circuit, a suitable filter can be used to bypass the noise of the tail current source [6]. The amount of distortion on the drain currents, which indicates where the amplitude of the oscillation will be self-stabilized, can be adjusted to a level that minimizes the phase noise. The use of short multi-finger transistors may reduce the g_m fluctuations related to the delay along the gate. It should also be noted that designing the circuit to fulfill the maximum frequency stability condition would naturally help to decrease the phase noise.

The effects of the modulation of the oscillation frequency with the noise can be discussed as follows:

From modulation theory, it is known that if the amplitude of a carrier is amplitude-modulated with a signal of f_m, two side-frequencies appear on the frequency spectrum, whose distances from f_0 are equal to f_m, and whose magnitudes are proportional to the amplitude of the modulating signal. Then we understand that, if an oscillator is amplitude modulated with "white" noise, two side-bands occur extending to infinity on both sides of f_0, whose envelope has an – almost – constant magnitude. If the noise in the system is not "white," due to the flicker (or $1/f$ noise), the magnitudes of the side

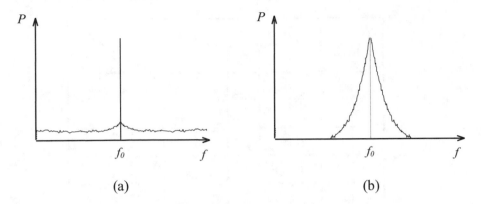

Fig. 6.25 (**a**) Side-bands of a carrier, amplitude modulated with noise (the increase in the vicinity of the carrier corresponds to the $1/f$ noise). (**b**) Side-bands of a carrier, frequency modulated with white noise

frequencies corresponding to low f_m values steadily increase toward the carrier, as shown in Fig. 6.25a.

The decreasing side-frequency components shown in Fig. 6.25b can be explained with the frequency modulation of f_0, i.e., the small Δf_0 fluctuations of the oscillation frequency around f_0. It is known from modulation theory that if the frequency of a carrier is modulated with a modulation signal, side frequencies occur on both sides of the carrier with a distance of f_m, $2 f_m$, $3 f_m$, etc. The magnitudes of these side frequencies relative to the magnitude of the carrier depend on the frequency modulation index, $\delta = \Delta f_0/f_m$, and can be expressed using Bessel functions of the first kind $J_1, J_2, ..., J_n$ of δ:

$$
\begin{aligned}
v = {}& J_0(\delta) \times A \sin \omega_0 t \\
&+ J_1(\delta) \times A [\sin (\omega_0 + \omega_m)t - \sin (\omega_0 - \omega_m)t] \\
&+ J_2(\delta) \times A [\sin (\omega_0 + 2\omega_m)t + \sin (\omega_0 - 2\omega_m)t] \\
&+ J_3(\delta) \times A [\sin (\omega_0 + 3\omega_m)t - \sin (\omega_0 - 3\omega_m)t] + \cdots
\end{aligned}
\tag{6.21}
$$

where A is the amplitude and v is the instantaneous value of the frequency modulated signal. Variations of the Bessel functions as a function of δ are given in the literature.[13]

Figure 6.26 shows the first three functions for small values of δ that correspond to a small Δf_0 fluctuation of the frequency around f_0. From these curves, it can be seen that the magnitudes of the side frequencies corresponding to f_m, $2 f_m$, $3 f_m$, etc. decrease with f_m, as shown in Fig. 6.24b. The rate of decrease of the RMS values of the side frequencies with f_m can be found as ≈ 20 dB/decade, in accordance with measurements.

The phase noise is the imperfection of an oscillator characterized in the frequency domain. Its counterpart in the time-domain is "jitter." Jitter means the random back-and-forth fluctuations of the zero-crossing points of a signal on the frequency axis. Jitter is – certainly – related to the phase noise and it has been shown that the RMS amplitude of jitter can be expressed as[14]

---

[13] For example [7]. But the easiest way is to refer to MatLab.

[14] Maxim Integrated, Application Note: 3359

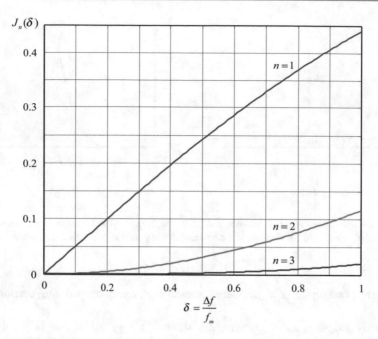

Fig. 6.26 Bessel functions for very small frequency deviations as a function of the frequency modulation index

$$J_{RMS}\big|_{(f_1 \text{ to } f_2)} = \frac{1}{2\pi f_0} \sqrt{2 \int_{f_1}^{\infty} 10^{\frac{L(f)}{10}} df} \qquad (6.22)$$

References

1. E.S. Yang, *Fundamentals of Semiconductor Devices* (McGraw-Hill, New York, 1978), pp. 100–103
2. J. Groszkowski, The interdependence of frequency variation and harmonic content, and the problem of constant-frequency oscillators. Proc. IRE **21**(7), 958–981 (1934)
3. A.B. Grebene, *Bipolar and MOS Analog Integrated Circuits* (Wiley, Hoboken, 1984), pp. 627–652
4. A. Hajimiri, T.H. Lee, A general theory of phase noise in electrical oscillators. IEEE J. Solid State Circuits **33**(2), 179–194 (1997)
5. A. Demir, A. Mehrorta, J. Roychowdhury, Phase noise in oscillators: a unifying theory and numerical methods for characterization. IEEE Trans. Circuits Systems I Fund. Theory Appl. **47**(5), 655–674 (2000)
6. P. Andreani, H. Sjöland, Tail current noise suppression in RF CMOS VCOs. IEEE J. Solid State Circuits **37**(3), 342–348 (2002)
7. H.B. Dwight, *Mathematical Tables* (Dover Publications, New York, 1958), pp. 144–167

Analog-Digital Interfaces

7

In the earlier chapters of this book, we introduced and examined the structure and operation of fundamental building blocks, or essential components, of high-frequency integrated circuits. The emphasis has been on transistor-level operation, the influence of device characteristics and parasitic effects, as well as the input-output behavior in time and frequency domains, with the intention to fill the gap between fundamental electronic circuits textbooks and more advanced RF IC design textbooks that mainly focus on the state-of-the-art. In this chapter, we will address a different domain and consider various aspects of "bringing together an entire system," especially for interfacing high-frequency analog components with corresponding digital processing blocks, using data converters. In the following, instead of considering specific system architectures, the main aspects of system design will be presented with as generic an approach as possible. The main philosophy is very similar to that of the earlier chapters: discussing design-oriented strategies and drawing attention to key issues that must be taken into account when combining analog and digital building blocks. For a detailed discussion of the system components and their circuit-level realizations, the reader is advised to consult any one of the excellent texts that already exist in this domain.[1]

7.1 General Observations

The vast majority of integrated systems used in communication applications today consist of analog as well as digital building blocks, combined within one package in close proximity to each other, or fabricated on a single chip substrate. The majority of the algorithmic signal processing and modulation is handled by the digital system blocks, while the high-frequency communication is delegated to the analog HF front-end. The task of translation between these two domains, i.e., the analog-digital interface, is the responsibility of the data converters: the *analog-to-digital converter* (ADC) for converting analog signals into discrete-time, quantized data, and the *digital-to-analog converter* (DAC) for converting the output of digital processing into modulated, continuous-time, analog signals that can be transmitted over larger distances.

The role of digital circuitry in conjunction with high-frequency analog components is not only limited to algorithmic signal processing. Due to their relatively smaller area overhead, digital circuits

---

[1] For a detailed treatment of data converter architectures and the circuit-level realization of key building blocks, refer to Maloberti [1], Razavi [2], or van Roermund [3].

© The Author(s), under exclusive license to Springer Nature Switzerland AG 2021
D. Leblebici, Y. Leblebici, *Fundamentals of High Frequency CMOS Analog Integrated Circuits*,
https://doi.org/10.1007/978-3-030-63658-6_7

Fig. 7.1 Simplified block diagram of a generic analog/digital system with data converters

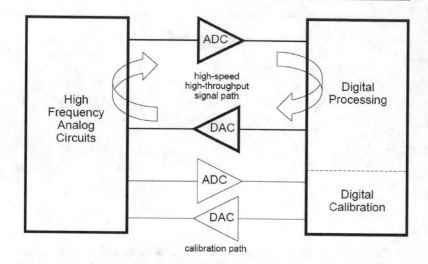

calibration path

such as simple controllers, lookup-tables, and coefficient registers are becoming very convenient means for handling the off-line and on-line calibration of sensitive analog components. Digital gain and offset calibration of amplifiers, for example, have become increasingly commonplace. Note that in the case of calibration, circuit speed is a secondary concern – whereas accuracy, measured in number of bits, is of paramount importance. Thus, when considering the relation between the analog and digital parts of a system, it is also increasingly necessary to make a distinction between the high-speed *signal path*, which must handle a high data throughput (i.e., high processing speed) in real-time, and the relatively low-speed *calibration path*, which must – by definition – have a higher bit resolution than the signal processing path (Fig. 7.1).

One particular observation is that in modern systems, the share of digital blocks is continuously increasing with respect to the overall system, and that the boundary between the analog and digital modules is becoming more ambiguous. The increasing share of digital system blocks is mostly reflected in terms of increasing transistor count and functional complexity – but not necessarily in terms of actual silicon area – which we have to address in the context of system-level design. In fact, the *proportion* of the silicon area that must be reserved for high-frequency analog blocks on a typical mixed-signal chip is rising with each new technology generation, even though the functional complexity of the remaining digital system blocks is increasing. This is mainly due to the fact that many of the active and passive components in high-frequency analog circuit blocks cannot be scaled down as easily as their digital counterparts, and this has to be taken into account in the overall system construction.

To give an example, an RF transceiver is usually defined as the entire system consisting of functional building blocks such as filters, amplifiers, frequency converters, modulator/demodulators, oscillators, synthesizers, data converters (ADC/DAC), switches, signal couplers, etc. The transceiver is not only formed by high-frequency (RF) components but also by intermediate frequency (IF) and analog baseband circuitry and devices. The ADC and DAC are often seen as the boundary between the high-frequency analog parts of the transceiver and its digital counterpart. However, this boundary is getting increasingly blurred with state-of-the-art data converters operating at higher sampling rates – in this sense, the ADC/DAC and the corresponding digital signal processors now take over more and more functions of the IF and even the RF blocks. Thus, high-speed data converters are increasingly considered as part of the high-frequency systems, opening up new possibilities for so-called direct conversion systems, especially for mobile applications. The tendency in system-level design is to shift an increasing portion of high-frequency functions (including some of the filtering and signal conditioning, etc.) onto the digital section, as higher sampling and processing speeds become available. It is safe to assume that this trend will continue in the future. The shifting of data converters closer toward the analog front-end ultimately decreases the proportion of the analog circuitry in the overall system, and increases the functional complexity of the digital back-end. While this may increase the overall robustness and the flexibility of the system (using programmable and/or reconfigurable digital signal processing), it also increases the demand for higher conversion bandwidth in the ADC and DAC units (Fig. 7.2).

The interface between analog and digital domains requires particular attention in system design – especially when defining the specifications on both sides. One significant issue that has to be taken into account is that the operational characteristics of digital circuits and systems are almost exclusively described in the *time domain*. On the other hand, most – not all – of the key characteristics of analog circuits are preferably described in the *frequency domain*, as we have seen extensively in the previous chapters of this book. Successful translation of key design specifications from one domain onto the other requires a good understanding of the fundamental aspects of sampling theory, which provides the bridge between the continuous-time analog world and the discrete-time digital world.

7.2 Discrete-Time Sampling

Sampling is the key transformation that is required in all data converters. The sampling of a continuous-time analog signal can be performed by successive sample and hold operations using a periodic sampling clock, as seen in Fig. 7.3. The output of the sampling process is a sequence of pulses whose amplitudes are derived from the input waveform samples. According to the *uniform sampling theorem*, a band-limited signal (i.e., a signal that does not have any spectral component beyond a certain frequency f_{max}) can be completely reconstructed from a set of uniformly spaced discrete-time samples, if these samples are obtained with a sampling rate (sampling frequency) of $f_s > 2 f_{max}$. Note that ideally, the output of the sampler is represented by a sequence of modulated Dirac deltas whose amplitude equals the amplitude of the sampled signal at the sampling times. Clearly, a practical sampling circuit does not generate a sequence of deltas, but pulses with finite duration. The pulses are intended to represent the input waveform only at the exact sampling instances, nT, where $T = 1 / f_s$. Figure 7.3a shows the block diagram of one such ideal sampler, which consists of a lossless, zero-delay switch, and an ideal unity-gain amplifier with infinite bandwidth. If the sampling clock pulse width δ is reduced to an infinitesimally short duration, the sampled output pulses will approach a sequence of Dirac deltas, as shown in Fig. 7.3b. In practical applications, it is usually preferable to apply a temporary memory (hold) function at the output of the sampler, so that the amplitude of the sampled signal is preserved until the next sampling instant. For sampled voltage waveforms, a linear capacitor at the output of the sampling switch can fulfill this memory function, as shown in Fig. 7.3c. This arrangement is commonly known as an ideal Sample-and-Hold circuit, with the assumption that the sampling pulse width δ can be made infinitesimally small. At the onset of each sampling clock pulse, the hold capacitor is instantaneously charged up to the level that corresponds to the input signal at that moment – and then preserves its charge until the next sampling instant. If the sampling clock pulse has a finite width, on the other hand, the same idealized circuit operates as a Track-and-Hold, where the output follows (tracks) the input during the period when the sampling clock is active, and then holds its value during the period when the sampling clock is inactive. To introduce some of the key aspects of sampling, we will assume ideal sampling with infinitesimally small pulse width, and no holding function, in the following.

We will present some of the fundamental properties associated with signal sampling and reconstruction, in order to provide further insight concerning the operation of data converters. Consider a

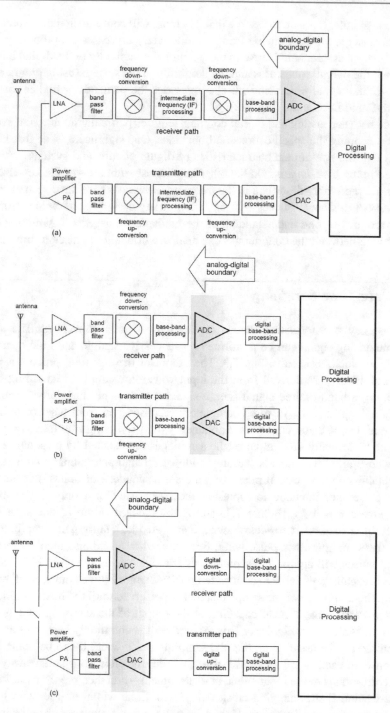

Fig. 7.2 (**a**) Example of a conventional super-heterodyne transceiver architecture where the incoming high-frequency signals are first down-converted to the intermediate frequency (IF) range, and then to the baseband range, in two steps. The up-conversion is also handled in two steps. The sampling and conversion from analog into digital are done in the baseband domain. (**b**) The so-called "zero-IF" architecture relies on a single frequency down-conversion step, with the ADC and DAC operating at a higher sampling rate. (**c**) The direct sampling approach relies on very high-speed converters next to the analog front-end, and the down-conversion (as well as the corresponding up-conversion) takes place completely in the digital domain

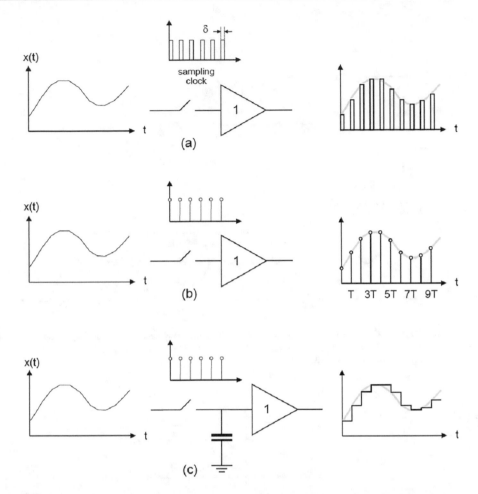

Fig. 7.3 (a) Sampling of the analog input signal with an ideal sampler circuit. (b) As the sampling clock pulse width is reduced to an infinitesimally small duration, the sampled output will approach a sequence of modulated Dirac deltas. (c) Ideal sample-and-hold operation performed with a linear capacitor at the output of the sampling switch

time-varying signal waveform which has a band-limited spectrum as shown in Fig. 7.4a. If this waveform is sampled with an ideal sampler, at a sampling frequency that is larger than $2 f_{max}$, the spectrum of the sampled signal can be reconstructed as depicted in Fig. 7.4b, consisting of frequency-domain superposition of an infinite number of replicas of the input spectrum. These replicas are centered at integer multiples of the sampling frequency f_s, i.e., with each *image* spectrum being shifted along the frequency axis by $n f_s$, where n is an integer. Hence, the spectrum repeats itself periodically in the frequency domain. As long as the highest frequency component of the signal spectrum is smaller than one half of the sampling frequency, the original signal can be reconstructed by using a low-pass filter to isolate the band-limited signal. If the sampling frequency is less than $2 f_{max}$, on the other hand, the replicas will partially overlap in the frequency domain (aliasing) and it will not be possible to reconstruct the original signal.

It should also be kept in mind that the sampling condition described above must hold for all spectral components of the sampled signal, including unwanted noise and interferences. Note that noise usually has an unpredictable spectrum and it can produce spectral components at any frequency.

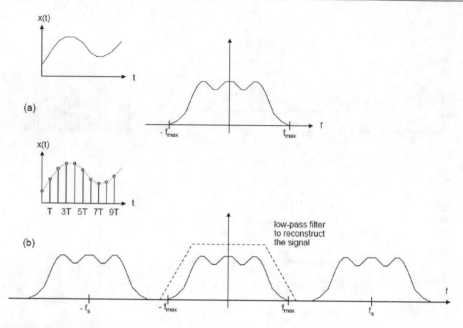

(a)

(b)

Fig. 7.4 (a) Time-varying signal waveform and its band-limited frequency spectrum, with f_{max} showing the highest frequency component. (b) Reconstruction of the sampled signal produces an infinite number of image spectra, each shifted from the origin by an integer multiple of f_s

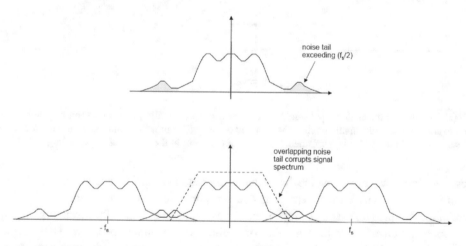

Fig. 7.5 Aliasing of noise tails can corrupt the reconstructed signal spectrum. To avoid this, the bandwidth of the input signal must be limited using an anti-alias filter, prior to sampling

If the input signal spectrum has an unexpected noise tail as shown in Fig. 7.5, the folded-shifted image (replica) of the same noise tail can corrupt the original signal in the sampled spectrum. To prevent this, it is advisable to filter out any unwanted spectral components that may lie beyond the intended limit frequency of f_{max} using an anti-aliasing filter, before sampling the signal.

Fig. 7.6 Timing jitter on the sampling clock will result in erroneous sample values, depending on the amount of timing error and the instantaneous slope of the signal

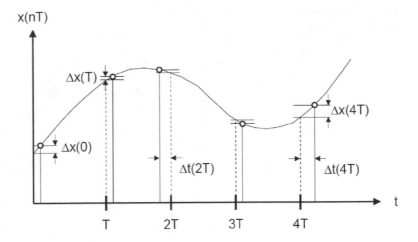

7.3   Influence of Sampling Clock Jitter

Up until this point, we have considered the conversion of continuous-time analog signals into discrete-time sampled signals, and their reconstruction back into continuous-time domain, assuming ideal components and ideal sampling conditions. Under realistic conditions, however, the sampling process is affected by the uncertainty of the clock edges in the time domain. The unpredictable variation of the clock edge, mainly due to the thermal noise in the clock generator and the uncertainty of the logic delay, is called the *jitter* in the actual sampling instants. Sampling of a time-domain signal waveform in the presence of jitter is illustrated in Fig. 7.6. Note that the magnitude as well as the sign (direction) of jitter $\Delta t(nT)$ at each sampling instant will change according to a random distribution function, and independently of the previous sampling periods. The resulting sampling errors $\Delta x(nT)$ are also indicated in the figure. The amount of the sampling error depends on the magnitude and sign of the clock jitter, and also on the magnitude and sign of the time derivative of the sampled input signal. For a sinusoidal input signal with amplitude A and angular frequency ω_0, the sampling error at each sampling instant can be found as

$$\Delta x(nT) = A\omega_0 \Delta t(nt) \cos(\omega_0 nT) \tag{7.1}$$

From the power of the jitter error, the resulting signal-to-noise ratio due to clock jitter can be found as a function of input frequency and the mean clock jitter.

$$SNR_{jitter} = -20 \log(\langle DJ(t)\rangle \omega_0) \ \ \text{dB} \tag{7.2}$$

Here, we assume that the instantaneous jitter amount $\Delta t(nT)$ is found by sampling the random variable $DJ(t)$, which is dictated by a white noise spectrum.

7.4   Quantization Noise

Regarding the conversion of a continuous-time analog signal into a digital data stream, the discrete-time sampling process examined above is not the only source of systematic error. In order to process the sampled signal algorithmically by using binary arithmetic operators, the samples have to be quantized with a finite number of bits into discrete levels. This quantization process inevitably

Fig. 7.7 Quantization of samples with a finite number of bits which results in quantization error

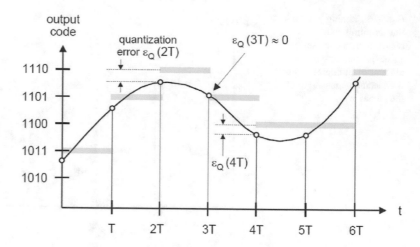

introduces a finite resolution for all sampled signals, which manifests itself as a *quantization error* ε_Q in the system – i.e., as an amplitude difference between the sampled signal and its level-quantized representation based on a limited number of bits. Under certain conditions, the quantization error can be represented in the form of *quantization noise*, which is reflected upon the signal-to-noise ratio of the data converter.

The main problems associated with the quantization process are illustrated with a simplified example in Fig. 7.7. Here, only a small section of the full signal range is shown, with discrete quantization levels described by four bits. It can be seen that for each discrete-time sampled signal, the quantizer has to perform a thresholding decision and assign the sample to the nearest discrete amplitude level. Unless the original sampled signal amplitude exactly corresponds to one of the quantization levels, the outcome of this thresholding decision is bound to produce an error term at every sample point. If the dynamic range of the sampled signal can exercise all quantization levels with equal probability, and if the time-domain variation of the signal causes frequent code transitions to decorrelate successive samples, then the resulting quantization error can be treated as quantization noise with a widely spread spectrum. Assuming a sinusoidal input with the amplitude corresponding to the full dynamic range of the quantization interval, the time average power of the quantization noise can be calculated as a function of the maximum possible quantization error at each sample, i.e., the number of bits (n) used for quantization. Thus, the maximum achievable signal-to-noise ratio that is due to quantization of samples with n bits can be expressed as:

$$SNR_{quantization} = (6.02n + 1.78) \ \text{dB} \tag{7.3}$$

This means that each additional bit of resolution can improve the SNR of the data converter by 6.02 dB. Note, however, that the expression (7.3) only accounts for the quantization noise due to a finite number of quantization steps assigned to the samples. In a real data converter, several other factors – such as thermal noise, bandwidth limitations of dynamic components, settling time limitations, etc. – bring about further errors that can also be viewed as additional noise components. Consequently, the expression given above can be rewritten to define the *equivalent number of bits* (ENOB), with SNR_{total} representing the combined signal-to-noise ratio that accounts for all noise sources influencing the signal band of the data conversion system, including, but not only limited to, the quantization noise.

$$ENOB = (SNR_{total} - 1.78)/6.02 \tag{7.4}$$

This number is usually specified for a particular sampling frequency, and for a maximum bandwidth of the sampled signal. While the bit-resolution of a data converter is given as an integer number n, the effective number of bits derived from the expression above can be a real number, which is always less than n itself. The ENOB nevertheless provides a good indication of the overall accuracy of the data conversion system, under the specified operating conditions.

7.5   Converter Specifications

This brief review of discrete-time sampling, and the limitations imposed by timing jitter and quantization noise, bring us to the fundamental specifications of data converters. The following discussion is kept at a sufficiently general level, with the aim of providing the reader an introductory understanding of the key issues that relate to their functional characteristics. The specifications introduced here do not depend on the specific circuit realizations of the data converters. In the following, we will categorize the specifications in two broad classes, namely, as static and dynamic specifications.

7.5.1   Static Specifications

The static specifications are exclusively defined on the input-output characteristics of the data converter. The ideal input-output characteristic of an analog-to-digital converter (ADC) is a staircase as depicted in Fig. 7.8, where the horizontal axis corresponds to the analog input level and the vertical axis corresponds to the quantized output codes. Note that each step width is equal to the quantization interval $\Delta = X_{full-range}/(2^n-1)$, which corresponds to one *least-significant-bit* (LSB) in the output code and thereby defines the analog *resolution* of the converter. In an n-bit converter, the output codes range from 0 to (2^n-1) and consequently, the input full-scale range $X_{full-range}$ is divided into (2^n-1) quantization intervals, with the decision threshold for each quantization interval placed in the

Fig. 7.8 Ideal input-output characteristic of the analog-to-digital converter

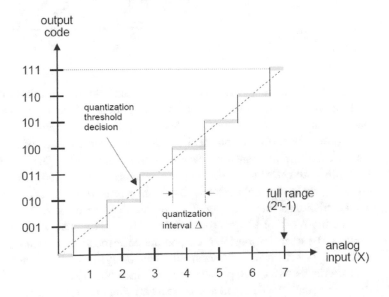

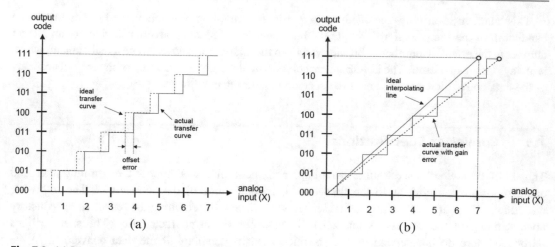

Fig. 7.9 (a) Input-output characteristic with offset error. (b) Input-output characteristic with gain error

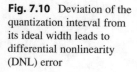

Fig. 7.10 Deviation of the quantization interval from its ideal width leads to differential nonlinearity (DNL) error

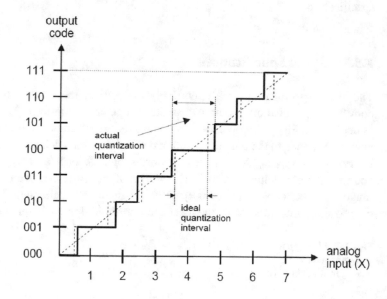

mid-point. Thus, the quantization error resulting from this thresholding decision ranges from $-(\Delta/2)$ to $+(\Delta/2)$, becoming equal to zero in the midpoint of each step. Deviations from this ideal input-output characteristic will manifest themselves in various ways.

Figure 7.9a shows a non-ideal input-output characteristic of an ADC with *offset error*, which causes a shift of all quantization steps by an equal amount. The offset error can also be defined for a DAC. In both cases, the input-output characteristic is shifted, while remaining parallel to the ideal one. A *gain error*, on the other hand, manifests itself as a change in the slope of the transfer characteristic, as shown in Fig. 7.9b. In the presence of offset and/or gain error, it is expected that the validity range of all quantization intervals (and output codes) will change, and in extreme cases, some of the quantization intervals and the corresponding output codes may not be exercised at all. Gain and offset errors can occur simultaneously, in which case their effects on the input-output characteristics will overlap. Nevertheless, since both gain and offset errors are linear non-idealities, they can usually be detected and corrected, using calibration techniques.

The *differential nonlinearity* (DNL) error is defined as the deviation of the quantization step (interval) width from that of an ideal step width (Fig. 7.10). This deviation can occur due to systematic non-idealities in circuit elements such as the nonlinear characteristics of transistors, and also as a result of random variations in the size and value of replicated components such as the resistors in a voltage-divider chain. The DNL error is usually specified in terms of percentage of the full scale, or in terms of LSB. Since the error of each quantization interval is measured separately, the stated DNL value is usually the maximum error among all quantization intervals of the converter. Regardless of its origin, the DNL error cannot be corrected completely by calibration, due to its inherently nonlinear nature. In a data converter with no other non-idealities, a DNL error with a maximum magnitude of less than 0.5 LSB is usually tolerable, since the resulting loss of accuracy, in this case, will always remain less than one LSB. Figure 7.11 shows the DNL variation of a typical 12-bit ADC, over 4096 individual quantization levels. The maximum deviation remains less than 0.5 LSB, which is (1/8192) of the full range.

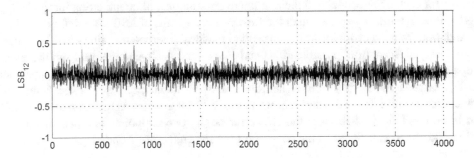

Fig. 7.11 DNL variation of a 12-bit ADC over the entire quantization range

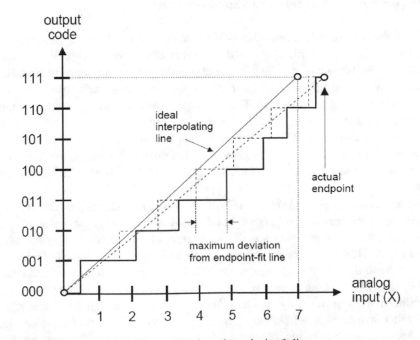

Fig. 7.12 Deviation of the input-output characteristic from the endpoint-fit line

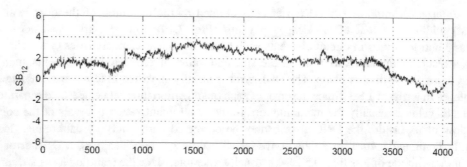

Fig. 7.13 INL variation of a 12-bit ADC, over the entire quantization range

The *integral nonlinearity* (INL) error is defined as the deviation of the input-output characteristic from the endpoint-fit line, which practically connects the starting and ending points of the transfer function (Fig. 7.12). As in the DNL case, the error is calculated individually for each quantization interval (or for each output code) and the maximum error among all quantization intervals is cited as the DNL value, in terms of percentage of the full scale or in terms of LSB. The definition of the error as the deviation from the endpoint-fit line instead of the ideal interpolating line effectively decouples the nonlinearity measure from possible offset and gain errors. A large deviation from the endpoint-fit line corresponds to harmonic distortion, which also manifests itself on the dynamic specifications. Similarly, a large DNL term is seen as an additional noise term that impacts the overall SNR, together with quantization noise. In the example shown in Fig. 7.13 for a 12-bit ADC, the maximum INL error is seen to be less than 4 LSB. Note the INL error on the two endpoints is equal to zero since by definition, the INL is calculated as the deviation from the endpoint-fit line.

7.5.2 Frequency-Domain Dynamic Specifications

The *signal-to-noise ratio* (SNR) is one of the key dynamic characteristics of data converters. As we have already seen, the upper limit of the SNR of a data converter is primarily determined by its bit-resolution, i.e., the number of bits used to quantize and represent the sampled signals. In addition to this quantization noise, the SNR of a data converter accounts for the influence of all noise sources, within the entire signal frequency interval. The nonlinear distortion terms generated by the sinusoidal input, however, are *not* included in this calculation.

The *total harmonic distortion* (THD) is the ratio of the root-mean-square value of the fundamental signal to the mean value of the root-sum-square of its harmonics. Note that, generally, only the first five harmonics are considered to be significant. The THD of an ADC is specified with the input signal at full-scale.

The *signal-to-noise-and-distortion ratio* (SNDR or SINAD) is defined as the ratio between the root-mean-square of the fundamental signal and the root-sum-square of all harmonic components, plus all noise components. In the literature, this measure is sometimes also called the *total harmonic distortion plus noise* (TDH + N). Being one of the quantitative specifications that takes into account all possible noise and distortion components, the SNDR is typically used to estimate the *equivalent number of bits* (ENOB) of a data converter – by replacing SNR_{total} in expression (7.4). The SNDR is usually specified at a given sampling rate, as a function of the input signal frequency, and also at a given input signal frequency, as a function of the sampling frequency. Since the linear non-idealities of circuit components (such as bandwidth and settling time limitations) tend to manifest themselves at

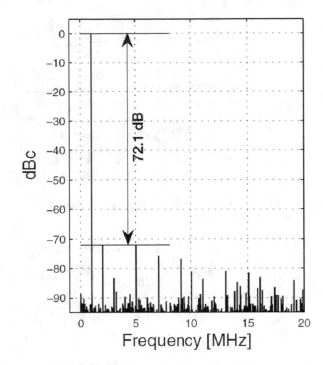

Fig. 7.14 Calculation of SFDR based on a sample frequency spectrum

higher operating frequencies, the SNDR usually degrades with frequency. The sampling frequency at which the SNDR drops by 3 dB determines the maximum sampling rate of the converter.

The *spurious-free dynamic range* (SFDR) is the ratio of the root-mean-square value of the signal to the root-mean-square value of the worst spurious signal, regardless of where this spur is located in the frequency spectrum. The worst spur may or may not be a harmonic of the original signal. SFDR is an important specification in communications systems because it represents the smallest value of signal that can be distinguished from a large interfering signal. Figure 7.14 shows the calculation of the SFDR based on a sample frequency spectrum, where the distance between the signal peak and the largest spur is shown to be 72.1 dB. As in the case of SNDR, the SFDR values are usually calculated for the entire input frequency spectrum at a given sampling clock frequency, and also for a range of sampling clock frequencies at a given input frequency. Figure 7.15 shows the variation of the SNDR and the SFDR of a pipelined ADC as a function of the sampling frequency, at an input frequency of 20 MHz. It can be seen that both SNDR and SFDR curves tend to roll-off as the sampling frequency is increased, and they both drop by more than 3 dB beyond the usable sampling speed limit of the ADC.

It is interesting to note that the three important frequency-domain specifications discussed here, namely SNR, THD, and SNDR, are linked in a very straightforward manner. These three measures are defined as the numerical ratios of (S/N), (S/D), and $(S/(N + D))$, respectively:

$$SNR = 20 \log (S/N)$$
$$THD = 20 \log (S/D) \tag{7.5}$$
$$SNDR = 20 \log [S/(N + D)]$$

With simple manipulation, it can be shown that any one of these three specifications can be derived using the following expressions, as long as the remaining two are known:

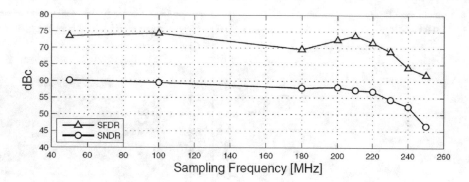

Fig. 7.15 Variation of SNDR and SFDR as a function of the sampling clock frequency, for a high-speed pipelined analog-to-digital converter

$$SNR = -10\log\left(10^{-SNDR/10} + 10^{-THD/10}\right)$$

$$THD = -10\log\left(10^{-SNDR/10} + 10^{-SNR/10}\right) \tag{7.6}$$

$$SNDR = -10\log\left(10^{-SNR/10} + 10^{-THD/10}\right)$$

7.6 Overview of ADC Architectures

ADCs can be classified into two major categories: oversampling ADCs and Nyquist-rate ADCs. The latter sample the input signal at the same rate as digitized values are generated. The most important Nyquist-rate ADC topologies are pipelined, SAR, and flash ADCs, which are outlined below. Less prominent with fewer applications are architectures such as folding, interpolating, and time-to-digital ADCs.

Oversampling ADCs, mainly $\Delta\Sigma$, run at a sampling speed much higher than the required frequency band to shape noise out of the frequency band of interest. In industry, $\Delta\Sigma$ ADCs are widely used for high-resolution, pipelined ADCs for medium-resolution, and flash ADCs for low-resolution. The resolution range of SAR ADCs partially overlaps with pipelined and flash ADCs.

7.6.1 Flash ADC

The principle of a flash ADC is shown in Fig. 7.16. The input voltage V_i is sampled onto a capacitor, which is connected to a number of comparators. Each comparator has a different offset voltage at the second input, against which the sampled voltage is compared. The offset voltages $V_1\ldots V_3$ are often derived from a resistive ladder formed by R_{bh}, R_{bl}, and $R_1\ldots R_2$. All comparators with an offset voltage above the sampled voltage will trigger high. This allows the ADC to determine the digital output by converting the thermometer-coded comparator output to a binary-coded output. Because of comparator noise and offset it can happen that there is a zero between the ones or a one between the zeros in the thermometer code, which is called a bubble. Most flash ADCs include bubble correction to prevent wrong output coding.

Flash ADCs need $2^n - 1$ comparators, making them power-inefficient for higher resolution n. These ADCs are therefore found mainly for resolutions up to 6 bits. The main advantages of flash ADCs are

Fig. 7.16 Flash ADC

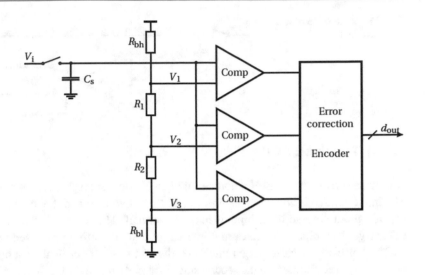

Fig. 7.17 SAR ADC

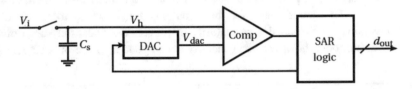

the high conversion rate, the simple design, and the short input-output latency. High conversion rates are possible due to the one-stage logic. Flash ADCs require less or no interleaving for sampling frequencies in the GHz range. The short input-output delay is of concern for closed-loop designs where a high loop-bandwidth is required.

7.6.2   SAR ADC

A SAR ADC converts one bit at a time by executing a binary search, starting with the most significant bit (MSB) and ending with the least significant bit (LSB). Its basic architecture is shown in Fig. 7.17. The sampled voltage V_h is compared against a voltage V_{dac} controlled by the DAC. For the MSB decision, V_{dac} is set to the mid-point voltage, i.e., to half the full-scale input voltage. Depending on the output of the comparator, V_{dac} is set to either 1/4 or 3/4 of the input voltage range to determine the MSB-1. An n-bit SAR ADC is approximately n times slower than a flash ADC, but only takes n comparator decisions to determine the output voltage, compared to $2^n - 1$ for flash ADCs, resulting in a higher energy efficiency for medium resolution ADCs at 6 to 16 bits than with other ADC topologies.

SAR ADCs require logic to count from MSB to LSB, to determine the DAC level and to calculate the output. Depending on the architecture, this logic can be very simple. As SAR ADCs do not require gain stages and consist only of a comparator as an analog building block, they are highly suitable for modern digital nanometer-scale CMOS processes and low supply voltages.

Fig. 7.18 One stage of a
pipelined ADC

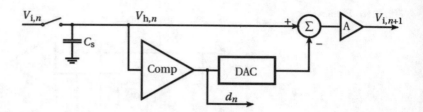

7.6.3   Pipelined ADC

A pipelined ADC consists of a number of equal or similar stages; one of them is shown in Fig. 7.18. The input voltage is compared against the mid-point voltage and the residue is formed by subtracting the mid-point voltage if the input is in the lower half. The residue is amplified and sampled onto the next stage. In a pipelined ADC, multiple samples are therefore processed at different stages.

The noise and linearity requirements on the first stage are critical. In a binary pipelined ADC, the noise power requirement in the second stage is already relaxed by a factor of four compared to that of the first stage. For power optimization, it is therefore often sufficient to design the first two stages and replicate the second stage for the remaining bits. Thanks to the flexibility on resolution, high conversion speed and design maturity of pipelined ADCs, they are often used by industry. Their power numbers, however, are generally not as good as those of SAR ADCs.

7.7   Further Details on SAR ADCs

SAR ADCs have been known for a few decades already. The concept was first proposed by McCreary in 1975 [4] as a MOS-only implementation of ADCs. For quite some time, the architecture was not popular because of the good gain and linearity of analog circuits in CMOS at that time. In particular, high-precision operational amplifiers pushed the performance of ADCs. With technology scaling and a focus on low-power design, the supply voltage dropped rapidly and made transistor stacking more challenging because of the lack of voltage headroom. SAR ADCs inherently do not depend on analog gain stages, which often require the stacking of transistors for good performance. The architecture is highly digital and contains only one critical analog block, the comparator. Digital blocks such as switches and logic cells can easily be implemented in a low-voltage design. Matching in newer CMOS technology nodes for minimum-sized structures is more random than systematic, and layout symmetry provides less matching advantage than in older CMOS nodes. On the other hand, digital switching speed increases with smaller technology sizes and digital logic power consumption drops. Together, these effects render the SAR ADC topology very attractive for medium-resolution ADCs with newer CMOS technology nodes.

7.7.1   Conventional SAR ADC

The conventional algorithm for SAR converters is shown in Fig. 7.19 for a single-ended implementation. The input voltage is first sampled while Φ_s is high, followed by a number of conversion cycles. The conversion sequence starts with the MSB, where $V_{ref}/2$ is added to V_{dac} and compared against zero. If V_{dac} is below zero, the voltage is kept, otherwise $V_{ref}/2$ is subtracted from V_{dac}. The conversion of MSB-1 is initiated by adding $V_{ref}/4$ to V_{dac}, compared against zero and kept if below zero. This sequence continues until the LSB conversion is complete. If V_{dac} is below zero, the comparator output is 1, otherwise it is 0. The conventional SAR algorithm involves one or two capacitive switching events for each conversion.

Fig. 7.19 Conventional
SAR capacitive array and
its operation

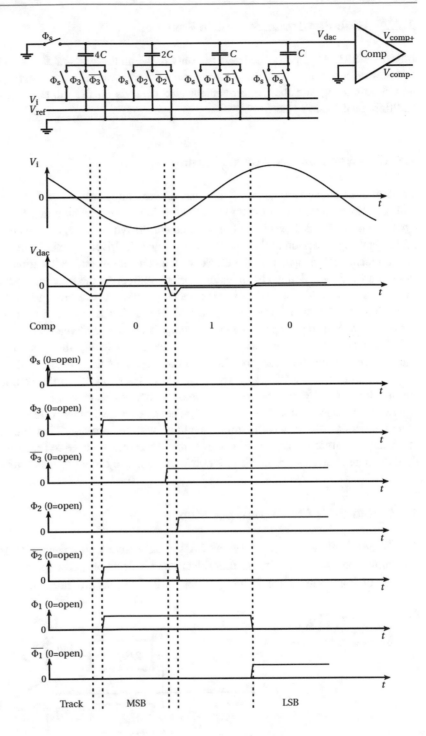

7.7.2 High-Performance Design Features

Improved SAR ADC performance is not only based on newer technology nodes but also on innovative new design approaches. They improve either power consumption, conversion speed, precision, or area requirement, and their improvement also depends on the implementation and potentially on additional calibration effort.

7.7.3 Asynchronous Internal Timing

One full SAR conversion generally consists of a sampling window, a number of comparison cycles with associated DAC feedback, and an offset calibration cycle including DAC reset. The calibration cycle is optional and depends on the architecture. In synchronous designs of SAR ADCs, the clock initiating the comparison cycles and calibration cycle is driven from an external periodic clock. Some of the recent designs generate the clock for the comparator and calibration internally, and only the sampling window is defined by an external clock source. As the internal clock is not synchronized to an external clock, this method of internal clock generation is referred to as asynchronous.

With asynchronous internal timing, the MSB decision is triggered by the external clock, which defines the end of the sampling window. Comparators usually have two outputs, where one of them switches state when the comparison is completed. To detect the completion of the comparator decision and trigger the next asynchronous conversion cycle, a decision detection block is used. It can be as simple as a NOR gate, where the output goes low when the comparison is completed. A comparator generally consists of a pre-amplification stage with a constant gain and delay, and a regenerative stage with exponential gain. The regenerative stage often consists of cross-coupled inverters with a path to bias the initial condition with the pre-amplified signal. The gain required to reach digital output levels depends on the input voltage. The time to reach this gain depends on the time the regenerative stage needs to amplify the signal to digital output levels.

7.7.4 Multi-Bit Conversion per Step

A very powerful way to increase the SAR ADC conversion speed is to invoke more than one comparator per step, resulting in more than one-bit decision per step, as shown in Fig. 7.20. The number of comparators for N-bits-per-step is $2^N - 1$, as in flash ADCs. This means that the

Fig. 7.20 A two-bit per conversion step SAR with 3 comparators

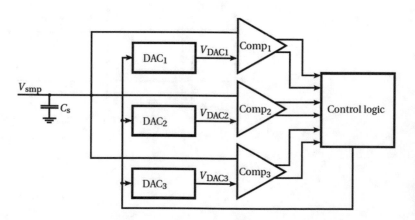

conversion for n bits will be finished in n/N cycles. The number of comparator decisions increases with larger N and equates to $n \times (2^N - 1)/N$. As each comparator decision is associated with a certain energy consumption, power increases with larger N. Minimum power consumption is reached at $N = 1$. As the number of conversion cycles in the SAR decreases with N, the benefit of asynchronous clock logic is lower than in a one-bit-per-step ADC.

If multi-bit conversion per step is used with a DAC configuration, where the DAC is equal to the sampling capacitor, as in the Set-And-Down DAC, usually one DAC has to be instantiated per comparator. The resulting input capacitance is then $2^N - 1$ higher for the same unit capacitor size or the unit capacitor has to be scaled down. Therefore, it can be beneficial to separate the input voltage from the DAC. This requires a comparator with two inputs or two differential input pairs in the case of differential input sampling. Generally, multi-bit conversion per step is beneficial for speed, but not for power and area.

7.8   Application Case Examples

7.8.1   ITU-OTU4

The optical communications standard ITU-OTU4 defines 100 Gb/s over a single fiber for long haul, i.e., up to a few thousand kilometers. A receiver implementation is shown in Fig. 7.21. The OTU4 channel consists of four sub-channels, an I/Q pair each for two orthogonal polarizations, which implements dual-polarization quadrature phase-shift keying (DP-QPSK). QPSK combines two orthogonal non-return to zero (NRZ) symbols in the complex domain (in-phase and quadrature-phase). DP-QPSK with coherent detection, i.e., with known phase and amplitude, is implemented. The baud rate is specified from 28 to 32 GS/s, with four bits per symbol from DP-QPSK and some margin for forward error correction (FEC).

It is generally suggested to run the ADC at two times the baud rate for coherent detection, resulting in a sampling rate requirement of 56 to 64 GS/s. Chromatic dispersion and polarization mode dispersion impact the signal integrity of long optical fibers. Both effects have to be compensated by the DSP shown in Fig. 7.21. Taking into account the channel imperfection and signal constellation with the bit-error-rate requirement of BER $< 10^{-12}$, about 5.5 to 6 ENOB at DC and 5 to 5.5 ENOB at 15 to 20 GHz input frequency are required with a 3 dB bandwidth >16 GHz. The $64\times$ interleaved 8 b ADC published in [5] primarily targets ITU-OTU4. A CMOS-only implementation provides significant benefits for power saving, particularly if the ADC is implemented in a modern CMOS process, where the large digital DSP can be implemented on the same chip, benefiting from the lower power consumption of modern CMOS processes.

7.8.2   IEEE 802.3bj

The IEEE 802.3bj defines the standard for 100 GS/s Ethernet over backplane or copper cable up to a length of 7 m, but is still in draft mode at the time of writing. Data is suggested to be transmitted electrically over four lanes with 25 Gb/s each. NRZ at 25 GS/s and 4 pulse amplitude modulation (4-PAM) are planned to be supported. NRZ and 4-PAM result in a Nyquist frequency of 12.5 GHz and 6.25 GHz, respectively. 4-PAM is more complex and entails more challenges in equalization, but provides better transmission properties because of the lower Nyquist frequency. As both NRZ and 4-PAM have to be supported, speed is given by the NRZ coding and precision is limited by 4-PAM,

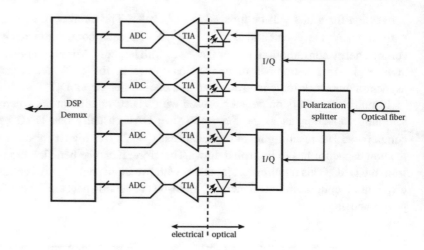

Fig. 7.21 Receiver of an OIF scalable SERDES Framer Interface

resulting in ADC requirements >25 GS/s at approximately 5 ENOB. The 32× interleaved 6 b ADC published in [6] targets these specifications at optimized power and area [7].

7.9 Additional Observations on Noise in High-Frequency ICs

We have already seen that the continuing trends in technology are enabling the integration of complete systems on a single die, which may include a combination of RF transceivers, analog processing, A/D, and D/A conversion as well as complex digital functions and memory on a single chip. However, when sensitive analog parts are combined with complex digital blocks operating at very high switching frequencies, the noise generated by the digital parts is inevitably transmitted to the analog blocks, predominantly through the common substrate, resulting in a reduction of the dynamic range, or reduction of the accuracy of the analog circuits. While discussing the coexistence of high-frequency, high-sensitivity analog blocks with high-speed digital blocks, this fact also has to be taken into account.

Noise in digital CMOS circuits is mainly generated by the rapid voltage variations caused by the switching of logic states, and the related charge-up/charge-down currents. In a conventional CMOS logic gate, the rapid change of voltage in internal nodes is coupled to the substrate through junction or wiring capacitances, causing charges to be injected into the substrate. Eventually, these substrate currents cause voltage drops that can perturb analog circuits through capacitive coupling and through variation of the threshold voltage due to body effect. Additionally, the high instantaneous currents needed to rapidly charge or discharge parasitic capacitances add up to large current spikes in the supply and ground distribution networks, a phenomenon known as simultaneous switching noise (SSN). These current spikes cause voltage noise primarily through the inductance of off-chip bondwires and on-chip power-supply rails. Ground supply networks are usually connected to the substrate, resulting in a direct coupling of the noise, and power networks are typically connected to very large N-well areas resulting in a consequently very large parasitic coupling capacitance to the substrate. Therefore, power and ground distribution networks are very noisy in CMOS circuits, and at the same time ideal mediums for the noise coupling to the substrate. Signal nets can also couple to the substrate, through diffusion and wiring capacitances, and signals with high energy and switching activity are thus critical from a noise perspective. This is the case especially for clock networks, which carry the most active signals and dissipate large amounts of power.

Two effective techniques to reduce the noise generation in digital circuits are the reduction of the voltage swings and the cancellation of transient currents during switching events. Specialized logic circuit families (single-ended and differential) that generate less noise than classical CMOS logic can be implemented, and are thus suitable for integration in a mixed-mode environment as a replacement or as a complement of CMOS logic.

Experimental studies have shown that single-ended logic families achieve only marginal improvement over regular CMOS in terms of noise. While differential logic families are the most promising candidates that offer improved noise reduction, traditional automation tools and design flows fail to accommodate many aspects associated with their differential nature. For this reason, large-scale implementation of digital circuits with low-noise differential logic families and their integration with high-sensitivity analog circuits remains a difficult task, which poses significant challenges for future systems design.

References

1. F. Maloberti, *Data Converters* (Springer, Dordrecht, 2007)
2. B. Razavi, *Principles of Data Conversion System Design* (Wiley, New York, 1994)
3. A. van Roermond, *Analog Circuit Design* (Springer, New York, 2011)
4. J. McCreary, P.R. Gray, A high-speed, all-MOS, successive-approximation weighted capacitor A/D conversion technique, in *ISSCC Dig. Tech. Papers, vol. XVIII, pp. 38–39*, (1975)
5. L. Kull, T. Toifl, M. Schmatz, P.A. Francese, C. Menolfi, M. Braendli, M. Kossel, T. Morf, T.M. Andersen, Y. Leblebici, *A 90GS/s 8b 667mW 64× interleaved SAR ADC in 32nm digital SOI CMOS* (ISSCC Dig. Tech, Papers, 2014)
6. L. Kull, J. Pliva, T. Toifl, M. Schmatz, P.A. Francese, C. Menolfi, M. Brändli, M. Kossel, T. Morf, T.M. Andersen, Y. Leblebici, Implementation of low-power 6–8 b 30–90 GS/s time-interleaved ADCs with optimized input bandwidth in 32 nm CMOS. IEEE J. Solid State Circuits **51**, 636–648 (2018)
7. L. Kull, *High-Speed CMOS ADC Design for 100Gb/s Communication Systems*, PhD Thesis, Ecole Polytechnique Fédérale de Lausanne (EPFL), 2014.

Correction to: Fundamentals of High Frequency CMOS Analog Integrated Circuits

Correction to:
D. Leblebici, Y. Leblebici, *Fundamentals of High Frequency CMOS Analog Integrated Circuits,* **https://doi.org/10.1007/978-3-030-63658-6**

The original version of this 2nd edition book was unfortunately published with errors in Figure 1.34, Figure 1.41, Figure 3.39, Figure 3.40, Figure 3.41, Figure 3.42, Figure 3.54, Figure 4.2, Figure 4.30, Figure 4.39, Figure 6.1, Figure 6.14 and Expression 5.20. This has now been corrected and an erratum to this book has been published.

The updated online versions of these chapters can be found at
https://doi.org/10.1007/978-3-030-63658-6_1
https://doi.org/10.1007/978-3-030-63658-6_3
https://doi.org/10.1007/978-3-030-63658-6_4
https://doi.org/10.1007/978-3-030-63658-6_5
https://doi.org/10.1007/978-3-030-63658-6_6

The corrected figures and expressions are given below.

Chapter 1

Fig. 1.34 (page 41)

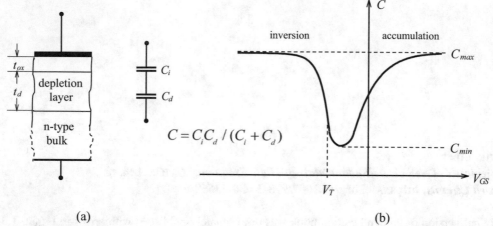

$$C = C_i C_d / (C_i + C_d)$$

(a)   (b)

Fig. 1.41 (page 47)

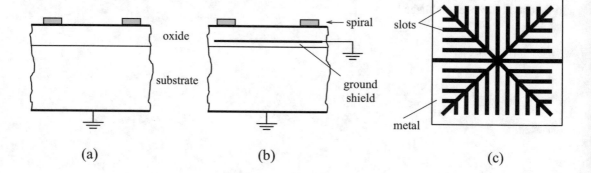

(a)   (b)   (c)

Chapter 3

Fig. 3.39 (page 146)

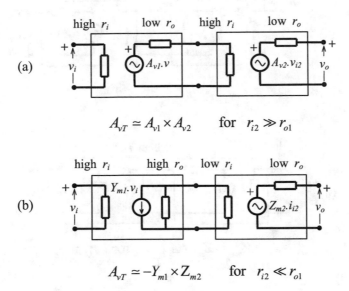

(a)

$$A_{vT} \simeq A_{v1} \times A_{v2} \qquad \text{for} \quad r_{i2} \gg r_{o1}$$

(b)

$$A_{vT} \simeq -Y_{m1} \times Z_{m2} \qquad \text{for} \quad r_{i2} \ll r_{o1}$$

Fig. 3.40 (page 147)

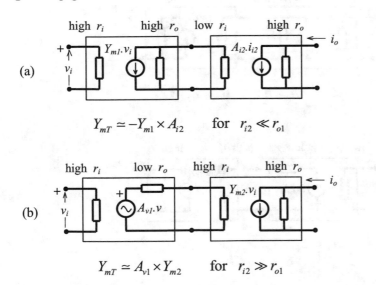

(a)

$$Y_{mT} \simeq -Y_{m1} \times A_{i2} \qquad \text{for} \quad r_{i2} \ll r_{o1}$$

(b)

$$Y_{mT} \simeq A_{v1} \times Y_{m2} \qquad \text{for} \quad r_{i2} \gg r_{o1}$$

Fig. 3.41 (page 147)

(a)

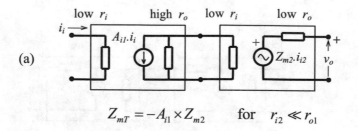

$$Z_{mT} = -A_{i1} \times Z_{m2} \qquad \text{for} \quad r_{i2} \ll r_{o1}$$

(b)

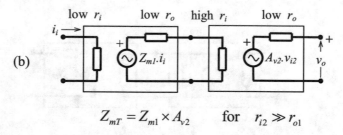

$$Z_{mT} = Z_{m1} \times A_{v2} \qquad \text{for} \quad r_{i2} \gg r_{o1}$$

Fig. 3.42 (page 148)

(a)

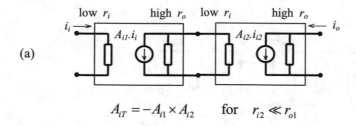

$$A_{iT} = -A_{i1} \times A_{i2} \qquad \text{for} \quad r_{i2} \ll r_{o1}$$

(b)

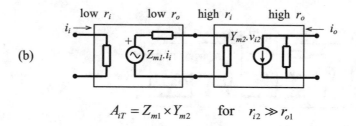

$$A_{iT} = Z_{m1} \times Y_{m2} \qquad \text{for} \quad r_{i2} \gg r_{o1}$$

Fig. 3.54 (page 161)

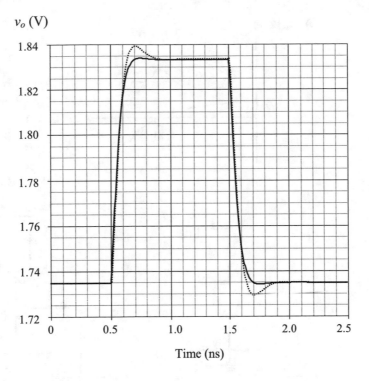

Chapter 4

Fig. 4.2 (page 168)

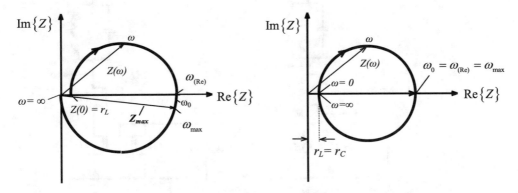

Fig. 4.30 (page 206)

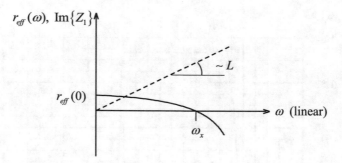

Fig. 4.39 (page 215)

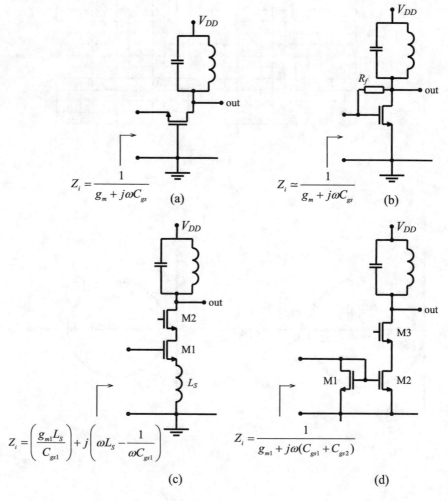

Chapter 6

Fig. 6.1 (page 270)

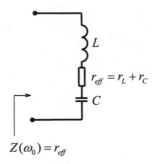

$$Z(\omega_0) = r_{\mathit{eff}}$$

Fig. 6.14 (page 284)

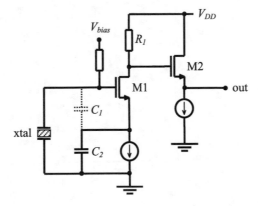

Chapter 5

Expression (5.20) (page 224)

$$\bar{i}_{ndT} = \bar{i}_{nd} \frac{1}{(1+g_m R_S) + \dfrac{R_D + R_S}{r_{ds}}} \cong \bar{i}_{nd} \frac{1}{(1+g_m R_S)} \tag{5.20}$$

Appendix

Appendix A

Mobility Degradation Due To the Transversal Field

In Chap. 1, it was mentioned that, especially for small geometry (thin gate oxide) MOS transistors, the electron and hole mobilities dramatically decrease due to the transversal electric field on the channel region. This effect is investigated in several publications and modeled in all advanced transistor models.

In this appendix, the degradation curves of electron and hole mobilities of surface channel NMOS and PMOS transistors are given for certain typical technologies.

The curves are based on the expressions given in Ref. [1.3] and [1.4].

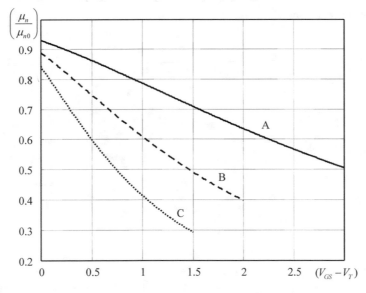

Fig. A.1 Relative variation of the channel electron mobility of typical NMOS transistors as a function of the gate overdrive voltage. Curve A: For a typical 0.35 micron technology (V_{TN} $-$ 0.5 V, $Tox = 7.5$ nm); Curve B: For a typical 0.18 micron technology ($V_{TN} = 0.35$ V, $Tox = 4$ nm); Curve C: For a typical 0.13 micron technology ($V_{TN} = 0.25$ V, $Tox = 2.3$ nm)

© The Author(s), under exclusive license to Springer Nature Switzerland AG 2021
D. Leblebici, Y. Leblebici, *Fundamentals of High Frequency CMOS Analog Integrated Circuits*,
https://doi.org/10.1007/978-3-030-63658-6

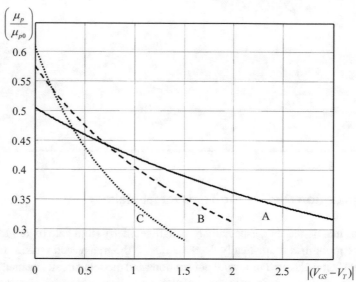

Fig. A.2 Relative variation of the surface channel hole mobility of typical PMOS transistors as a function of the gate overdrive voltage. Curve A: For a typical 0.35 micron technology ($V_{TP} = -1$ V, $Tox = 7.5$ nm); Curve B: For a typical 0.18 micron technology ($V_{TP} = -0.35$ V, $Tox = 4$ nm); Curve C: For a typical 0.13 micron technology ($V_{TP} = -0.20$ V, $Tox = 2.3$ nm)

Appendix B

The Typical Characteristic Curves and Basic Parameters for Hand Calculations of AMS 0.35 Micron NMOS and PMOS Transistors

AMS 035 NMOS TRANSISTORS

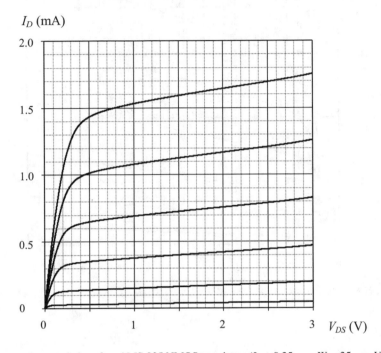

Fig. B.1 The output characteristics of an AMS 035 NMOS transistor. ($L = 0.35\ \mu$m, $W = 35\ \mu$m. $V_{GS} = 0.6$ to 1.1 V)

Basic Parameters for Hand Calculations

(Obtained from BSIM3v3 parameters and rounded)

$V_T(0) = 0.5\ [\text{V}],\ \mu_{n0} = 475\ \left[\text{cm}^2/\text{V.s}\right]$

$T_{ox} = 7.6\ [\text{nm}], C_{ox} = 4.54 \times 10^{-7}\ \left[\text{F}/\text{cm}^2\right] = 4.54\left[\text{fF}/\mu\text{m}^2\right]$

$C_{\text{DGO}} = C_{\text{SGO}} = 1.2 \times 10^{-10}\ [\text{F/m}] = 0.12\ [\text{fF}/\mu\text{m}], C_{\text{GBO}} = 1.1 \times 10^{-10}\ [\text{F/m}] = 0.11\ [\text{fF}/\mu\text{m}]$

$C_j(0) = 9.4 \times 10^{-4}\ \left[\text{F}/\text{m}^2\right] = 0.94\left[\text{fF}/\mu\text{m}^2\right],\ \ C_{\text{jsw}}(0) = 2.5 \times 10^{-10}\ [\text{F/m}] = 0.25[\text{fF}/\mu\text{m}]$

$\text{RDSW} = 345[\text{ohm}/\mu\text{m}], \text{RSH} = 75[\text{ohm}/\text{square}]$

$\lambda = 0.073\ [\text{V}^{-1}]$ (derived from output curves) (Fig. B.1)

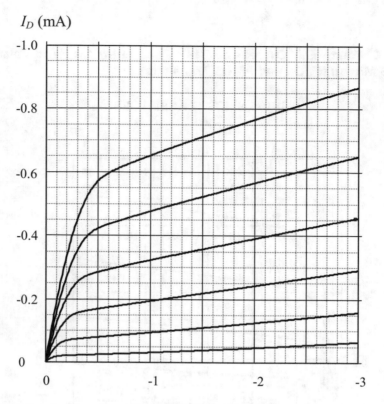

Fig. B.2 The output characteristics of an AMS 0.35 PMOS transistor. ($L = 0.35\ \mu$m, $W = 35\ \mu$m. $V_{GS} = -0.6$ to -1.3 V)

Basic Parameters for Hand Calculations

(Obtained from BSIM3v3 parameters and rounded)

$V_T(0) = -0.7$ [V], $\mu_{p0} = 148\ \left[\text{cm}^2/\text{V.s}\right]$

$T_{ox} = 7.6$ [nm], $C_{ox} = 4.54 \times 10^{-7}\ \left[\text{F/cm}^2\right] = 4.54\left[\text{fF}/\mu\text{m}^2\right]$

$C_{DGO} = C_{SGO} = 8.6 \times 10^{-11}[\text{F/m}] = 0.086$ [fF/μm], $C_{GBO} = 1.1 \times 10^{-10}$ [F/m] = 0.11 [fF/μm]

$C_j(0) = 1.36 \times 10^{-3}\ \left[\text{F/m}^2\right] = 1.36\left[\text{fF}/\mu\text{m}^2\right]$, $C_{jsw}(0) = 3.2 \times 10^{-10}$ [F/m] = 0.25[fF/μm]

RDSW = 1033 [ohm/μm], RSH = 130 [ohm/square]

$\lambda = 0.2$ [V^{-1}] (derived from output curves) (Fig. B.2)

Appendix C

BSIM3-v3 Parameters of AMS 0.35 Micron NMOS and PMOS Transistors[1]

```
.MODEL MODN NMOS LEVEL=7
* --------------------------------------------------------------------
*********** SIMULATION PARAMETERS*********
* --------------------------------------------------------------------
* format    : PSPICE
* model     : MOS BSIM3v3
* process   : C35
* revision  : 3.1;
* extracted : B10866 ; 2002-12; ese(487)
* doc#      : ENG-182 REV_3
* --------------------------------------------------------------------
*              TYPICAL MEAN CONDITION
* --------------------------------------------------------------------
*
*      *** Flags ***
+MOBMOD =1.000e+00          CAPMOD =2.000e+00
+NOIMOD =3.000e+00

*      *** Threshold voltage related model parameters ***
+K1 = 5.0296e-01
+K2 = 3.3985e-02           K3 = -1.136e+00          K3B = -4.399e-01
+NCH = 2.611e+17           VTH0 = 4.979e-01
+VOFF = -8.925e-02         DVT0 = 5.000e+01         DVT1 = 1.039e+00
+DVT2 = -8.375e-03         KETA = 2.032e-02
+PSCBE1 = 3.518e+08        PSCBE2 = 7.491e-05
+DVT0W = 1.089e-01         DVT1W = 6.671e+04        DVT2W = -1.352e-02

*      *** Mobility related model parameters ***
+UA = 4.705e-12            UB = 2.137e-18           UC = 1.000e-20
+U0 = 4.758e+02

*      *** Subthreshold related parameters ***
+DSUB = 5.000e-01          ETA0 = 1.415e-02         ETAB = -1.221e-01
+NFACTOR = 4.136e-01
```

---

[1] Note that usually there are several versions of device parameters for transistors with the same lithographic dimensions. The parameters given in this appendix are related to one of the conventional 0.35-micron devices fabricated by Austria Micro Systems AG (AMS). For other versions (for example, high voltage (5V) version or RF version) and for detailed technology-related knowledge and for design rules, it is necessary to apply to the manufacturing company or academic service organizations like EUROPRACTICE.

```
*       *** Saturation related parameters ***
+EM = 4.100e+07              PCLM = 6.948e-01
+PDIBLC1 = 3.571e-01         PDIBLC2 = 2.065e-03        DROUT = 5.000e-01
+A0 = 2.541e+00              A1 = 0.000e+00             A2 = 1.000e+00
+PVAG = 0.000e+00            VSAT = 1.338e+05           AGS = 2.408e-01
+B0 = 4.301e-09              B1 = 0.000e+00             DELTA = 1.442e-02
+PDIBLCB = 3.222e-01

*       *** Geometry modulation related parameters ***
+W0 = 2.673e-07              DLC = 3.0000e-08
+DWB = 0.000e+00             DWG = 0.000e+00
+LL = 0.000e+00             LW = 0.000e+00            LWL = 0.000e+00
+LLN = 1.000e+00             LWN = 1.000e+00            WL = 0.000e+00
+WW = -1.297e-14             WWL = -9.411e-21           WLN = 1.000e+00
+WWN = 1.000e+00

*       *** Temperature effect parameters ***
+AT = 3.300e+04              UTE = -1.800e+00
+KT1 = -3.302e-01            KT2 = 2.200e-02            KT1L = 0.000e+00
+UA1 = 0.000e+00             UB1 = 0.000e+00            UC1 = 0.000e+00
+PRT = 0.000e+00

*       *** Overlap capacitance related and dynamic model parameters   ***
+CGDO = 1.200e-10            CGSO = 1.200e-10           CGBO = 1.100e-10
+CGDL = 1.310e-10            CGSL = 1.310e-10           CKAPPA = 6.000e-01
+CF = 0.000e+00              ELM = 5.000e+00
+XPART = 1.000e+00           CLC = 1.000e-15            CLE = 6.000e-01

*       *** Parasitic resistance and capacitance related model parameters ***
+RDSW = 3.449e+02
+CDSC = 0.000e+00            CDSCB = 1.500e-03          CDSCD = 1.000e-03
+PRWB = -2.416e-01           PRWG = 0.000e+00           CIT = 4.441e-04

*       *** Process and parameters extraction related model parameters ***
+TOX = 7.575e-09             NGATE = 0.000e+00
+NLX = 1.888e-07

*       *** Substrate current related model parameters ***
+ALPHA0 = 0.000e+00     BETA0 = 3.000e+01

*       *** Noise effect related model parameters ***
+AF = 1.507e+00              KF = 2.170e-26             EF = 1.000e+00
+NOIA = 1.121e+19            NOIB = 5.336e+04           NOIC = -5.892e-13

*       *** Common extrinsic model parameters ***
+LINT = -5.005e-08           WINT = 9.4030e-08          XJ = 3.000e-07
+RSH = 7.000e+01             JS = 1.000e-05
+CJ = 9.400e-04              CJSW = 2.500e-10
+CBD = 0.000e+00             CBS = 0.000e+00            IS = 0.000e+00
+MJ = 3.400e-01              N = 1.000e+00              MJSW = 2.300e-01
+PB = 6.900e-01              TT = 0.000e+00
+PBSW = 6.900e-01
```

• ---

MODEL MODP PMOS LEVEL=7
* --
********** SIMULATION PARAMETERS *********
* --
* format   : PSPICE
* model   : MOS BSIM3v3
* process : C35
* revision : 3.1;
* extracted : C64685 ; 2002-12; ese(487)
* doc#   : ENG-182 REV_3
* --
*   TYPICAL MEAN CONDITION
* --
*
*   *** Flags ***
+MOBMOD = 1.000e+00   CAPMOD = 2.000e+00
+NOIMOD = 3.000e+00

*   *** Threshold voltage related model parameters ***
+K1 = 5.9959e-01
+K2 = -6.038e-02   K3 = 1.103e+01   K3B = -7.580e-01
+NCH = 9.240e+16   VTH0 = -6.915e-01
+VOFF = -1.170e-01   DVT0 = 1.650e+00   DVT1 = 3.868e-01
+DVT2 = 1.659e-02   KETA = -1.440e-02
+PSCBE1 = 5.000e+09   PSCBE2 = 1.000e-04
+DVT0W = 1.879e-01   DVT1W = 7.335e+04   DVT2W = -6.312e-03

*   *** Mobility related model parameters ***
+UA = 5.394e-10   UB = 1.053e-18   UC = 1.000e-20
+U0 = 1.482e+02

*   *** Subthreshold related parameters ***
+DSUB = 5.000e-01   ETA0 = 2.480e-01   ETAB = -3.917e-03
+NFACTOR = 1.214e+00

*   *** Saturation related parameters ***
+EM = 4.100e+07   PCLM = 3.184e+00
+PDIBLC1 = 1.000e-04   PDIBLC2 = 1.000e-20   DROUT = 5.000e-01
+A0 = 5.850e-01   A1 = 0.000e+00   A2 = 1.000e+00
+PVAG = 0.000e+00   VSAT = 1.158e+05   AGS = 2.468e-01
+B0 = 8.832e-08   B1 = 0.000e+00   DELTA = 1.000e-02
+PDIBLCB = 1.000e+00

*   *** Geometry modulation related parameters ***
+W0 = 1.000e-10   DLC = 2.4500e-08
+DWB = 0.000e+00   DWG = 0.000e+00
+LL = 0.000e+00   LW = 0.000e+00   LWL = 0.000e+00
+LLN = 1.000e+00   LWN = 1.000e+00   WL = 0.000e+00
+WW = 1.894e-16   WWL = -1.981e-21   WLN = 1.000e+00
+WWN = 1.040e+00

```
*       *** Temperature effect parameters ***
+AT = 3.300e+04              UTE = -1.300e+00
+KT1 = -5.403e-01            KT2 = 2.200e-02              KT1L = 0.000e+00
+UA1 = 0.000e+00            UB1 = 0.000e+00             UC1 = 0.000e+00
+PRT = 0.000e+00


*       *** Overlap capacitance related and dynamic model parameters   ***
+CGDO = 8.600e-11           CGSO = 8.600e-11            CGBO = 1.100e-10
+CGDL = 1.080e-10           CGSL = 1.080e-10            CKAPPA = 6.000e-01
+CF = 0.000e+00             ELM = 5.000e+00
+XPART = 1.000e+00          CLC = 1.000e-15             CLE = 6.000e-01


*       *** Parasitic resistance and capacitance related model parameters ***
+RDSW = 1.033e+03
+CDSC = 2.589e-03           CDSCB = 2.943e-04           CDSCD = 4.370e-04
+PRWB = -9.731e-02          PRWG = 1.477e-01            CIT = 0.000e+00


*       *** Process and parameters extraction related model parameters ***
+TOX = 7.754e-09            NGATE = 0.000e+00
+NLX = 1.770e-07


*       *** Substrate current related model parameters ***
+ALPHA0 = 0.000e+00         BETA0 = 3.000e+01


*       *** Noise effect related model parameters ***
+AF = 1.461e+00             KF = 1.191e-26              EF = 1.000e+00
+NOIA = 5.245e+17           NOIB = 4.816e+03            NOIC = 8.036e-13


*       *** Common extrinsic model parameters ***
+LINT = -7.130e-08          WINT = 3.4490e-08           XJ = 3.000e-07
+RSH = 1.290e+02            JS = 9.000e-05
+CJ = 1.360e-03             CJSW = 3.200e-10
+CBD = 0.000e+00            CBS = 0.000e+00             IS = 0.000e+00
+MJ = 5.600e-01             N = 1.000e+00               MJSW = 4.300e-01
+PB = 1.020e+00             TT = 0.000e+00
+PBSW = 1.020e+00
*   -----------------------------------------------------------------------
```

Appendix D

Current Sources and Current Mirrors

Current sources/mirrors are among the most important building-blocks of analog CMOS circuits and are being extensively used as DC current sources, DC current mirrors, and AC current mirrors. For DC and AC current mirrors, the mirroring factor can be unity, or more than unity; in other words, a current mirror can be considered and used as a current amplifier.

In this appendix, first the DC behavior of the DC current sources and current mirrors will be analyzed. Later, the frequency characteristics of AC current mirrors, which have severe effects on the overall performance of many circuits, will be investigated in some detail.

D.1. DC Current Sources

The simplest MOS current source is a MOS transistor working in the saturation region, i.e., $V_{DS} \geq (V_{GS}-V_T)$, (Fig. D.1a). Here "load" represents the circuit to be fed by this current source. The current is mainly determined by the gate-source voltage V_{GS}, but due to the channel length modulation effect, also has a weak dependence on V_{DS}. We know from (1.14a) that it can be expressed as

$$I_L = \frac{1}{2}\mu \cdot C_{ox} \frac{W}{L}(V_{GS} - V_T)^2 \left(\frac{1+\lambda \cdot V_{DS}}{1+\lambda \cdot (V_{GS}-V_T)} \right) \tag{D.1}$$

In this circuit, due to the quadratic character of the load current-control voltage relation, the sensitivity is high. Therefore, to obtain a certain desired load current it is necessary to adjust and maintain V_{GS} at an appropriate value with precision.

Another way to bias the current-source transistor is shown in Fig. D.1b. In this circuit, the gate bias of the current source transistor M2 is determined as the gate-source voltage of the diode-connected transistor M1, biased with a constant drain bias current, or "reference current," I_r. Figure D.1c shows the multiple output version of the circuit, sharing the same reference transistor.

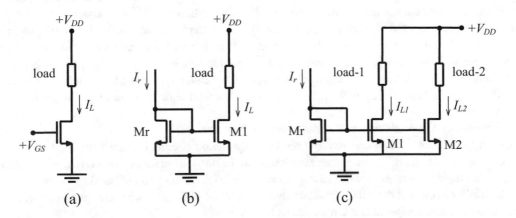

Fig. D.1 (a) MOS transistor as a DC current source. (b) single-output, (c) multiple-output basic MOS current mirrors

From (1.14a), the drain currents of M1 and M2 in Fig. D.1b can be written as

$$I_r = \frac{1}{2}\mu_n C_{ox}\frac{W_1}{L}(V_{GS} - V_T)^2\left(\frac{1 + \lambda V_{GS}}{1 + \lambda(V_{GS} - V_T)}\right) \tag{D.2}$$

$$I_L = \frac{1}{2}\mu_n C_{ox}\frac{W_2}{L}(V_{GS} - V_T)^2\left(\frac{1 + \lambda V_{DS2}}{1 + \lambda(V_{GS} - V_T)}\right) \tag{D.2a}$$

From these expressions I_L can be solved in terms of the reference current I_r:

$$I_L = \frac{W_2}{W_1}\frac{1 + \lambda V_{DS2}}{1 + \lambda V_{GS}}I_r = BI_r \tag{D.3}$$

(D.3) shows that for this circuit the load current is proportional to the reference current. The mirroring coefficient is

$$B = \frac{I_L}{I_r} = \frac{W_2}{W_1}\frac{1 + \lambda.V_{DS2}}{1 + \lambda.V_{GS}} \tag{D.4}$$

and mainly depends on the ratio of the widths. The secondary effects of λ and V_{DS2} on B are negligible only for long-channel-length transistors. The sensitivity of the load current with respect to the variations of V_{DS} (in other words the variations of the DC resistance of the load circuit) can be calculated as

$$\frac{dI_L}{dV_{DS2}} = \frac{W_2}{W_1}I_r\frac{\lambda}{1 + \lambda.V_{GS}} \tag{D.5}$$

which is also equal to the small-signal output conductance of M2, in other words, the internal conductance of the current source.

To approach an ideal current source, which has a zero internal conductance, λ must be as small as possible, in other words long channel transistors must be used, at the expense of increasing the parasitics. There are several alternative solutions to obtain a very low internal conductance. One of them is the cascode current mirror and is shown in Fig. D.2. Although the low-frequency small signal output conductance of this circuit is considerably lower than that of the basic circuit given in Fig. D.1b, it has a severe drawback. In this circuit there is more than one transistor on the load current path and each additional transistor "steals" from the supply voltage budget, which is getting increasingly limited with diminishing supply voltages. Consequently, for small geometry-low voltage circuits, using the basic circuit with sufficiently longer channel lengths is more advantageous.

D.2. Frequency Characteristics of Basic Current Mirrors

Current mirrors are being extensively used in many analog circuits to mirror the AC current of a branch to another branch. The mirroring coefficient can be unity, as in the active loads of a differential pair, or can have a value greater than unity, as in some OTA circuits investigated in Chap. 3.

The basic circuit and the small-signal equivalent with its main parasitics are given in Fig. D.3a. The parasitic capacitances in this circuit are as follows:

Fig. D.2 Cascode current mirror

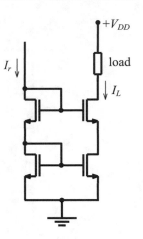

- The total gate-source capacitance of M1, which is the sum of the gate capacitance and the gate-source overlap capacitance:

$$C_{gs1} = C_{ox}W_1L\left(\frac{2}{3} + \frac{\text{CGSO}}{C_{ox}L}\right) = k_{ol}C_{ox}W_1L$$

- The total gate-source capacitance of M2, which is the sum of the gate capacitance and the gate-source overlap capacitance:

$$C_{gs2} = C_{ox}W_2L\left(\frac{2}{3} + \frac{\text{CGSO}}{C_{ox}L}\right) = k_{ol}C_{ox}W_2L$$

- The gate-drain capacitance of M2:

$$C_{dg2} = \text{CDGO} \times W_2 \cong \text{CGSO} \times W_2$$

- The total drain junction capacitance of M1:

$$C_{jd1} \cong W_1X_1C_j\left[1 + \frac{2}{X_1}\frac{C_{jsw}}{C_j}\right] = W_1X_1C_j'$$

where X_1 denotes the second dimension of the drain area, C_j the bottom junction capacitance per unit area, and C_{jsw} the side-wall capacitance per unit length (c.f. Chap. 1).

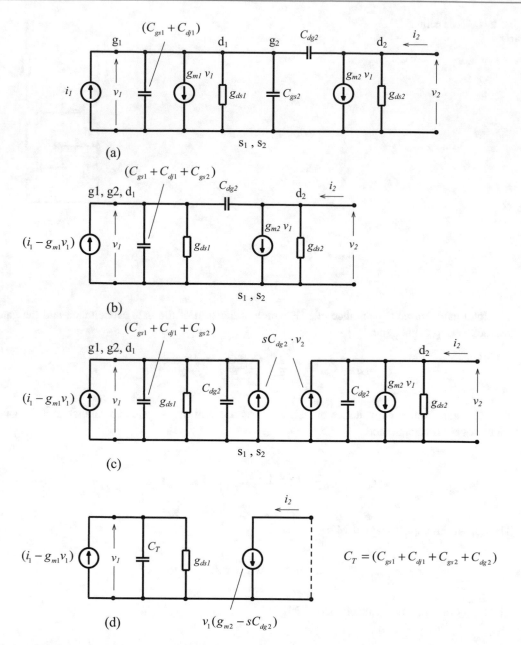

Fig. D.3 (a) The small-signal equivalent circuit of the basic current source. (b) and (c) Simplification and application of the modified Miller transformation. (d) Simplified circuit for $|Y_L| \gg |g_{ds2} + j\omega C_{dg2}|$ to calculate B

The current transfer ratio of the mirror can be easily calculated from Fig. D.3d:

$$B = \frac{i_2}{i_1} = B_0 \frac{s_p}{s_0} \frac{(s - s_0)}{(s - s_p)} \tag{D.6}$$

where

$$s_0 = +\frac{g_{m2}}{C_{dg2}}, \quad s_p = -\frac{g_{m1} + g_{ds1}}{C_T} \cong -\frac{g_{m1}}{C_T} \tag{D.7}$$

and the low-frequency value of B

$$B_0 = \frac{g_{m2}}{(g_{m1} + g_{ds1})} \cong \frac{g_{m2}}{g_{m1}} \cong \frac{W_2}{W_1} \tag{D.8}$$

The magnitude of the zero frequency is obviously very high compared to that of the pole; the pole strongly dominates and the 3 dB frequency is determined by the pole:

$$f_{(3dB)} \cong \frac{1}{2\pi} \frac{g_{m1}}{C_T} \tag{D.9}$$

In this expression, g_{m1} depends on the geometry, the DC operating conditions, and the mode of operation. It is known that the transistors can operate either in the normal saturation region (the pre-velocity saturation region) or in the velocity saturation region, depending on the channel length and the DC operating point. The value of the gate source capacitance that is the dominant component of C_T is also different for non-velocity-saturated and velocity-saturated regimes. Therefore, the 3 dB frequency of a current mirror must be investigated separately for these two cases.

D.2.1. Frequency Characteristics for Normal Saturation

If the transistors of the current mirror are operating in the pre-velocity-saturation mode, the transconductance of the reference transistor and the total capacitance of the input node are

$$g_{m1} \cong \mu C_{\text{ox}} \frac{W_1}{L} (V_{\text{GS}} - V_T)$$

and

$$\begin{aligned} C_T &= C_{\text{gs1}} + C_{\text{gs2}} + C_{\text{dg2}} + C_{\text{dj1}} \\ &= k_{ol} C_{\text{ox}} L W_1 + k_{ol} C_{\text{ox}} L W_2 + \text{CDGO} \cdot W_2 + X_1 C'_j W_1 \end{aligned}$$

which can be arranged as

$$C_T = W_1 \left[k_{ol} C_{\text{ox}} L (1 + B_0) + B_0 \cdot \text{CDGO} + X_1 C'_j \right] \tag{D.10}$$

The 3 dB frequency of the gain function can be arranged assuming $g_{m1} \gg g_{dg1}$ and $C_{GSO} = C_{GDO}$, which is reasonable for most cases, as

$$f_{(3dB)} = \frac{\mu (V_{\text{GS}} - V_T)}{\left[k_{ol} L^2 (1 + B_0) + B_0 \frac{L \cdot \text{CDGO}}{C_{\text{ox}}} + L X_1 \frac{C'_j}{C_{\text{ox}}} \right]} \tag{D.11}$$

From this expression, it is possible to examine the effects of any one of the parameters, and to reach valuable design hints. It can be easily seen that the 3 dB frequency:

- Decreases with B_o (or with the number of outputs)
- Strongly decreases with the channel length
- Is independent from the gate-width
- Increases with mobility (NMOS is better)
- Increases with the gate overdrive, or the DC current of the source transistor

The parameters in these expressions are directly given in the model parameter lists, except μ. Especially for gate oxide thicknesses smaller than 10 nm, the value of μ must be calculated in terms of the gate voltage (c.f. Chap. 1). The C_j and C_{jsw} values given in the model parameter lists correspond to zero junction voltage. The actual values of these capacitances are smaller due to the reverse bias of the junction.

D.2.2. Frequency Characteristics Under Velocity Saturation

We know that for small gate lengths – and especially for NMOS transistors – the operating mode must be checked to see if the transistor is in the velocity-saturation mode, or not. If the transistor is operating in the velocity saturation mode, the transconductance must be calculated from (1.34) as $g_m = kC_{ox}Wv_{sat}$. In addition, the gate-source capacitance of a velocity-saturated transistor is $C_{gs} = C_{ox}WL + (CGSO \times W)$ as derived in Problem 1.1.

Using these parameters, the 3 dB frequency can be found as

$$f_{(3dB)} = \frac{kv_{sat}}{\left[L(1 + B_0) + B_0 \frac{\text{CDGO}}{C_{ox}} + X_1 \frac{C'_j}{C_{ox}} \right]} \tag{D.12}$$

This expression can be interpreted as follows:

- The 3 dB frequency of a transistor does not change with the operating point, provided that the transistor remains in velocity saturation.
- The gate-length dependence of the 3 dB frequency of a velocity-saturated transistor is not as severe as that of a non-velocity saturated transistor.
- The 3 dB frequencies of NMOS and PMOS transistors are approximately equal, provided that both are in velocity saturation.

Appendix E

S-Parameters

Throughout the design and simulation procedures of an electronic circuit, we can calculate or "measure" the voltage of any node and the current on any branch in the circuit without any restrictions and then calculate the gains, impedances, etc. At the end, the designed circuit can be characterized in terms of terminal voltages and currents. For this purpose, several small-signal parameter sets are developed, such as y-parameters, z-parameters, etc. It is known that the y-parameters are the most suitable set at high frequencies.

After the realization of a circuit, to check if the design goals are fulfilled, it is necessary to determine terminal voltages and currents by direct measurements. Except for the low-end of the RF, there arise some severe problems:

1. At RF frequencies, due to the finite (non-infinite) input impedances of voltmeters and non-zero input impedances of ammeters, it is not possible to determine the voltages and currents without interfering with the circuit.
2. For the y-parameter measurements it is necessary to short-circuit the output or input port, which becomes more and more difficult (even impossible) as the frequency increases to the deep GHz region.
3. The cables (transmission lines) connecting the signal source to the input port of the circuit and the output port to the load are another problem. It is known from transmission line theory that the voltage at the output of a line depends not only on the voltage at the input of the line, but also on the length, the frequency, and the terminating impedance of the line. This problem can be eliminated only if the line is terminated with its characteristic impedance[2].

For the characterization of electronic circuits at radio frequencies, the "s-parameters" technique was borrowed from microwave circuits and systems, and adapted to RF electronic circuits and a class of measuring instruments; "network analyzers" were developed based on this approach.

In this approach, the basic definition of the s-parameters of a two-port is expressed with this equations set:

$$b_1 = s_{11}a_1 + s_{12}a_2$$
$$b_2 = s_{21}a_1 + s_{22}a_2$$

(E.1)

where a_1 is the incident voltage wave toward the input port, b_1 is the reflected voltage wave from the input port, both normalized to the characteristic impedance. Similarly a_2 is the normalized voltage wave toward the output port, and b_2 is the normalized voltage wave toward the load. Note that due to the phase differences of the related waves, all of these parameters are complex quantities. These parameters can be expressed in terms of the amplitudes of the incident and reflected voltage waves[3]:

---

[2] The worldwide accepted standard value for the coaxial transmission lines is 50 ohm. There are several other standard values (60 ohm, 75 ohm, etc.) but they are not as widespread as 50 ohm.

[3] H-P Application Note 95-1.

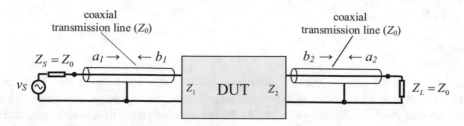

Fig. E.1 The measurement set for the s_{11} and s_{21} parameters

$$a_1 = \frac{V_{i1}}{\sqrt{Z_0}}, \quad a_2 = \frac{V_{i2}}{\sqrt{Z_0}}, \quad b_1 = \frac{V_{r1}}{\sqrt{Z_0}}, \quad b_2 = \frac{V_{r2}}{\sqrt{Z_0}} \tag{E.2}$$

This means that the corresponding powers[4] are equal to

$$P_{i1} = \frac{V_{i1}^2}{2Z_0}, \quad P_{i2} = \frac{V_{i2}^2}{2Z_0}, \quad P_{r1} = \frac{V_{r1}^2}{2Z_0}, \quad P_{r2} = \frac{V_{r2}^2}{2Z_0} \tag{E.2a}$$

In Fig. E.1, the block diagram of an s-parameters measurement set for s_{11} and s_{21} is shown. The internal impedance of the signal source and the load impedance are both equal to the characteristic impedance of the lines. To eliminate (or minimize) reflections from the input and output ports, the Z_1 and Z_2 must be[5] equal or as close as possible to Z_0.

From the first line of expression (E.1), we see that if $Z_L = Z_0$ there is no reflection from the load, therefore, $a_2 = 0$. In this case, the s_{11} parameter can be found as

$$s_{11} = \frac{b_1}{a_1}\Big|_{a_2=0} \tag{E.3}$$

which is the ratio of the reflected to incident waves of the input port and is also called the input reflection coefficient (Γ_1). This parameter is a measure of the proximity of the input impedance to the targeted Z_0 value. s_{11} is usually expressed in dB[6]:

$$s_{11}(\text{dB}) = 20.\log|s_{11}| \tag{E.3a}$$

whose achievable or acceptable value is usually in the range of -10 dB to -50 dB.

From the second line of (E.1) s_{21} can be written as

$$s_{21} = \frac{b_2}{a_1}\Big|_{a_2=0} \tag{E.4}$$

---

[4] Originally, the s-parameters technique was developed based on the powers of the incident and reflected waves in the waveguides. This was because, during the early days of microwaves, the only measurable quantity of electromagnetic waves was the power.

[5] If the signal source and the load (for example antenna) is very close to the input and output ports of the circuit and no transmission lines are necessary, the input and output internal impedances do not have to be 50 ohm (see Sect. 5.4). But for the characterization of the circuit with s-parameters technique, the 50 ohm assumption is necessary. In such cases, it is possible to properly use external 50-ohm resistors to perform measurements and then to de-embed their effects to reach the original parameters of the circuit.

[6] Note that s-parameters are complex quantities, and therefore contain not only magnitude but phase information. s-parameters expressed in dB, however, contain only magnitude information.

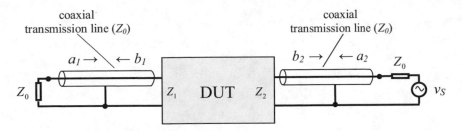

Fig. E.2 The measurement set for the s_{22} and s_{12} parameters

which is the ratio of the amplitude of the voltage wave going toward the load, to the amplitude of the voltage wave applied to the input when the output is matched. This means that this is the "voltage gain." It can be expressed in dB[7] as

$$s_{21}(\mathrm{dB}) = 20.\log|s_{21}| \qquad (\text{E.4a})$$

According to the second line of (E.1)

$$s_{22} = \frac{b_2}{a_2}\Big|_{a_1=0} \qquad (\text{E.5})$$

which is the reflection coefficient of the output port (Γ_2). The s_{22} parameter in dB is

$$s_{22}(\mathrm{dB}) = 20.\log|s_{22}| \qquad (\text{E.5a})$$

To fulfill $a_1 = 0$, there must be no incident wave to the input, i.e., the input must be matched and no signal must be applied. The corresponding measurement set is shown in Fig. E.2.

By definition, the s_{12} parameter is

$$s_{12} = \frac{b_1}{a_2}\Big|_{a_1=0} \qquad (\text{E.6})$$

which expresses the effect of an output signal on the input when the input port is matched and there is no signal applied to the input. In other words, it is the voltage transfer ratio in the reverse direction and also usually expressed in dB.

It is obvious that the s-parameters can be converted to y-parameters and vice-versa. Although this is usually not a necessity for an IC designer, the conversion expressions between s-parameters and y-parameters (or any other 2-port small-signal parameter set) can be found in the literature.[8]

---

[7] Note that since the powers are proportional to the squares of the amplitudes of the voltage waves, the numerical value of the power gain in dB is also equal to (E.3a).

[8] For example, H-P Application Note 95-1.

Index

© The Author(s), under exclusive license to Springer Nature Switzerland AG 2021
D. Leblebici, Y. Leblebici, *Fundamentals of High Frequency CMOS Analog Integrated Circuits*,
https://doi.org/10.1007/978-3-030-63658-6

Printed in the United States
by Baker & Taylor Publisher Services